PRINCIPLES OF
AQUATIC CHEMISTRY

PRINCIPLES OF AQUATIC CHEMISTRY

François M. M. Morel

Massachusetts Institute of Technology

A Wiley-Interscience Publication

JOHN WILEY & SONS

New York • Chichester • Brisbane • Toronto • Singapore

Library of Congress Cataloging in Publication Data:

Morel, François, 1944–
  Principles of aquatic chemistry.

  "A Wiley-Interscience publication."
  Includes index.
  1. Water chemistry. I. Title.
GB855.M67 1983   546′.22   83-6840
ISBN 0-471-08683-5

Printed in the United States of America

10 9 8 7 6 5 4 3

Ce livre, à Chantal qui comprendra un jour
     qu'un oursin n'y était pas nécessaire,
   à Sébastien qui est presque assez grand
   pour savoir qu'il n'y est pas superflu,
   à Nicole, partenaire de jeu et de travail,
   qui assura presque seule notre survie quotidienne.

S.M.

PREFACE

Because the hydrologic cycle powers all others, and because life on earth is so intimately tied to water, aquatic chemistry is a central link between the cycles of elements at the surface of the earth and the workings of biological systems. Life adapts to its environment and modifies it. As noted by Alfred Redfield some 50 years ago, the composition of natural waters reflects both the geochemistry of the planet earth and the biochemical requirements of its inhabitants. These are delicately in tune, constantly modifying each other, linked by a few billion years of "co-evolution," and both ultimately responding to the same fundamental laws of chemistry.

This central position of aquatic chemistry in the natural sciences gives it an increasing popularity in science and engineering curricula; it also makes it a difficult topic to teach for it requires exploring some aspects of almost all sciences. This text is intended for first year graduate students and advanced undergraduates. I have assumed only a general college chemistry background and all incursions into geology, microbiology, oceanography, chemical kinetics, and other associated fields are kept strictly elementary. The first three chapters which deal with conservation of mass, thermodynamics and kinetics, and chemical equilibrium calculations, form effectively a separate entity. I intend them as a self-contained introduction to solution chemistry. In these three chapters, which provide the general background indispensible for the rest of the book, thermodynamics is the major stumbling block. Because typical student backgrounds differ widely, I have chosen a treatment of thermodynamics that is pragmatic and intuitive, to serve as a first introduction for some and as a review for others.

In the next five chapters of the text the subject of aquatic chemistry proper is developed following the accepted divisions of acid-base, precipitation-dissolution, coordination, redox, and surface chemistry. (A possible ninth chapter on organic aquatic chemistry never got written; I think it would require a text and a course of its own.) Chapters 5 and 6 focus more on chemical principles and less on the intricate realities of natural waters which are introduced more fully in the later chapters. At the end of each chapter the more advanced topics, such as kinetics, are treated in an elementary fashion. Sections that are either digressive or not essential to the overall comprehension of the text are indicated by three squares at the beginning and at the end.

□ □ □

The expression "aquatic chemistry" made its official entry into scientific language with the publication of the text by Stumm and Morgan in 1970. The title of this new text is meant in part as a general reference and acknowledgment to that classic book, as a plain expression of the debt I owe to its authors who started a scientific and pedagogic tradition.

My primary goal in writing this text was to satisfy the demands of a particular assemblage of students. For ten years I have taught aquatic chemistry to students from MIT engineering and science departments, from the MIT-WHOI Joint Program in Oceanography, and from Harvard College and the School of Public Health. During those years I developed class notes that are the basis for this text. Much has been added and reworked to be sure, but the skeleton of these original class notes remains, still reflecting a wistful and very Gallic ambition for organization and coherence.

From those notes also comes an emphasis on the geochemical and geobiological cycles of elements in natural waters. This central conceptual link between environmental chemists and oceanographers is used as a thread throughout the text. I have also tried to meet the students' demand for mathematical rigor without mathematical sophistication. The price sometimes is a cumbersomely detailed algebraic and numerical development. I hope to have at least generally resisted a natural propensity for reducing lively chemical concepts to cold mathematics.

While writing this text I often found myself squeezed between the lofty goals of pedagogy and the sheer boredom of a systematic and repetitious treatment. My pen attempted to escape and I had to fight bouts of lyricism, jocularity, and even obscurantism. (Such wonderful depth there is in the murky corners of the mind!) No help to students, of course, and most of it has been bravely edited away. Perhaps the scattered remains will entertain those who wander through; may they find the text more enjoyable where it is less orderly.

In the end, of course, this book is a compromise, a truce in a quixotic battle against the modern realities of academic life. Still, there is much satisfaction just in getting here and, as I sigh in relief, I must also thank all those who helped along the way. I shall mention only a few: W. Fish, P. M. Gschwend, T. D. Waite, and O. Zafiriou who contributed some of their expertise to particular sections; R. E. McDuff, J. J. Morgan, and J. C. Westall who commented on the complete manuscript; R. Selman and B. A. Hanrahan who transformed my hieroglyphic handwriting into print; D. A. Dzombak and S. J. Tiffany who helped simplify and anglicize my prose and freed me of countless detailed tasks; S. W. Chisholm whose enlightening influence permeates the whole text. To these and many unnamed others I express my gratitude. Still, it is all the students in Aquatic Chemistry at MIT over the past ten years who deserve the greatest credit: they insisted on their right to understand. If this text helps clarify the subject in some way, the next generations of students will owe it to their elders, and I shall derive no small satisfaction from it.

FRANÇOIS M. M. MOREL

Lexington, Massachusetts
September 1983

CONTENTS

CHAPTER ONE

CONSERVATION
PRINCIPLES

A poet physicist once calculated how many atoms from Plato's body each of us has appropriated as our own—probably in excess of 10^{10}. Behind the vertigo of the multiple reincarnations and the comforting thought of all the great people each of us has been lies the profound and fundamental principle of conservation of mass.

The idea that the total mass of a closed system should remain constant regardless of internal changes is surely one of our most dearly held notions, and the classical chemical version of this principle, according to which the total mass of an element is invariant throughout all its physical and chemical transformations, is almost as deeply engraved in our minds. In a paradoxical way our understanding of these fundamental principles has benefited from the discovery of their fallacy as demonstrated by the theory of relativity and the discovery of nuclear reactions. Although closed systems can change mass, and transmutation of elements does occur, we retain in practice the useful approximation that mass conservation and elemental permanence are absolute truth.

So easily intuitive seems the idea of mass conservation that its application into workable mathematical expressions can be ironically frustrating. In the study of aquatic transport phenomena the difficulty lies in writing—and then solving—the differential equations expressing the conservation of solvent and solutes when they are subjected to various motions and changes of volume. In this chapter we study how conservation principles are applied to aqueous chemical systems in which multiple reactions, but no transport, are taking place. The underlying mathematics are more in the field of linear algebra than that of differential analysis. The key concept to be mastered is that of *components*, the fundamental entities to which the conservation principle is applied in the form of mole balance equations.

1

It may admittedly be excessive to devote a whole chapter to a subject that is often taken as intuitively obvious. Indeed much of what follows will seem terribly rudimentary to many. Yet it is our purpose to develop a formalism that can handle even the most complex situations—those where intuition typically seems to break down—and that provides a thread throughout the text. To this end, it may prove useful to look as deeply as we can behind the appearance of simplicity, even if in the process we seem to complicate or even to obscure the most transparent concepts.

1. PRELIMINARY NOTION

To introduce the concept of mole balance equations which express mass conservation in chemical systems, let us start with a simple concrete example. Consider an aqueous solution containing a strong base, NaOH, and a strong acid, HCl ("strong" = completely dissociated in water).

Recipe $(H_2O)_T = 55.4\ M$ (molarity of water in water at $25°C$:
                          $\qquad 997\ g\ l^{-1}/18.01\ g\ mol^{-1} = 55.4\ M$)
                          $(NaOH)_T = 10^{-4}\ M$
                          $(HCl)_T = 10^{-3}\ M$

Reaction   The only reaction considered to take place in this system is that of the dissociation of water, symbolized as

$$H_2O = H^+ + OH^- \tag{1}$$

(The reactions of dissociation of $NaOH : NaOH \rightarrow Na^+ + OH^-$, and $HCl : HCl \rightarrow H^+ + Cl^-$, are considered to have gone to completion, and are not properly "taking place" in the system. There is no equilibrium between reactants and products. These reactions are implicitly included in the description of the system by considering the species Na^+, OH^-, H^+, and Cl^- to exist in the system but not NaOH or HCl.)

Species $H_2O,\ H^+,\ OH^-,\ Na^+,\ Cl^-$.

[The molar concentrations of the species are symbolized by parentheses: (H_2O), (H^+), etc.]

We can obtain a mathematical description of mass conservation in this system by expressing that transmutation does not occur: namely, the concentration of each element present in the various species in the system is equal to the concentration of the element present in the compounds that enter in the recipe of the system:

$$TOTH = 2(H_2O) + (H^+) + (OH^-) = 2(H_2O)_T + (HCl)_T + (NaOH)_T \tag{2}$$

$$\cong 110.8\ M$$

$$TOTO = (H_2O) + (OH^-) = (H_2O)_T + (NaOH)_T \tag{3}$$

$$\cong 55.4\ M$$

$$TOTNa = (Na^+) = (NaOH)_T \tag{4}$$

$$= 10^{-4}\ M$$

$$TOTCl = (Cl^-) = (HCl)_T \tag{5}$$

$$= 10^{-3}\ M$$

In this system the mole balance Equations 2, 3, 4, and 5, which are based on the principle of elemental conservation, provide a perfectly good description of mass conservation.

Suppose, however, that we add $10^{-4}\ M$ of dissolved oxygen to our basic recipe and that there is no further reaction. The recipe is modified by adding $(O_2)_T = 10^{-4}\ M$, and the species list by adding $O_2(aq)$. The conservation equations for H, Na, and Cl (Equations 2, 4, and 5) are obviously unchanged, while that for O becomes

$$TOTO = (H_2O) + (OH^-) + 2(O_2 \cdot aq) = (H_2O)_T + (NaOH)_T + 2(O_2)_T \tag{6}$$

$$\cong 55.4\ M$$

Although Equations 2, 4, 5, and 6 are still perfectly valid, they no longer provide an adequate expression of conservation in this system. Since the added oxygen has not reacted, it must be conserved, and the equation

$$(O_2 \cdot aq) = (O_2)_T = 10^{-4}\ M \tag{7}$$

must also be applicable. Equation 7 is independent of the four equations of elemental conservation, Equations 2, 4, 5, and 6, which are thus insufficient to describe mass conservation in the new system.

Conversely, suppose that we add instead $10^{-4}\ M$ hydrogen cyanide to our basic recipe (without O_2). HCN is a weak acid that dissociates into H^+ and CN^-:

Additional ingredient in recipe $\quad (HCN)_T = 10^{-4}\ M$

Additional reaction $\quad\quad\quad\quad\quad HCN = H^+ + CN^-$

Additional species $\quad\quad\quad\quad\quad\quad HCN,\ CN^-$

The elemental conservation equations for O, Na, and Cl (3, 4, and 5) are obviously unchanged, while that for H becomes

$$TOTH = 2(H_2O) + (H^+) + (OH^-) + (HCN) = 2(H_2O)_T + (HCl)_T \tag{8}$$

$$+ (NaOH)_T + (HCN)_T \cong 110.8\ M$$

In addition we can write conservation equations for C and N:

$$TOTC = (HCN) + (CN^-) = (HCN)_T = 10^{-4}\ M \tag{9}$$

$$TOTN = (HCN) + (CN^-) = (HCN)_T = 10^{-4}\ M \tag{10}$$

In this case Equations 3, 4, 5, 8, 9, and 10 do provide a complete description of mass conservation in the system, but they do not constitute a proper set of mathematical expressions since Equations 9 and 10 are identical. In this system, expressions of conservation for carbon and nitrogen are not independent because these two elements only occur together as the combination (CN) in the cyanide species.

As seen in this simple example, conservation of elements provides a first intuitive notion for conservation equations in chemical systems: mole balance equations are obtained by writing that what is present in the system (species concentrations) is somehow equal to what is put into the system (recipe). In some cases, however, the equations resulting from elemental conservation are insufficient to describe properly the conservation principle, while in others they are in fact redundant. There is thus more generality and subtlety to mole balance equations than mere elemental conservation. To improve our understanding of mass conservation in chemical systems, let us then forego elemental symbolism for a while and make the effort of working through a purely abstract example.

2. DEFINITION OF COMPONENTS

Consider a hypothetical system containing only three species, A, B, and C with only one possible variation that can occur, the progress to the right or to the left of the chemical reaction

$$2A + 3B = C \tag{11}$$

The numbers 2, 3, and 1 (implicit for C) are known as the *stoichiometric coefficients* of the reaction, and (A), (B), and (C) are the number of moles of A, B, and C in the system, respectively. Such a system that does not exchange matter with its surroundings is called a *closed system*. We wish to formulate mathematically the conservation principle applicable to this system, to obtain an algebraic expression of the idea that what is put into the system must remain there regardless of the progress of the chemical reaction.

Suppose first that the system is made up by mixing a given number of moles of A (A_T) with a given, smaller, number of moles of B (B_T). After the reaction has proceeded to the point where the system contains (C) moles of C, using up in the process $2(C)$ moles of A and $3(C)$ moles of B, the following conservation equations must apply:

$$TOTA = (A) + 2(C) = A_T \tag{12a}$$

$$TOTB = (B) + 3(C) = B_T \tag{12b}$$

These two equations are known as *mole balance equations*, and the two species A and B whose conservation is expressed in these equations are the *components* of the system. The components are effectively "reference species," and on the basis of these all other species can be formulated. The two mole balance Equations 12a and 12b simply reflect the stoichiometric formula of C as a function of A and B, which according to Reaction 11 is

$$C = (A)_2(B)_3$$

Consider now another situation in which the system is made up by mixing a certain number of moles of A (A'_T) with a certain number of moles of C (C'_T), leaving the initial concentration of B at zero. It is apparent that this second system would have the same overall composition as the first one if we choose A'_T and C'_T so that

$$A'_T + 2C'_T = A_T \tag{13a}$$

$$3C'_T = B_T \tag{13b}$$

After the dissociation of C has proceeded to the point where (B) moles of B have been formed along with $\frac{2}{3}(B)$ additional moles of A, while $\frac{1}{3}(B)$ moles of C have been used up, the conservation equations for the system can be written

$$TOTA = (A) - \tfrac{2}{3}(B) = A'_T \tag{14a}$$

$$TOTC = (C) + \tfrac{1}{3}(B) = C'_T \tag{14b}$$

We have two new mole balance equations corresponding to the new components (reference species) A and C and expressing the stoichiometry of B as a function of A and C:

$$B = (C)_{1/3}(A)_{-2/3}$$

These mole balance equations are not independent of the previous ones, and we can verify that they are mathematically equivalent when the system has the same overall composition: a linear combination of Equations 14a and 14b according to 13a and 13b leads to Equations 12a and 12b. Both sets of components, A & B and A & C, are equally appropriate and lead to mathematically equivalent sets of mole balance equations.

To provide an intuitive interpretation of the notion of components, so far we have chosen our components to be the chemical compounds that make up the system, A and B or A and C. This may in fact be quite misleading as components are only accounting entities and need not correspond to the compounds that enter in the recipe of a system. Consider, for example, that the system is made up by mixing A''_T, B''_T, and C''_T moles of A, B, and C, respectively. First, choosing A and B as our components, and reasoning as before, we find that the mole balance Equations 12a and 12b still apply, with a modified right-hand side:

$$TOTA = (A) + 2(C) = A''_T + 2C''_T$$

$$TOTB = (B) + 3(C) = B''_T + 3C''_T$$

A similar modification of the mole balances given by Equations 14a and 14b is obtained if we choose A and C as our components:

$$TOTA = (A) - \tfrac{2}{3}(B) = A''_T - \tfrac{2}{3}B''_T$$

$$TOTC = (C) + \tfrac{1}{3}(B) = C''_T + \tfrac{1}{3}B''_T$$

Regardless of the recipe of the system, the two components sets, A & B or A & C, provide equally valid accounting currencies, and the corresponding mole balance equations are mathematically equivalent expressions of conservation in the system.

It is important to note that the components of a chemical system need not even represent actual chemical compounds. To emphasize this abstract nature of components, consider the quantities $TOTX$ and $TOTY$ defined by a linear combination of Equations 12a and 12b:

$$TOTX = \tfrac{2}{3}TOTA - \tfrac{1}{3}TOTB = \tfrac{2}{3}(A) - \tfrac{1}{3}(B) + \tfrac{1}{3}(C)$$

$$= \text{constant} \tag{15a}$$

$$TOTY = -\tfrac{1}{3}TOTA + \tfrac{2}{3}TOTB = -\tfrac{1}{3}(A) + \tfrac{2}{3}(B) + \tfrac{4}{3}(C)$$

$$= \text{constant} \tag{15b}$$

Given that this system of equations is again equivalent to those of 12 and 14, what are the components X & Y whose conservation is expressed by 15a and 15b? The chemical species A which has a coefficient of $\tfrac{2}{3}$ in 15a must "contain" $\tfrac{2}{3}X$, whatever X may be. In the same way it must contain $-\tfrac{1}{3}Y$. The stoichiometric expression of A as a function of the components X & Y is then

$$A = (X)_{2/3}(Y)_{-1/3}$$

Following the same reasoning, we obtain the expressions of B and C:

$$B = (X)_{-1/3}(Y)_{2/3}$$

$$C = (X)_{1/3}(Y)_{4/3}$$

Vice versa, the stoichiometry of X and Y is derived by appropriate combination of these expressions:

$$X = (A)_2(B)_1$$

$$Y = (A)_1(B)_2$$

The exact nature of these two components, X and Y, may seem fairly mysterious at this point since they do not correspond to compounds used in the recipe of the system nor even to species considered to exist in the system. Nonetheless, X and Y are perfectly acceptable "reference species"; they define a proper set of accounting currencies, as would any two entities that provide a unique stoichiometric formula for each species in the system.

With the previous example in mind, we now define components as *a set of chemical entities that permits a complete description of the stoichiometry of the system*. Two equivalent concepts are contained in this definition:

1   The mole balances corresponding to components must provide a complete expression of conservation for the chemical system (our goal in this chapter).

2   Components must provide a complete and unique stoichiometric formula for each chemical species in the system.

3. PROPERTIES OF COMPONENTS

Although the example of the preceding section provides an initial handle on the notion of components and illustrates some of their key properties, our final definition is nonetheless somewhat fuzzy. In lieu of developing a systematic and somewhat formidable algebraic formalism to remedy this situation, let us examine in greater detail the exact nature of components, what they are, and are not, in order to sharpen the concept and make it more useful in practice.

3.1 Components Are Not (Necessarily) Elements

The paradigm of the elemental composition of matter, as formalized in universally accepted chemical formulae for all chemical species, is so central to modern chemistry that it is often identified with the field of chemistry itself. Yet this paradigm may well be the greatest conceptual obstacle to our present purpose: the understanding of components and mass conservation in chemical systems. Because "balancing" chemical reactions is *prima facie* an expression of the conservation of elemental mass, chemical elements are seen intuitively as the fundamental conservative entities. As chemical accountants we take grams or moles of H, C, N, O, Fe and so on, as our basic currencies. Therein lies the first misconception. To dispel it requires almost an unlearning of our clearest chemical concepts, a backward use of our intellectual time machine, viewing chemistry (as did J. W. Gibbs) without a complete knowledge of elements or elemental composition. For example, when we write the chemical reaction

$$3H^+ + PO_4^{3-} = H_3PO_4 \tag{16}$$

we demonstrate more knowledge than the reaction itself is supposed to formalize. If asked, at this point, for the "formula" of phosphoric acid, we would undoubtedly write

phosphoric acid = 3 hydrogen atoms + 1 phosphorus atom

+ 4 oxygen atoms

This is a direct translation of our symbolic chemical formalism, a piece of information truly extraneous to the writing of the chemical reaction itself which expresses only that 3 hydrogen ions and 1 phosphate ion combine to form 1 phosphoric acid. The corresponding formula is

phosphoric acid = 3 hydrogen ions + 1 phosphate ion

that is,

$$H_3PO_4 = (H^+)_3(PO_4^{3-})_1$$

The correct stoichiometry for the reaction expresses the conservation of H^+ and PO_4^{3-}, not that of H, P, and O. We need not even know that these elements exist. With the proper names for each symbol the reaction

$$3X + Y = Z \tag{17}$$

expresses the same chemical reality. Organic compounds provide perhaps more familiar examples. The reaction of formation of an organic acid

$$H^+ + Ac^- = HAc \qquad (18)$$

clearly expresses the conservation of the hydrogen ion and of the organic radical regardless of the exact formula of the latter. By using a nonelemental symbol for compactness, we have made the true stoichiometric nature of the reaction more apparent.

In any closed chemical system (defined as a set of chemical species and a set of chemical reactions among these species) at fixed temperature and pressure, all possible changes are those in the concentration of species due to the progress of one or several reactions. The conservative quantities in such a system are thus highly dependent upon the set of reactions considered. The chemical components that we can choose as basic accounting currencies, to insure that mass is conserved, are then strictly subordinate to the possible reactions among the species, not to the elemental composition of the species. Only if the formation or dissociation of the species from or into their elements were considered as possible reactions in the system would the elements themselves automatically become possible choices for components. These distinctions between elements and components may appear now as pointless nit-picking, however, we shall see later (Section 4) that they are essential. The point being made here can be expressed simply: *All the stoichiometric information relevant to mass conservation in a chemical system is contained in the stoichiometry of the possible reactions, no more, no less.*

It is also important to emphasize that, with respect to mass conservation, chemical reactions are strictly stoichiometric relationships: so many moles of each reactant yield so many moles of each product. The actual reaction process, the path of the reaction, might be very different from that implied by its symbolic representation. For example, one or several intermediary compounds not explicitly shown may be formed during the reaction. As a consequence chemical reactions may be added together, subtracted from each other, and subjected to all linear operations to yield new reactions equally valid for the chemical system under consideration (validity being determined from the point of view of mass conservation and energetic principles).

3.2 Basic Components

Having demolished our foundation of familiar chemical concepts, can we replace it by an equivalent one of the same compelling intuitiveness? Given the set of species and reactions considered in aquatic chemistry, can we define a set of simple building blocks to replace the repudiated elements? The correct answer is no, in principle, as the set of reactions considered to take place, and hence the possible sets of conservative components, are dependent on the particular system being studied. Nonetheless, we shall see that most species considered in aquatic chemistry (and aqueous chemistry) are in fact combinations of simple building blocks, such as

$$H^+, Na^+, K^+, Ca^{2+}, Mg^{2+}, Cu^{2+}, Co^{2+}, Fe^{3+}, Al^{3+} \quad \text{(metals)}$$

and

$$OH^-, Cl^-, NH_3, SO_4^{2-}, NO_3^-, PO_4^{3-}, SiO_3^{2-}, Ac^- \quad \text{(ligands)}$$

In many cases these building blocks can be taken as *basic components* for the system, and a complete expression of mass conservation will be obtained as the corresponding set of mole balance equations. The major exceptions to this rule are encountered in the study of redox reactions.

For example, consider an aqueous system containing the species Hg^{2+}, Cl^-, $HgCl_3^-$, and $HgCl_4^{2-}$ where only the following two reactions are assumed to take place:

$$Hg^{2+} + 3Cl^- = HgCl_3^- \tag{19}$$

$$Hg^{2+} + 4Cl^- = HgCl_4^{2-} \tag{20}$$

Choosing Hg^{2+} and Cl^- as the basic components for this system, we obtain the stoichiometric formulae:

$$Hg^{2+} = (Hg^{2+})_1$$

$$Cl^- = (Cl^-)_1$$

$$HgCl_3^- = (Hg^{2+})_1(Cl^-)_3$$

$$HgCl_4^{2-} = (Hg^{2+})_1(Cl^-)_4$$

The corresponding mole balance equations expressing the conservation of the components Hg^{2+} and Cl^- are

$$TOTHg^{2+} = (Hg^{2+}) + (HgCl_3^-) + (HgCl_4^{2-}) = \text{constant} \tag{21}$$

$$TOTCl^- = (Cl^-) + 3(HgCl_3^-) + 4(HgCl_4^{2-}) = \text{constant} \tag{22}$$

where each of the quantities above represents a concentration in moles per liter. A compact and convenient way to represent the relationship between components, species, formulae, and mole balance equations is to organize the stoichiometric data in the form of a "tableau," for example, as shown in Tableau 1.1.

TABLEAU 1.1

Components →	Hg^{2+}	Cl^-
Species ↓		
Hg^{2+}	1	0
Cl^-	0	1
$HgCl_3^-$	1	3
$HgCl_4^{2-}$	1	4

The stoichiometric formulae of the species as a function of the components are written horizontally (rows); the coefficients of the mole balance equations are then given vertically (columns), as can be seen by comparing the first column with Equation 21 and the second column with Equation 22.

3.3 Independence of Components

If, in the previous example, we had replaced the second reaction by an equivalent one showing the stepwise coordination of the mercuric ion with chloride:

$$Hg^{2+} + 3Cl^- = HgCl_3^- \tag{23}$$

$$HgCl_3^- + Cl^- = HgCl_4^{2-} \tag{24}$$

we might have been tempted to look no further than the list of reactants and to choose Hg^{2+}, Cl^-, *and* $HgCl_3^-$ as our components for the system. We would have been faced, however, with the dilemma of expressing $HgCl_3^-$ either as $(HgCl_3^-)_1$ or as $(Hg^{2+})_1(Cl^-)_3$. The stoichiometric representation of the system as a function of our components would not have been unique, and we would not have had a proper accounting system to express conservation of mass.

Although this necessity of "independence" of components does not always take such an obvious form, we can make it a general rule: *no component should be expressible as a formula of other components*, or equivalently, *a set of components should yield a unique formula for each species in the system.**

Note also that, if we had considered all three possible reactions simultaneously,

$$Hg^{2+} + 3Cl^- = HgCl_3^- \tag{25}$$

$$Hg^{2+} + 4Cl^- = HgCl_4^{2-} \tag{26}$$

$$HgCl_3^- + Cl^- = HgCl_4^{2-} \tag{27}$$

our choice of components and the corresponding mole balance equations in Tableau 1.1 would still be valid. Indeed the third equation provides no independent stoichiometric information; it is merely obtained by subtracting the first two.

3.4 Components Are Defined as a Set

In the second section of the chapter we have seen that several alternative choices are possible—and a priori equivalent—for the components of a system. The resulting sets of mole balance equations are merely linear transformations of each other.

If we consider a given component, say, Cl^- in our current example, the corresponding mole balance equation is strictly dependent on the choice of the other component(s). Choosing then the three sets of components: Cl^- & Hg^{2+},

---

* For mathematically inclined readers it becomes clear at this point that we are defining a set of components as a basis in the vector space of chemical species. This linear algebraic nature of chemical systems will be transparent throughout this chapter and the next.

TABLEAUX 1.2

	(a)		(b)		(c)	
	Cl^-	Hg^{2+}	Cl^-	$HgCl_3^-$	Cl^-	$HgCl_4^{2-}$
Cl^-	1	0	1	0	1	0
Hg^{2+}	0	1	-3	1	-4	1
$HgCl_3^-$	3	1	0	1	-1	1
$HgCl_4^{2-}$	4	1	1	1	0	1

TABLEAUX 1.3

	(a)		(b)	
	Cl^-	Hg^{2+}	$HgCl_3^-$	Hg^{2+}
Cl^-	1	0	1/3	$-1/3$
Hg^{2+}	0	1	0	1
$HgCl_3^-$	3	1	3/3	0
$HgCl_4^{2-}$	4	1	4/3	$-1/3$

Cl^- & $HgCl_3^-$, Cl^- & $HgCl_4^{2-}$, we can form Tableaux 1.2. By inspection of the first column of each tableau, we can see that the only coefficient in common among the three mole balances for Cl^- is that of Cl^- itself.

Conversely, two completely different sets of components can yield an identical mole balance equation. For example, the two sets of components Cl^- & Hg^{2+} and $HgCl_3^-$ & Hg^{2+} result in the two Tableaux 1.3. The mole balances for $TOTCl^-$ in the first case (first column of a) and for $TOTHgCl_3^-$ in the second case (first column of b) are identical, after adjustment by a factor of three.

Components and their corresponding mole balance equations are thus defined as a set, *not* as individuals, and any individual species can always be made a component.

3.5 Components Need Not Be Species

So far, to stay within what is chemically acceptable, we have chosen components that were also existing species in the system (except of course in Section 1 where we considered the conservation of elements). Such a limitation is not in fact necessary, and a chemical entity such as $HgCl_2$, which we have not considered as an existing species in the system, can quite properly be chosen as a component, as shown in Tableau 1.4. To go further, there is in fact no need for components to have any sort of chemical significance as seen in Tableau 1.5. The components Hg_3Cl^{5+} and $HgCl_{100}^{98-}$ correspond to no chemical reality whatsoever, yet they define a perfectly acceptable accounting currency for the species in our system. Because an accounting method only requires information relating to the number of moles, not to actual chemical processes, the components and the stoichiometric formulation of the species need not have any

TABLEAU 1.4

	Cl^-	$HgCl_2$
Hg^{2+}	-2	1
Cl^-	1	0
$HgCl_3^-$	1	1
$HgCl_4^{2-}$	2	1

TABLEAU 1.5

	Hg_3Cl^{5+}	$HgCl_{100}^{98-}$
Hg^{2+}	$100/299$	$-1/299$
Cl^-	$-1/299$	$3/299$
$HgCl_3^-$	$97/299$	$8/299$
$HgCl_4^{2-}$	$96/299$	$11/299$

structural or chemical meaning. Note that the four "reactions" corresponding to each line of the tableau have no more chemical significance than the components that they contain. However, by linear combinations of these four reactions, the original reactions may be readily obtained.

3.6 Number of Components

It is rather intuitive at this point that the number of components (N_C) for a given chemical system is independent of the particular set of components chosen and is in fact equal to the number of species (N_S) minus the number of (independent) chemical reactions (N_R) among these species. Indeed, since each reaction defines a stoichiometric relationship among species and allows the expression of one of them as a formula of the others, the minimum number of species necessary to formulate all others is precisely $N_C = N_S - N_R$. There is one caveat to this rule: the reactions must be stoichiometrically independent. Even if they correspond to separate chemical processes taking place simultaneously in a system, reactions or sets of reactions that contain the same stoichiometric information are redundant from the point of view of mass conservation. For example, of the seven reactions

$$CO_2(g) = CO_2(aq) \tag{28}$$

$$CO_2(aq) + H_2O = H_2CO_3 \tag{29}$$

$$CO_2(aq) + OH^- = HCO_3^- \tag{30}$$

$$CO_2(aq) + OH^- = CO_3^{2-} + H^+ \tag{31}$$

$$H_2CO_3 = HCO_3^- + H^+ \tag{32}$$

$$HCO_3^- = CO_3^{2-} + H^+ \tag{33}$$

$$H_2O = H^+ + OH^- \tag{34}$$

only five are independent, yielding three components for this system (eight species). Note that in this example, if we did not consider the reaction between the gas and the aqueous phase to take place $[CO_2(g) = CO_2(aq)]$, say, for reasons of slow kinetics, then the system would include four components, not three. The gaseous species $CO_2(g)$ would not be related to any of the aqueous species, and *two separate* mole balance equations would have to be written for carbon in the gas and the aqueous phases.

4. EXAMPLES OF CHEMICAL ACCOUNTING

After pondering in a general way the notions of components and mole balances, let us now come to grips with their practical application and study how the conservation principle is applied to real, rather than purely conceptual chemical systems. For this purpose, and for consistency with the following chapters, we consider two examples of the chemical equilibrium problem posed in its most classical form:

What is the equilibrium composition of a system, given (1) a complete recipe for how the system has been "made up," (2) a complete list of possible chemical species, and (3) a complete set of possible chemical reactions (and their equilibrium constants)?

We do not intend to solve such problems at this stage but merely to express conveniently a complete set of conservation equations.

4.1   Example 1

Before working through the details of a more suitable example, let us simply see how the methodology can be applied to our initial example (Section 1).

	A	B	C
Recipes	$(H_2O)_T = 55.4\ M$	$(H_2O)_T = 55.4\ M$	$(H_2O)_T = 55.4\ M$
	$(NaOH)_T = 10^{-4}\ M$	$(NaOH)_T = 10^{-4}\ M$	$(NaOH)_T = 10^{-4}\ M$
	$(HCl)_T = 10^{-3}\ M$	$(HCl)_T = 10^{-3}\ M$	$(HCl)_T = 10^{-3}\ M$
		$(O_2)_T = 10^{-4}\ M$	$(HCN)_T = 10^{-4}\ M$
Species	$H_2O, H^+, OH^-,$	$H_2O, H^+, OH^-,$	$H_2O, H^+, OH^-, Na^+,$
	Na^+, Cl^-	$Na^+, Cl^-, O_2(aq)$	Cl^-, HCN, CN^-
Reactions	$H_2O = H^+ + OH^-$	$H_2O = H^+ + OH^-$	$H_2O = H^+ + OH^-$
			$HCN = H^+ + CN^-$

TABLEAU 1.6a

		Components			
		H^+	OH^-	Na^+	Cl^-
Species	H_2O	1	1		
	H^+	1			
	OH^-		1		
	Na^+			1	
	Cl^-				1
Recipe	H_2O	1	1		
	NaOH		1	1	
	HCl	1			1

For the first system, A, the simple "building blocks" H^+, OH^-, Na^+, and Cl^- provide a suitable basic component set from which we obtain Tableau 1.6a (leaving blanks for zero). At the bottom of this tableau we have included the list of chemicals used in the recipe of the system and written in each corresponding line their stoichiometric coefficients as a function of the chosen components. It is indeed imperative that the same currencies be employed throughout the accounting process, that the same components be used to express both the compounds that make up the system and the species present in the system. The equations of conservation between what is introduced into the system—the recipe—and what is found at equilibrium can then be obtained from the columns of the tableau as mole balances:

$$TOTH^+ = (H_2O) + (H^+) = (H_2O)_T + (HCl)_T \cong 55.4\ M \tag{35}$$

$$TOTOH^- = (H_2O) + (OH^-) = (H_2O)_T + (NaOH)_T \cong 55.4\ M \tag{36}$$

$$TOTNa^+ = (Na^+) = (NaOH)_T = 10^{-4}\ M \tag{37}$$

$$TOTCl^- = (Cl^-) = (HCl)_T = 10^{-3}\ M \tag{38}$$

We can easily verify that this set of Equations 35 through 38 is equivalent to the original Equations 2 through 5.

In the same way the component sets H^+, OH^-, Na^+, Cl^-, O_2, and H^+, OH^-, Na^+, Cl^-, CN^-, lead to Tableaux 1.6b and 1.6c. The two corresponding sets of mole balance Equations, 39 through 43 and 44 through 48, obtained from the columns of each of these tableaux

$$TOTH^+ = (H_2O) + (H^+) = (H_2O)_T + (HCl)_T \cong 55.4\ M \tag{39}$$

$$TOTOH^- = (H_2O) + (OH^-) = (H_2O)_T + (NaOH)_T \cong 55.4\ M \tag{40}$$

$$TOTNa^+ = (Na^+) = (NaOH)_T = 10^{-4}\ M \tag{41}$$

$$TOTCl^- = (Cl^-) = (HCl)_T = 10^{-3}\ M \tag{42}$$

$$TOTO_2 = (O_2 \cdot aq) = (O_2)_T = 10^{-4}\ M \tag{43}$$

TABLEAU 1.6b

		Components				
		H^+	OH^-	Na^+	Cl^-	O_2
Species	H_2O	1	1			
	H^+	1				
	OH^-		1			
	Na^+			1		
	Cl^-				1	
	$O_2(aq)$					1
Recipe	H_2O	1	1			
	NaOH		1	1		
	HCl	1			1	
	O_2					1

TABLEAU 1.6c

		Components				
		H^+	OH^-	Na^+	Cl^-	CN^-
Species	H_2O	1	1			
	H^+	1				
	OH^-		1			
	Na^+			1		
	Cl^-				1	
	HCN	1				1
	CN^-					1
Recipe	H_2O	1	1			
	NaOH		1	1		
	HCl	1			1	
	HCN	1				1

and

$$TOTH^+ = (H_2O) + (H^+) + (HCN)$$

$$= (H_2O)_T + (HCl)_T + (HCN)_T \cong 55.4 \ M \tag{44}$$

$$TOTOH^- = (H_2O) + (OH^-) = (H_2O)_T + (NaOH)_T \cong 55.4 \ M \tag{45}$$

$$TOTNa^+ = (Na^+) = (NaOH)_T = 10^{-4} \ M \tag{46}$$

$$TOTCl^- = (Cl^-) = (HCl)_T = 10^{-3} \ M \tag{47}$$

$$TOTCN^- = (HCN) + (CN^-) = (HCN)_T = 10^{-4} \ M \tag{48}$$

can be verified to be equivalent to the two original sets of Equations 2, 4, 5, 6, 7, and 3, 4, 5, 8, 9, respectively. These sets of mole balance equations are necessary and sufficient expressions of mass conservation in these systems.

4.2   Example 2

Suitable Choices of Components

To illustrate the writing of mole balances in more detail, let us choose an example that is important in aquatic systems—a sulfate and hydrogen sulfide bearing water.

Posing the Problem

Recipe   0.0341 g of H_2S have been bubbled into 1 liter of water containing 1.42 g of Na_2SO_4.

Species Na^+, SO_4^{2-}, HS^-, H_2S, S^{2-}, H^+, OH^-, H_2O (all aqueous).

Reactions

$$S^{2-} + H_2O = HS^- + OH^- \tag{49}$$

$$HS^- + H_2O = H_2S + OH^- \tag{50}$$

$$H^+ + OH^- = H_2O \tag{51}$$

$$HS^- + H^+ = H_2S \tag{52}$$

$$S^{2-} + 2H^+ = H_2S \tag{53}$$

$$S^{2-} + H^+ = HS^- \tag{54}$$

The first choice in an accounting system has to be that of units. We shall systematically use the molar scale and express all concentrations in moles per liter of solution. Considering the molecular weight of the various chemicals and the density of water at 25°C, we obtain:

$$(H_2O)_T = 55.4 \ M;$$

$$(H_2S)_T = \ 0.001 \ M;$$

$$(Na_2SO_4)_T = \ 0.010 \ M$$

The chemical reactions that are given are clearly not independent. For example, Reactions 49 and 50 can be obtained by subtracting Reaction 51 from 54 and 52, respectively. In the same way Reaction 52 can be obtained by subtracting 54 from 53. After some thoughtful elimination we are left with

$$H^+ + OH^- = H_2O \tag{51}$$

$$S^{2-} + 2H^+ = H_2S \tag{53}$$

$$S^{2-} + H^+ = HS^- \tag{54}$$

Use of Basic Components

Given a list of eight species and three independent reactions, we are seeking a set of five components that can be used to express all species in a unique way. A rather obvious choice is given by the *basic components*: H^+, OH^-, Na^+, SO_4^{2-}, and S^{2-} which yield Tableau 1.7a and the corresponding mole balances from each column of the tableau:

$$TOTH = (H^+) + (HS^-) + 2(H_2S) + (H_2O) \tag{55}$$

$$= (H_2O)_T + 2(H_2S)_T = 55.4\ M + 2(0.001\ M)$$

$$= 55.4\ M \text{ (approximately)}$$

$$TOTOH = (OH^-) + (H_2O) = (H_2O)_T = 55.4\ M \tag{56}$$

$$TOTNa = (Na^+) = 2(Na_2SO_4)_T = 2(0.01\ M) = 0.02\ M \tag{57}$$

$$TOTSO_4 = (SO_4^{2-}) = (Na_2SO_4)_T = 0.01\ M \tag{58}$$

$$TOTS = (S^{2-}) + (HS^-) + (H_2S) = (H_2S)_T = 0.001\ M \tag{59}$$

For simplicity, the charges are now omitted in the designation of the mole balance equations: $TOTH$, $TOTSO_4$, and so on. It is worth pointing out once more how different these mole balances are from expressions of elemental conservation. Note, for example, that conservation of sulfur is expressed in two separate expressions, one for sulfate (Equation 58) and one for sulfide (Equation 59). There are in fact five components and only four elements. Note also that in the $TOTH$ expression (Equation 55), the coefficient of (H_2S) is 2 while that of (H_2O) is 1, illustrating again that the coefficients of the mole balance

TABLEAU 1.7a

		Components				
		H^+	OH^-	Na^+	SO_4^{2-}	S^{2-}
	Na^+			1		
	SO_4^{2-}				1	
	H^+	1				
	OH^-		1			
Species	S^{2-}					1
	HS^-	1				1
	H_2S	2				1
	H_2O	1	1			
	H_2O	1	1			
Recipe	H_2S	2				1
	Na_2SO_4			2	1	

equations are not the stoichiometric coefficients of the elements but those of the components.

A More Convenient Choice of Components

Although the set of Equations 55 through 59 is a complete and coherent expression of conservation for our system, and any other set of mole balance equations has to be mathematically equivalent to it, it is not automatically the most convenient set, neither practically nor numerically. Consider, for example, Equations 55 and 56. Reasoning that none of the sulfide species can exceed 0.001 M (the total sulfide concentration in the system), and arguing that the hydrogen and hydroxyl ion concentrations, (H^+) and (OH^-), are much smaller than that of water (in anticipation of future chapters), we can neglect the small concentrations and simplify both equations to

$$(H_2O) = 55.4 \ M$$

This is hardly a surprising result for a dilute aqueous solution. However, in the process of making these approximations, which here are perfectly appropriate and result simply from rounding off errors, critical information has been lost regarding the conservation of protons (= hydrogen ions) in the system. The reason is seen by rearranging Equation 55:

$$(H^+) + (HS^-) + 2(H_2S) = 55.4 - (H_2O)$$

Such an expression which equates a small number to a difference between two large numbers typically leads to poor numerical behavior. One way out of this difficulty is to subtract Equation 56 from 55:

$$TOTH - TOTOH = (H^+) + (HS^-) + 2(H_2S) - (OH^-) = 2(H_2S)_T = 0.002 \ M$$

thus obtaining a numerically convenient equation.

To eliminate the possibility of errors, it would be helpful to obtain this equation directly without having to examine the mole balance equations in detail and to manipulate them. Since we want to eliminate the species (H_2O)—whose high concentration swamps all other—from all mole balance equations but one, a simple solution is to choose H_2O itself as one of the components. The formula of H_2O is then simply $H_2O = (H_2O)_1$ and, in the corresponding tableau, the species H_2O has a zero coefficient in all columns except that of the component H_2O. Here lies the true elegance of the concept of components: any component set provides a simple and systematic way of writing a complete set of mole balance equations; in addition a judicious choice of components leads directly to a *convenient* set of such mole balances. Let us, for example, choose H^+, H_2O, Na^+, SO_4^{2-}, and S^{2-} as our new components. From Tableau 1.7b we obtain the following mole balance equations:

$$TOTH = (H^+) - (OH^-) + (HS^-) + 2(H_2S) = 2(H_2S)_T = 0.002 \ M \quad (60)$$

$$TOTH_2O = (OH^-) + (H_2O) = (H_2O)_T = 55.4 \ M \quad (61)$$

TABLEAU 1.7b

		Components				
		H^+	H_2O	Na^+	SO_4^{2-}	S^{2-}
Species	Na^+			1		
	SO_4^{2-}				1	
	H^+	1				
	OH^-	-1	1			
	S^{2-}					1
	HS^-	1				1
	H_2S	2				1
	H_2O		1			
Recipe	H_2O		1			
	H_2S	2				1
	Na_2SO_4			2	1	

$$TOTNa = (Na^+) = 2(Na_2SO_4)_T = 0.02\ M \tag{62}$$

$$TOTSO_4 = (SO_4^{2-}) = (Na_2SO_4)_T = 0.01\ M \tag{63}$$

$$TOTS = (S^{2-}) + (HS^-) + (H_2S) = (H_2S)_T = 0.001\ M \tag{64}$$

The $TOTH$ equation now has the desired form. Modification of the set of components results in new mole balance expressions that are more convenient linear combinations of the old ones.

As we shall be dealing with dilute solutions, the reasoning carried out for this example will always be applicable and H_2O will *always be chosen as a component*. Since the resulting mole balance equation has a rather trivial solution $[(H_2O) = 55.4\ M]$, and since we shall see that the degree of hydration of the various species (i.e., their stoichiometric coefficient for H_2O) need not be considered for calculating the composition of the system, H_2O will be altogether omitted from the tableaux. (It is good practice, however, to include it explicitly until one is familiar with the use of components.) In other words, an arbitrary number of (H_2O)'s will be understood as included in the formulae of the various species. For example, we shall write

$$OH^- = (H^+)_{-1}$$

instead of

$$OH^- = (H^+)_{-1}(H_2O)_1$$

Together with this choice of H_2O as an "understood" component, H^+ will also be chosen as a component systematically, explicitly, and strictly by convention. The corresponding *proton conservation equation, TOTH*, will be the focus of much of our attention. As a first exercise in this matter, let us consider the situation of charge balance (electroneutrality) in our example.

The Electroneutrality Condition

Since the system has been made up of neutral chemical constituents, and no electrical charge can be created or lost in any of the reactions, the electroneutrality equation expressing the balance of positive and negative charges in the system has to be verified:

$$(Na^+) + (H^+) = 2(SO_4^{2-}) + (OH^-) + 2(S^{2-}) + (HS^-) \qquad (65)$$

If our set of mole balance equations, Equations 60 through 64, is truly complete, it must somehow contain Equation 65. This is verified by adding Equations 60 and 62 and subtracting twice both Equations 63 and 64.

By the same type of manipulation we used in obtaining Equation 60, we should be able to make a choice of components that would directly yield the electroneutrality expression as one of the mole balance equations. Keeping with our convention to use H^+ as a component, we can insure that each species will have a coefficient in the H^+ column equal to its electrical charge by choosing all other components (besides H^+) to be neutral. Under these conditions the *TOT*H equation should then logically be the electroneutrality expression. Let us choose, for example, the components H^+, NaOH, H_2SO_4, H_2S (and H_2O!) as shown in Tableau 1.7c. The corresponding *TOT*H expression is indeed the sought-for electroneutrality condition:

$$TOTH = (Na^+) - 2(SO_4^{2-}) + (H^+) - (OH^-) - 2(S^{2-}) - (HS^-) = 0$$

Note that in this example the chemical formulae of the various species as given by the lines of the tableau do not strictly correspond to chemical reactions since the components are not species actually present in the system.

TABLEAU 1.7c

		Components				
		H^+	NaOH	H_2SO_4	H_2S	(H_2O)
	Na^+	1	1			(-1)
	SO_4^{2-}	-2		1		
	H^+	1				
Species	OH^-	-1				(1)
	S^{2-}	-2			1	
	HS^-	-1			1	
	H_2S				1	
	H_2O					(1)
	H_2O					(1)
Recipe	H_2S				1	
	Na_2SO_4		2	1		(-2)

5. NOTATION, SYMBOLS, UNITS, RULES, AND TERMINOLOGY

Throughout the preceding three sections we have used particular units and introduced a few symbols. We have also stated a certain number of rules and have given the word "component" a special meaning. Let us compile here these various conventions while extending and explaining them a little more.

5.1 Equal Signs

It must be emphasized that the equal sign is used here for three different purposes:

1 To equate algebraic or arithmetic quantities.
2 To define stoichiometric formulae of compounds.
3 As a shorthand notation for chemical reactions (replacing $\rightleftarrows$).

For all three types of expressions written with an equal sign the numerical or chemical entities on each side of the corresponding "equation" can be freely subjected to all usual linear operations, such as multiplication by a scalar and addition of equations. See, for example, Table 1.1.

TABLE 1.1

Chemical Reactions	Equivalent Formulae
$1 \times CO_2 + H_2O = H_2CO_3$	$H_2CO_3 = (CO_2)_1(H_2O)_1$
$-1 \times CO_3^2 + 2H^+ = H_2CO_3$	$H_2CO_3 = (CO_3^{2-})_1(H^+)_2$
$2 \times H^- + OH^- = H_2O$	$H_2O = (H^+)_1(OH^-)_1$
$CO_2 + 2OH^- = CO_3^{2-} + H_2O$	$CO_3^{2-} = (CO_2)_1(OH^-)_2(H_2O)_{-1}$

5.2 Types of Concentration

Consider a system made up of a three millimolar concentration of NaCl, a one millimolar concentration of NaOH, and a two millimolar concentration of HCl, yielding the species H^+, OH^-, Na^+, Cl^- (and H_2O). Four different types of concentrations involving sodium may be defined for such a system.

1. $(NaCl)_T = 3 \times 10^{-3} M$ and $(NaOH)_T = 10^{-3} M$
These are simply the molar concentrations of the chemical compounds used in the recipe of the system. Neither NaCl nor NaOH is present as a species in this system.

2. (Na^+)

This is the molar concentration of the species Na^+ (sodium ion), and its numerical value is not known a priori. Although it is not the case here, other species involving sodium could in principle be present in the system (e.g., $NaCO_3^-$, $NaSO_4^-$).

3. *TOT*Na

This is the value of the mole balance equation for Na^+, if Na^+ is chosen as a component. As before, the symbol for the charge is omitted for simplicity. *TOT*Na is strictly dependent upon the particular choice for the other components and may be negative or null, although it has dimensions of moles per liter. For example, with the choices of components H^+, Na^+, Cl^-, or H^+, Na^+, NaCl, two vastly different values are obtained for *TOT*Na as seen in Tableaux 1.8:

$$(TOTNa)_a = (Na^+) = (NaCl)_T + (NaOH)_T = 3 \times 10^{-3} \ M + 10^{-3} \ M$$
$$= 4 \times 10^{-3} \ M \tag{66}$$

$$(TOTNa)_b = (Na^+) - (Cl^-) = (NaOH)_T - (HCl)_T = 10^{-3} \ M - 2 \times 10^{-3} \ M$$
$$= -10^{-3} \ M \tag{67}$$

Note that the mole balance equations consist of two equal quantities: the left-hand side (LHS), which is a sum of concentrations of species present in the system (the value of each of these is usually unknown); and the right-hand side (RHS), which is a sum of concentrations of chemical compounds used in the recipe of the system.

4. Na_T

This is the *analytical concentration* of sodium, not previously introduced in this chapter. The charge is again omitted for simplicity; a parenthesis may be used for clarity whenever the chemical formula is complicated, for example, $(CO_3)_T$. This concentration, sometimes called the *total concentration*, is defined as that which would be measured were we to analyze the system for sodium (or car-

TABLEAUX 1.8

	(a) H^+	Na^+	Cl^-	(b) H^+	Na^+	NaCl
H^+	1	0	0	1	0	0
OH^-	-1	0	0	-1	0	0
Na^+	0	1	0	0	1	0
Cl^-	0	0	1	0	-1	1
NaCl	0	1	1	0	0	1
NaOH	-1	1	0	-1	1	0
HCl	1	0	1	1	-1	1

bonate). It is equal to the sum of the concentrations of all species containing sodium—in our case only Na^+—and to the sum of the concentrations of the chemicals containing sodium used in the recipe of the system, if the system is so defined:

$$Na_T = (Na^+) = (NaCl)_T + (NaOH)_T = 4 \times 10^{-3} \, M \qquad (68)$$

From a comparison of Equations 66 and 68, it is clear that some of the more common choices for components yield the equality:

$$TOTNa = Na_T$$

This is true when none of the other components contain the species or element of interest. For example, in Chapter 4, we will use chiefly three alternative sets of components to study the carbonate system:

H^+, CO_3^{2-}, + other components containing no carbonate.
H^+, HCO_3^-, + other components containing no carbonate.
H^+, H_2CO_3, + other components containing no carbonate.

Then, for any given system,

$$TOTCO_3 = TOTHCO_3 = TOTH_2CO_3 = (CO_3)_T$$

Note that Concentrations **1** and **4** are conceptually related and the corresponding notations are similar. The numerical values of the other concentrations may sometimes be the same, as illustrated by Equations 66 and 68; however, they are based on different concepts and are designated by clearly different notations.

5.3   Concentration Units

So far we have expressed all concentrations on the molar scale; we shall continue to do so. It is further necessary to be aware of other commonly used concentration scales and to understand their interrelationships.

Weight Fraction

Symbols   % (percent); ‰ (per mil = parts per thousand); ppm (parts per million); ppb (parts per billion).

Dimensions   None.

Definition   The weight of the species or element of interest per total weight of the system. This is a commonly used analytical scale, particularly useful for solids (dry weight). In water its relation to the molar scale is a function of the molecular weight of the species in question and the density of the solution. In using this scale, it is critical to specify what particular species are being considered. Contemplate, for example, the not uncommon and truly ambiguous situation in which, say, a 3 ppm concentration of phosphate is reported. As P? As PO_4? As H_3PO_4? Such a unit that does not provide direct information on the number of chemical entities (atoms, molecules, ions) is inherently less chemically relevant than a unit based on the number of moles.

Volume Fraction

Symbols $\%$, etc.
Dimensions   None.
Definition   The volume of the species of interest per total volume of the system. Used chiefly for liquid mixtures such as concentrated acid solutions, this concentration scale is of little interest in aquatic chemistry.

Molal Concentration

Symbols m, mm, μm, etc.
Dimensions   mol kg^{-1}.
Definition   The number of moles of the species of interest per kilogram of *solvent*.

Molar Concentration

Symbols M, mM, μM, etc.
Dimensions   mol $liter^{-1}$.
Definition   The number of moles of the species of interest per liter of *solution*. The molal scale is in principle thermodynamically preferable to the molar scale because it is independent of the effects of temperature and solutes on the density of the solution or its molar volume.* However, the molar scale is analytically more convenient, and in aquatic chemistry, which deals mostly with dilute solutions over a small range of temperatures, the two scales are almost equivalent: at 20°C, with a total salt content of less than 3%, molar and molal concentrations differ by less than 1%. For convenience, in this text the symbol M is often omitted, particularly in the derivation of equations. The notation pX is used to indicate the negative logarithm of the molar concentration of X:

$$pX = -\log_{10}(X)$$

---

* In oceanography a hybrid unit, mole per kg of solution (seawater), is convenient and widely used.

Atom Concentrations

Symbols	g-at liter^{-1}, mg-at liter^{-1}, μg-at liter^{-1}, etc.
Dimensions	gram-atom liter^{-1}.
Definition	The number of gram-atoms of an element per liter of solution. This scale is equivalent to the molar scale but normalizes the concentration to the number of atoms of an element of interest rather than to the number of moles of species. For example, 1 M urea $[CO(NH_2)_2] = 2$ g-at liter^{-1} N. This scale is widely used by biologists interested in elemental ratios.

Equivalent Charge Concentration

Symbols	eq liter^{-1}, $m$eq liter^{-1}, μeq liter^{-1}, etc.
Dimensions	Equivalent liter^{-1} ($=96,500$ coulombs liter^{-1}).
Definition	The number of equivalent charges of a given ion per liter of solution. Again, this scale is similar to the molar scale, but here the molar concentration of a given ion is multiplied by the absolute charge number of the ion. For example, 1 M H$^+$, OH$^-$, HCO$_3^-$, or NO$_3^- = 1$ eq liter^{-1}; 1 M SO$_4^{2-}$, Ca^{2+}, or CO$_3^{2-} = 2$ eq liter^{-1}.

Mole Fraction

Symbols	%, etc.
Dimensions	None.
Definition	The number of moles of the species of interest per total number of moles in the system. This is the thermodynamic scale *par excellence*. Its occasional use is largely restricted—as it is here—to thermodynamic developments.

Other concentration units used for historical reasons or for particular applications are not widely encountered in aquatic chemistry.

5.4   Rules for Choosing Components

Let us summarize here the various rules we have discussed for choosing components for a particular chemical system.

General Requirements

A proper component set must be such that:

1   All species can be expressed stoichiometrically as a function of the components, the stoichiometry being defined by the chemical reactions.

2 Each species has an unique stoichiometric expression as a function of the components.

A necessary, but not sufficient, condition to fulfill these requirements is that the number of components be equal to the number of species minus the number of *independent* reactions considered to take place in the system.

Practical Rules

For obtaining a convenient set of mole balance equations:

3 H_2O should always be chosen as a component. Because the corresponding mole balance equation and the corresponding column in the tableau yield no useful information, in this text H_2O will be omitted from all tableaux. It is, however, always included implicitly in the component set, and it may be good practice for beginners to include it explicitly in the tableaux.

4 H^+ should always be chosen as a component.

For the other components, unless there are reasons to do otherwise, the simplest "building blocks" (basic components) can be chosen. In addition, if we wish for some reason to isolate a particular concentration in one equation, it is sufficient to choose the corresponding species as a component. Other practical rules will be derived in Chapter 3.

☐ ☐ ☐

5.5 A Matter of Terminology and History

There is some confusion in the literature regarding the meaning of the word "component." It is fairly common practice to limit the use of the word to designate uncharged species, those that can exist as individual salts or pure solutions. In this tradition, components are considered to be the minimum set of chemicals that have to be taken off the shelf to duplicate the system under consideration. We shall not follow this practice which appears to originate from a confusion regarding the meaning that Gibbs intended for the word "component."

To illustrate the differences between these two approaches, consider the simplest of all aqueous solutions, pure water. Three species, H_2O, H^+, and OH^-, and one reaction have to be considered:

$$H_2O = H^+ + OH^-$$

According to our definition, two components must be chosen, and following our newly set rules, these are H_2O and H^+ as shown in Tableau 1.9. In our approach,

TABLEAU 1.9

	H$^+$	(H$_2$O)
H$_2$O	0	(1)
H$^+$	1	(0)
OH$^-$	-1	(1)

two independent mole balance equations can be written, the second one being of little interest:

$$TOTH = (H^+) - (OH^-) = 0 \qquad (69)$$

$$TOTH_2O = (H_2O) + (OH^-) = 55.4 \ M \qquad (70)$$

If we limited the choice of components to chemicals necessary to make up the system, only one component would and could be chosen: H_2O. In order to obtain Equation 69, we would then impose the electroneutrality condition as an additional constraint on the system, independent of the conservation constraints. This appears somewhat illogical since electroneutrality results necessarily from the fact that the system is originally made up of an electrically neutral compound: H_2O. In fact there are of course chemical systems or subsystems that are not electrically neutral. Some conservation equation for electrical charge must then pertain in such systems while electroneutrality does not. Conservation of charge (i.e., electroneutrality in a neutral system) is included implicitly in the mole balance equations when the component set is complete and accounts for the electrical charge of the ions.

The historical justification for our use of the word "components" rests with the writings of the man who originally coined its chemical meaning. Although they may be difficult to fully understand out of their context, the following three quotes from J. W. Gibbs, which establish his own view of what components are, appear consistent with our use of the word.[1]

1.   All species must be expressible as a function of the components. (See Section 6 for the relationship between completeness of the component set and completeness of the set of possible variations.)

... The substances S_1, S_2 ... S_n of which we consider the mass composed, must of course be such that the values of the differentials dm_1, dm_2 ... dm_n shall be independent, and shall express every possible variation in the composition of the homogeneous mass considered, including those produced by the absorption of substances different from any initially present. It may therefore be necessary to have terms in the equation relating to component substances which do not initially occur in the homogeneous mass considered, provided, of course, that these substances, or their components, are to be found in some part of the whole given mass.

If the conditions mentioned are satisfied, the choice of the substances which we are to regard as the components of the mass considered, may be determined entirely by convenience, and independently of any theory in regard to the internal constitution of the mass. The number of components will sometimes be greater, and sometimes less, than the number of chemical elements present.

2. It is not necessary for a component to exist as a homogeneous chemical.

... In fact, we may give a definition of a potential which shall not presuppose any choice of a particular set of substances as the components of the homogeneous mass considered.

Definition. If to any homogeneous mass we suppose an infinitesimal quantity of any substance to be added, the mass remaining homogeneous and its entropy and volume remaining unchanged, the increase of the energy of the mass divided by the quantity of the substance added is the *potential* for that substance in the mass considered. (For the purpose of this definition, any chemical element or combination of elements in given proportions may be considered a substance, whether capable or not of existing by itself as a homogeneous body.)

3. Ions can be chosen as components.

... It will be observed that the choice of the substances which we regard as the components of the fluid is to some extent arbitrary, and that the same physical relations may be expressed by different equations of the form (682), in which the fluxes are expressed with reference to different sets of components. If the components chosen are such as represent what we believe to be the actual molecular constitution of the fluid, those of which the fluxes appear in the equation of the form (682) are called the *ions*, and the constants of the equation are called their *electro-chemical equivalents*. For our purpose, which has nothing to do with any theories of molecular constitution, we may chose such a set of components as may be convenient, and call those *ions*, of which the fluxes appear in the equation of the form (682), without farther limitation.

6. A DIFFERENTIAL APPROACH TO THE PROBLEM

As mentioned before, it is possible to develop rigorously the concept of chemical components in a mathematical framework. Although we do not wish to be so theoretical, it is instructive to examine the essential aspects of such a development.

In mathematical terms, a conservative quantity is a parameter that remains constant while all *possible variations* are considered in the system. In other words, it is a quantity whose total differential is identically null. The only possible variations in a closed chemical system at constant pressure and temperature are changes in the concentrations of the species due to the advancement of one or several reactions. Our purpose is then simply to formulate chemical quantities that are invariant with respect to the advancement of the reactions in a given system.

Consider our example of Section 2: three hypothetical species, A, B, C, and only one reaction:

$$2A + 3B = C \tag{11}$$

We can reason that the formation or the dissociation of C is the only possible variation in this system. The conservation principle, as symbolized by Reaction 11, is then readily written in differential form:

$$\frac{d(A)}{d(C)} = -2, \quad \frac{d(B)}{d(C)} = -3$$

and thus

$$d(A) + 2d(C) = 0, \quad d(B) + 3d(C) = 0$$

These exact differentials which define two invariant quantities—say, $dTOTA = 0$ and $dTOTB = 0$—are then integrated into familiar mole balance equations:

$$(A) + 2(C) = \text{constant} = TOTA$$

$$(B) + 3(C) = \text{constant} = TOTB$$

The chemical meaning of the invariants $TOTA$ and $TOTB$, and thus the definition of the components, is obtained by a reasoning similar to that presented in Section 2.

The situation is somewhat more complicated when more than one reaction is taking place. Consider our example of Section 3.2. For four species, Hg^{2+}, Cl^-, $HgCl_3^-$, and $HgCl_4^{2-}$, there were two reactions:

$$Hg^{2+} + 3Cl^- = HgCl_3^- \tag{71a}$$

$$Hg^{2+} + 4Cl^- = HgCl_4^{2-} \tag{71b}$$

The principle of conservation then needs to be expressed with partial derivatives relating the changes in the number of moles of each of the species to the advancement of each of the Reactions, 71a and 71b. This is most simply achieved by defining parameters ξ_1 and ξ_2 corresponding to the advancement, from the left to the right, of Reactions 71a and 71b, respectively:

$$\begin{aligned}
\frac{\partial(Hg^{2+})}{\partial\xi_1} = -1, \quad \frac{\partial(Cl^-)}{\partial\xi_1} = -3, \quad \frac{\partial(HgCl_3^-)}{\partial\xi_1} = +1, \quad \frac{\partial(HgCl_4^{2-})}{\partial\xi_1} = 0 \\
\frac{\partial(Hg^{2+})}{\partial\xi_2} = -1, \quad \frac{\partial(Cl^-)}{\partial\xi_2} = -4, \quad \frac{\partial(HgCl_3^-)}{\partial\xi_2} = 0, \quad \frac{\partial(HgCl_4^{2-})}{\partial\xi_2} = +1
\end{aligned} \tag{72}$$

The parameters ξ_1 and ξ_2 are actually defined here by the equations themselves and have units of mol liter^{-1}. Since Reactions 71a and 71b are the only ones considered to take place, the total differentials are written:

$$d(Hg^{2+}) = -d\xi_1 - d\xi_2 \tag{73a}$$

$$d(Cl^-) = -3d\xi_1 - 4d\xi_2 \tag{73b}$$

$$d(HgCl_3^-) = d\xi_1 \tag{73c}$$

$$d(HgCl_4^{2-}) = d\xi_2 \tag{73d}$$

We can combine these equations to eliminate the variables $d\xi_1$ and $d\xi_2$ as in the following:

$$(73a) + (73c) + (73d) = d(Hg^{2+}) + d(HgCl_3^-) + d(HgCl_4^{2-}) = 0$$

$$(73b) + 3(73c) + 4(73d) = d(Cl^-) + 3d(HgCl_3^-) + 4d(HgCl_4^{2-}) = 0$$

These exact differentials can now be integrated to yield the sought-for mole balance equations:

$$(Hg^{2+}) + (HgCl_3^-) + (HgCl_4^{2-}) = \text{constant} = TOTHg \qquad (74a)$$

$$(Cl^-) + 3(HgCl_3^-) + 4(HgCl_4^{2-}) = \text{constant} = TOTCl \qquad (74b)$$

Other combinations of Equations 73 to eliminate the variables $d\xi_1$ and $d\xi_2$ would result in different sets of mole balance equations that are linear combinations of Equations 74a and 74b.

The general problem involving a chemical system with N_S species, each with a concentration (S_i), and N_R reactions, each characterized by a degree of advancement ξ_j, would be obviously cumbersome to approach in this manner. A more elegant solution can be written with matrix algebra, but this is beyond the scope of this chapter. Here it is sufficient to note that we can always find $N_C = N_S - N_R$, and no more than N_C, independent ways to eliminate the $d\xi_j$ variables among the N_S total differentials of the species concentrations:

$$d(S_i) = \sum_j \frac{\partial(S_i)}{\partial \xi_j} d\xi_j$$

The coefficients of these N_C linear combinations to eliminate $d\xi_j$'s define the stoichiometry of the components. By integration, N_C mole balance equations are obtained that express fully the notion of mass conservation in the system.

Such a differential approach provides a fundamental mathematical definition of components and mole balance equations. However, we have seen that components of a chemical system can usually be defined intuitively and the stoichiometric relations between species, components, and mole balances are readily expressed in the form of a tableau. In effect this intuitive approach is no more than an efficient utilization of our chemical symbolism which expresses inherently the notion of mass conservation in the elemental stoichiometry of chemical species.

7. USE OF MOLE BALANCES IN TRANSPORT EQUATIONS

In natural waters chemical species are subjected to transport by advective and diffusive processes at the same time that they are reacting chemically. In order to help bridge the gap between aquatic chemists and hydrodynamicists, let us study how the concept of components can, in some instances, be used to simplify the problem of simultaneous transport and reactions of chemical species.

Notation

t	time
x	one-dimensional coordinate
U	advective velocity in the x direction
(S_k)	molar concentration of the kth species, $k = 1, 2, \ldots, N_S$
D_k	diffusion coefficient of S_k (turbulent diffusion coefficient if applicable)
R_{kj}	rate of production of S_k in the jth reaction, $j = 1, 2, \ldots, N_R$
$d\xi_j$	degree of advancement of the jth reaction
v_{jk}	stoichiometric coefficient of S_k in the jth reaction ($v > 0$ for products; $v < 0$ for reactants)
$TOTC_i$	mole balance equation for the component C_i, $i = 1, 2, \ldots, N_C$
α_{ki}	stoichiometric coefficient of S_k as a function of the component C_i

Tableau 1.10 is implied by this notation. The mole balance equation for the component C_i is thus written

$$TOTC_i = \sum_k \alpha_{ki}(S_k) \tag{75}$$

The condition of conservation for the component C_i in the jth chemical reaction symbolized by

$$0 = v_{j1}S_1 + v_{j2}S_2 + \cdots + v_{jN_S}S_{N_S}$$

results in the identity

$$\sum_k v_{jk}\alpha_{ki} = 0 \tag{76}$$

To keep the notation elementary, let us consider the one-dimensional transport problem and write the conservation equation:

$$\frac{\partial(S_k)}{\partial t} = -U\frac{\partial(S_k)}{\partial x} + D\frac{\partial^2(S_k)}{\partial x^2} + \sum_j R_{kj} \tag{77}$$

TABLEAU 1.10

	C_1	C_2	$\cdots$	C_i	$\cdots$
S_1	α_{11}	α_{12}	$\cdots$	α_{1i}	$\cdots$
S_2	α_{21}	α_{22}	$\cdots$	α_{2i}	$\cdots$
$\vdots$	$\vdots$	$\vdots$			
S_k	α_{k1}	α_{k2}		α_{ki}	
$\vdots$	$\vdots$	$\vdots$		$\vdots$	

The reaction rates (R_{kj}'s) can have complicated and often poorly known functionalities, so the solution to these N_S coupled differential equations can be very difficult to obtain even if the advective velocity field is obtained independently by neglecting the effects of the solutes on the solvent motion—a reasonable approximation in almost all cases.

From our definition of the degree of advancement, ξ_j, of the jth reaction, R_{kj} can be written more explicitly as

$$R_{kj} = \frac{\partial(S_k)}{\partial \xi_j} \frac{d\xi_j}{dt} \tag{78}$$

which results in

$$R_{kj} = v_{jk} \frac{d\xi_j}{dt} \tag{79}$$

Introducing Equation 79 in 77, multiplying by α_{ki} and summing over k,

$$\sum_k \alpha_{ki} \frac{\partial(S_k)}{\partial t} = -U \sum_k \alpha_{ki} \frac{\partial(S_k)}{\partial x} + \sum_k \alpha_{ki} D_k \frac{\partial^2(S_k)}{\partial x^2} + \sum_k \alpha_{ki} \sum_j v_{jk} \frac{d\xi_j}{dt} \tag{80}$$

Using Equations 75 and 76 and rearranging,

$$\frac{\partial TOTC_i}{\partial t} = -U \frac{\partial TOTC_i}{\partial x} + \sum_k D_k \frac{\partial^2(\alpha_{ki}S_k)}{\partial x^2} \tag{81}$$

The N_S equations describing the rate of change of the species concentrations have now been replaced by $N_C (<N_S)$ equations describing the transport of the component concentrations. The chemical reaction rates have been effectively eliminated from the equations. By judicious choice of components and reasonable approximations, Equation 81 can be made to have particularly simple forms and convenient boundary conditions.

It is, for example, possible to choose components so as to ensure that all species in a given mole balance equation are soluble. It is then sometimes a good approximation to take all diffusion coefficients as equal:

$$D_k = D \tag{82}$$

$$\frac{\partial TOTC_i}{\partial t} = -U \frac{\partial TOTC_i}{\partial x} + D \frac{\partial^2 TOTC_i}{\partial x^2} \tag{83}$$

Equation 83 (which could be obtained in three dimensions as well as in one) is surprisingly simple. It expresses the equation of transport of the component C_i as a *conservative* entity. The only condition for the applicability of such an equation is that the diffusion coefficients be approximately equal for all species in the corresponding mole balance equation.

Let us take as an example the dissolution of mercuric sulfide in a high chloride medium (say, 0.5 M NaCl). The reactions to be considered are

$$HgS(s) = Hg^{2+} + S^{2-}$$

$$S^{2-} + H^+ = HS^-$$

$$HS^- + H^+ = H_2S(aq)$$

$$H_2S(aq) = H_2S(g)$$

$$H_2O = H^+ + OH^-$$

$$Hg^{2+} + 2Cl^- = HgCl_2$$

$$HgCl_2 + Cl^- = HgCl_3^-$$

$$HgCl_3^- + Cl^- = HgCl_4^{2-}$$

This is a complicated situation with dissolution of a solid phase, loss of a gas phase, and a complex set of coordination reactions in the aqueous phase. Consider the choice of components H^+, Na^+, Cl^-, H_2S, HgS (and H_2O) in Tableau 1.11:

$$TOTH = (H^+) - (OH^-) - (HS^-) - 2(S^{2-}) + 2(Hg^{2+})$$

$$+ 2(HgCl_2) + 2(HgCl_3^-) + 2(HgCl_4^{2-}) = 0 \qquad (84)$$

$$TOTNa = (Na^+) = (NaCl)_T = 0.5\ M \qquad (85)$$

$$TOTCl = (Cl^-) + 2(HgCl_2) + 3(HgCl_3^-) + 4(HgCl_4^{2-})$$

$$= (NaCl)_T = 0.5\ M \qquad (86)$$

$$TOTH_2S = (H_2S) + (H_2S \cdot g) + (HS^-) + (S^{2-}) - (Hg^{2+})$$

$$- (HgCl_2) - (HgCl_3^-) - (HgCl_4^{2-}) = 0 \qquad (87)$$

$$TOTHgS = (HgS \cdot s) + (Hg^{2+}) + (HgCl_2) + (HgCl_3^-)$$

$$+ (HgCl_4^{2-}) = (HgS)_T = ? \qquad (88)$$

TABLEAU 1.11

	H^+	Na^+	Cl^-	H_2S	HgS
Na^+		1			
Cl^-			1		
H_2S				1	
$H_2S(g)$				1	
$HgS(s)$					1
H^+	1				
OH^-	-1				
HS^-	-1			1	
S^{2-}	-2			1	
Hg^{2+}	2			-1	1
$HgCl_2$	2		2	-1	1
$HgCl_3^-$	2		3	-1	1
$HgCl_4^{2-}$	2		4	-1	1
$NaCl$		1	1		
HgS					1

Equations 87 and 88 are of little interest to us inasmuch as we have no knowledge of the number of moles of H_2S in the gas phase or of HgS in the solid phase. Because HgS(s) is very insoluble, it is a safe approximation that chloride is in great excess of all the mercuric species. Thus Equations 85 and 86 yield

$$(Na^+) = (Cl^-) = 0.5 \ M$$

a reasonable if obvious result.

Equation 84 which expresses the proton balance with respect to the chosen components is thus the only interesting one. It is quite remarkable that, if all diffusion coefficients can be taken as approximately equal, this proton balance will remain null at all times and in all places regardless of the rate of dissolution of HgS(s), of the rate of escape of $H_2S(g)$, or of the rate of formation of the various mercuric chloride species, and regardless of the mixing and diffusive processes. With some assumption on the controlling reactions (i.e., the slow ones) and reasonably simple transport conditions, this equation permits one to obtain the concentrations of all species as a function of time and place.

This general approach to coupling transport and chemical transformation equations has been used in a few applications.[2,3] It may serve as a basis for future models that describe the fate of aquatic pollutants.

□ □ □

REFERENCES

1 J. W. Gibbs, *On the Equilibrium of Heterogeneous Substances, The Collected Works*, vol. 1, Yale Univ. Press, New Haven, 1906, pp. 63, 93, 332.
2 J. C. Westall, F. M. M. Morel, and D. N. Hume, *Anal. Chem.* **51**, 1792 (1979).
3 D. M. DiToro, in *Modeling Biochemical Processes in Aquatic Systems*, R. C. Canale, Ed., Ann Arbor Science, Ann Arbor, 1976.

PROBLEMS

1.1 Write the "formulae" for the species indicated in italics in the following chemical reactions using (H_2O), (H^+), and whatever other species involved in each reaction are necessary.

a. $NH_3 + H_2O = NH_4^+ + OH^-$
b. $H_2S = S^{2-} + 2H^+$
c. $Cl_2(g) + H_2O = HOCl + H^+ + Cl^-$
d. $Ca(OH)_2(s) = Ca^{2+} + 2OH^-$
e. $CaCO_3(s) + H_2CO_3 = Ca^{2+} + 2HCO_3^-$
f. $FeCO_3(s) + 2H^+ = Fe^{2+} + CO_2(g) + H_2O$
g. $Zn^{2+} + 3OH^- = Zn(OH)_3^-$

h. $NaAlSi_3O_8(s) + CO_2(g) + \frac{11}{2}H_2O = Na^+ + HCO_3^- + 2H_4SiO_4$
   $+ \frac{1}{2}Al_2Si_2O_5(OH)_4(s)$
i. $Fe^{3+} + 3OH^- = Fe(OH)_3(s)$
j. $Fe(OH)_3(s) = FeOH^{2+} + 2OH^-$
k. $Fe(OH)_3(s) = Fe(OH)_2^+ + OH^-$
l. $2Cu^{2+} + 2OH^- = Cu_2(OH)_2^{2+}$
m. $Cu_2(OH)_2CO_3(s) + 4H^+ = 2Cu^{2+} + 3H_2O + CO_2(g)$
n. $Pb^{2+} + 2OH^- = Pb(OH)_2(s)$

1.2 A system is made up of a solution of Na_2SO_4 and H_2S (see species and reactions in Section 4). Which of the following component choices is acceptable, even if not very practical?

a. $SO_4^{2-}, H_2S, Na^+, OH^-, (H_2O)$
b. $H_2SO_4, NaHS, NaOH, H^+, (H_2O)$
c. $Na_2SO_4, S^{2-}, Na_2S, NaOH, (H_2O)$
d. $Na_2SO_4, H_2SO_4, Na_2S, NaOH, (H_2O)$
e. $Na_2SO_4, H_2SO_4, Na_2S, Na^+, (H_2O)$
f. $SO_4^{2-}, H_2S, S^{2-}, Na^+, (H_2O)$
g. $Na_2SO_4, SO_4^{2-}, S^{2-}, Na^+, (H_2O)$
h. $S_8^0, O_2^0, Na^0, H^+, (H_2O)$
i. $S^{2-}, O_2^0, Na^+, H^+, (H_2O)$
j. $Na_2HSO_4^+, H_2SO_4, Na_3S^+, HS^-, (H_2O)$

1.3 a. Given the following definitions of aqueous chemical systems, write complete sets of mole balance equations for each system, $[(H_2O)_T = 55.4 \ M$ in all recipes, and H_2O is always a species$]$.

1 Recipe $(CH_2O)_T = (CO_2)_T = 10^{-4} \ M$
  Species $CH_2O, O_2(g), CO_2(g)$
  Reactions $CH_2O + O_2(g) = CO_2(g) + H_2O$

2 Recipe $(HNO_3)_T = 10^{-3} \ M; (NH_4Cl)_T = 10^{-4} \ M$
  Species $NH_4^+, NH_3, NO_3^-, Cl^-, H^+, OH^-$
  Reactions $NH_4^+ = NH_3 + H^+$
               $NH_4^+ + OH^- = NH_3 + H_2O$
               $H_2O = H^+ + OH^-$

3 Recipe $(CO_2)_T = 10^{-3} \ M; (NaHCO_3)_T = 10^{-4} \ M$
  Species $H_2CO_3, HCO_3^-, CO_3^{2-}, Na^+, H^+, OH^-$
  Reactions $H_2CO_3 = HCO_3^- + H^+$
               $HCO_3^- = CO_3^{2-} + H^+$
               $H_2CO_3 = CO_3^{2-} + 2H^+$
               $HCO_3^- + OH^- = CO_3^{2-} + H_2O$
               $H_2O = H^+ + OH^-$

4 Recipe $(NaHCO_3)_T = (Na_2CO_3)_T = (NaOH)_T$
               $= (CH_3COONa)_T = 10^{-3} \ M$

Species   Same as system 3 + CH_3COOH, CH_3COO^-
Reactions   Same as system 3 + CH_3COOH
$$= CH_3COO^- + H^+$$

b. Show that electroneutrality is satisfied by each system of mole balance equations in part a.

c. Suppose that half of the carbon is eliminated from systems 3 and 4 by bubbling some inert gas:

$$H_2CO_3 \rightarrow H_2O + CO_2(g))$$

Write new mole balance equations for 3 and 4 (for the aqueous phase only).

1.4 Given the system defined by

Recipe $(NaHCO_3)_T = 10^{-3}\ M$
$(CO_2)_T = 2 \times 10^{-3}\ M$

Species (H_2O), H^+, OH^-, Na^+, CO_2, H_2CO_3, HCO_3^-, CO_3^{2-}

Reactions $H_2O = H^+ + OH^-$
$CO_2 + H_2O = H_2CO_3$
$H_2CO_3 = HCO_3^- + H^+$
$HCO_3^- = CO_3^{2-} + H^+$

Choose components and write tableaux and mole balance equations, first, isolating HCO_3^- in one mole balance equation and, second, isolating H_2CO_3 in one equation. Verify that the two sets of equations are equivalent and that the electroneutrality condition is satisfied.

1.5 In systems containing solid and gas phases, it is usually convenient to isolate each solid or gaseous species in one equation. Do this for a chemical system similar to Problem 1.4, but add

Recipe $(CaSO_4)_T = 10^{-1}\ M$
$(NaOH)_T = 10^{-1}\ M$

Species Ca^{2+}, SO_4^{2-}, $CaCO_3(s)$, $Ca(OH)_2(s)$, $CaSO_4(s)$

Reactions $Ca^{2+} + CO_3^{2-} = CaCO_3(s)$
$Ca^{2+} + 2OH^- = Ca(OH)_2(s)$
$Ca^{2+} + SO_4^{2-} = CaSO_4(s)$

1.6 Consider a chemical system containing only three species, X, Y, and Z. Only one reaction takes place:

$$xX + yY = zZ$$

(x, y, and z are stoichiometric coefficients). Taking X and Y as components, write the mole balance equations for the system, first using molar concentration units and then mole fractions. Apply your result to the H_2O, H^+, OH^- system.

CHAPTER TWO

ENERGETICS AND KINETICS

With the ultimate purpose of understanding the chemical behavior of natural waters, we need to answer three questions: What is the chemical composition of the system of interest? What reactions take place in this system? How are these reactions progressing in time? Supposing the analytical and purely chemical information to be given (e.g., a list of total concentrations, a list of chemical species, and a list of reactions), we focus here on the third question, that of the advancement of chemical reactions.

If we are to insist on a strict kinetic description of a chemical system, and if this system comprises more than a few species as do all natural waters, our task is quite discouraging. The system's behavior is governed by the simultaneous and interacting advancement of all possible chemical reactions, each controlled by its individual rate law. We are then faced with a formidable barrier to achieving an adequate quantitative description of even moderately complex chemical systems. If we could write them, the governing differential equations would be very difficult; however, we do not even know most of the applicable rate laws. If only by ignorance, we are thus prompted to ask an easier question and consider not what the detailed time history of the system might be, but what its final composition should be—supposing that we wait long enough for the reactions to reach equilibrium.

The fundamental reason why equilibrium conditions are much easier to describe than kinetic behavior rests with our greater ability to study the energetics (the thermodynamics) rather than the dynamics of systems. Think, for example, of the difficulty in predicting the flight of a feather, while its equilibrium position on the floor is a foregone conclusion. To be sure, if the time course of the process is precisely what matters, then knowledge of the final equilibrium condition is of little help. However, chemical equilibrium often provides a good approximation of the composition of natural waters. Furthermore chemical thermodynamics is more than a study of ultimate composition; it is a study of energetics and as such provides information on the direction of spontaneous change, on the energy available from or required for a particular reaction, and, in some situations, even on the kinetics of chemical reactions.

Many chemical reactions taking place in natural waters are quite fast and can be considered to be at equilibrium. By ignoring very slow reactions, one obtains a *partial equilibrium model* of the real system which includes some kinetic information just by the choice of the reactions considered to take place. This notion of partial equilibrium is a chemical one, not a thermodynamic one. Once the system is defined with *all* its possible variations (i.e., all its chemical reactions), its equilibrium state is uniquely defined. If additional independent variations are considered, the system is thermodynamically a different one and has a different equilibrium state. For example, if our feather experiment is conducted in a tall building, we may wish to consider the *possibility* of the feather flying through the window. Its equilibrium state is then the street, not the floor.

A kinetic description of chemical reactions that are neither very fast nor very slow (compared to some time scale of interest, e.g., the residence time of the water) can, under certain conditions, be superimposed on an equilibrium model for the fast reactions, then called a *pseudoequilibrium* model.

Even when equilibrium is not reached, chemical thermodynamics tells us the direction of spontaneous change. From energetic considerations we can decide, for example, whether or not a particular species is at equilibrium. If it is not at equilibrium, we can estimate how much energy can be obtained from the existing disequilibrium state, or may be required to maintain it. We can then investigate the nature of this disequilibrium, which may be caused by kinetic hindrance or an energy-consuming process mediated by light or by the biota. Since microbial communities seem to have evolved to exploit the most energetically favorable reactions first, such energetic analysis allows us to predict, or at least rationalize, the sequence of biologically mediated chemical events in many aquatic systems.

Energetic principles are the cornerstone of the principal modern theory of chemical kinetics, the *transition state theory*. Understanding thermodynamics is thus a prerequisite to understanding kinetics. In addition there are many cases of correlation between the kinetic and the thermodynamic parameters describing particular families of chemical reactions. In the absence of precise kinetic data, equilibrium information can then serve to guide us in describing the kinetics of some chemical processes.

In this chapter we do not attempt to provide a compact classical presentation of chemical thermodynamics.[1,2,3] Although a good grasp of this wide subject is indispensable to an advanced study of the physical chemistry of solutions, our purpose here is only to provide an intuitive basis for the essential thermodynamic concepts used in aquatic chemistry. This particular point of view biases considerably the relative importance of the various aspects of the subject. For example, a familiarity with the nonideal behavior of solutions is more important to us than a thorough understanding of Carnot's principle. As a result of this rather pragmatic approach, the presentation is more intuitive than rigorous and is not intended as a substitute for the many excellent texts on chemical thermodynamics. Following our discussion of the thermodynamic equilibrium of chemical systems, a few elements of chemical kinetics are provided, concluding with a brief section on pseudoequilibrium conditions.

1. THERMODYNAMICS OF CHEMICAL SYSTEMS

As a first and major simplification for our study of chemical thermodynamics, we limit ourselves to the study of systems under given *fixed pressure and temperature* (e.g., 1 at; 25°C)*. Our short exploration into thermodynamics is thus limited to examining the energetic consequences of changes in the composition of a *closed chemical system*, a system that does not exchange matter with its surroundings. These changes in composition are brought about by the progress of various chemical reactions taking place in the system. Our goals are to be able to calculate the equilibrium composition of the system and to estimate the energy associated with a particular change in composition.

For a thermodynamic study of such limited scope we bypass the fundamental principles of thermodynamics and postulate directly that a free energy function can be defined for our system and that this function possesses some key properties.

1.1   The Energy of a System

For the purpose of our discussion we define the free energy of a system as the amount of work necessary to reproduce the system from an arbitrary reference state (supposing 100% efficiency). For example, when we talk of the gravitational energy of an object of mass m at some elevation z above ground as being mgz, we have implicitly defined the system to be the object plus the earth's gravitation (acceleration g) and the reference state ($=0$ energy) to be the object resting on the ground ($z = 0$). Following international conventions, we count positively the work done to or stored by the system of interest and negatively the work done by or extracted from the system.

There are three fundamental properties of this quantity called the free energy of a system that are necessary to our thermodynamic development and that we simply accept here as postulates.

Postulate 1. The free energy of a system is a state property; it is only a function of the state of the system (given the reference state). According to our definition the work required to reproduce the system from the reference state must thus be independent of the way we do it (independent of the path), as long

---

* Pressure and temperature effects are of course important in some aquatic systems such as deep ocean basins. They are dealt with in Section 1.8.

as we are perfectly efficient of course. This postulate is clearly important for defining the free energy of a system as a useful entity; it also allows us to consider the free energy of a system as a sum of energies of different types. For example, one way to reproduce the system may be to elevate and pressurize the reference state system sequentially and bring it into an electrical field. Thus the gravitational, mechanical, and electrical energies are regarded as additive components of the total free energy of the system.

Postulate 2. The free energy of a system is the sum of the free energies of its constituent parts. This postulate may be considered simply as a choice in the way we account for the free energy of a system. In particular, the work necessary merely to assemble the different parts of a system must, in some way or another, be reflected in the free energies of the individual parts. From a chemical point of view we shall of course consider individual molecules as the parts whose assemblage constitutes the system. Defining, respectively, n_i as the number of moles and μ_i as the (partial) molar free energy of the ith species, we can write the total free energy G as

$$G = \sum_i n_i \mu_i \tag{1}$$

Postulate 3. The equilibrium of a system (i.e., its stable resting state) is given by its lowest possible energy state. The word *possible* is important in this rather intuitive postulate, as it reminds us that the proper definition of a system includes a definition of the possible variations in the system. Foremost among those for chemical systems are the advancements of the various chemical reactions. Implicit in this postulate is the idea that, in the absence of an external energy source, a system proceeds spontaneously to a lower energy state. For example, chemical reactions must take place in such a way as to decrease the free energy of a system.

As is the case for the classical thermodynamic principles to which they are obviously related, the ultimate justification of these three postulates rests on intuitive correctness and experimental verification of their consequences. Pragmatically, the conceptual and mathematical framework with which they allow us to structure our observations must, and does, enhance the success of our scientific predictions.

1.2   Chemical and Concentration Energies

The principal types of energies we have to consider in chemistry are those associated with the chemical nature and with the concentrations of chemical species. For simplicity we distinguish these two and refer to them as "chemical" and "concentration" energies. Other energies that come into play on occasion in aquatic chemistry are electrical, mechanical, and, less often, gravitational energies. The other known types of macroscopic energies are not usually considered in the context of aquatic chemistry; for example, the effect of the earth's magnetic field on ocean chemistry is not a topic of active investigation.

Chemical Energy

Molecules can react together to form compounds, and such chemical reactions may release or absorb energy. Chemical species are assemblages of atoms of various elements and thus possess a different free energy from that of the pure elements. By defining as the reference state for the pure elements a chemical standard state corresponding to their most stable forms (solid, liquid, or gas) under standard pressure (1 atmosphere $= 1.0133 \times 10^5$ Pascals) and temperature ($25°C = 298.15°K$), we can define the chemical energy content of all chemical species under any condition. For example, the molar free energy of liquid water under arbitrary conditions of pressure (P) and temperature (T) is given by the work necessary ($=$ minus the energy released) to carry out the overall reaction

$$P = 1 \text{ at}, T = 25°C; \quad H_2(g) + \tfrac{1}{2}O_2(g) \rightarrow H_2O(l); \quad P, T$$

This work includes the energy of the gas phase reaction at standard pressure and temperature, the energy of condensation of water vapor, and the energy necessary to heat (or cool) and pressurize (or depressurize) one mole of water from standard conditions to the temperature and pressure considered.

The nature of the solvent is of course important in determining the free energy of a solute (energy of solvation), as is, in principle, the particular composition of the solution. In the case of ions it is necessary to define arbitrarily one additional reference state in order to obtain expressions for the free energy of individual ions; by convention, the free energy of formation of the hydrogen ion H^+ ($=$ hydrated proton) is taken to be zero. The free energy of an ion such as chloride can then be obtained from the energy required for the formation of a completely ionized hydrochloric acid solution:

$$\tfrac{1}{2}H_2(g) + \tfrac{1}{2}Cl_2(g)(+ \text{ water}) = H^+(aq) + Cl^-(aq)$$

Free energies for other ions can be consequently obtained by consideration of the work necessary for the formation of other electrolyte solutions; for example,

$$Na(l) + \tfrac{1}{2}Cl_2(g)(+ \text{ water}) = Na^+(aq) + Cl^-(aq).$$

Concentration Energy (Definition of Phases)

We know that even in the absence of any mixing, gases and solutes spontaneously diffuse to achieve a homogeneous concentration distribution. We have an intuitive feeling for diffusion of sugar in our coffee as we do for falling cups: both processes occur spontaneously and must thus be energetically favorable. Our purpose here is to obtain a mathematical expression for the concentration energy.

To this end we need, first, to define the possible variations, to decide what geometrical limits there are to the diffusion process. This is done through the familiar concept of *phases*, which distinguishes typically gas, solution, and solid phases, each referring to the particular physical state of the constitutive matter. Normally in aquatic chemistry we consider only one gas phase (air) and one solution phase (water); nonetheless, it is sometimes useful to consider others such as the intracellular solutions in aquatic microorganisms. Gases and solutes are free to diffuse throughout the air and water phases, respectively. For simplicity, we may consider at this point that all solid phases are pure phases (i.e., they are constituted of only one chemical species) and that no diffusion takes place in solid phases. This restriction leads to the obvious result that there is *no* concentration energy term for solids.

Defining the mole fraction (X_i) of a gas or solute as the ratio of the number of moles of the gas or solute to the total number of moles in the corresponding phase,

$$X_i = \frac{n_i}{\sum_j n_j} \tag{2}$$

The work necessary to go from a concentration X_i to X_i' is given by the formula

$$w = RT \ln \frac{X_i'}{X_i} \tag{3}$$

where R is the universal gas constant (8.3143 J deg^{-1} mol^{-1}) and T is the absolute temperature; ln designates the Naperian (natural) logarithm. ($2.3 RT \cong 5.7$ kJ mol$^{-1} \cong 1.36$ kcal mol^{-1} at $25°$C.) This expression can be justified on the basis of empirical knowledge. For example, it can be derived for gases by considering the corresponding mechanical pressure-volume work in terms of Boyle-Mariotte's law. The result can then be extended to solutions by application of Raoult's and Henry's laws for the equilibrium between concentrations of solvent and solutes and their vapor pressures. Many other experimental justifications of Equation 3 are provided by verification of its consequences, for example, the mass law expression for reactions among solutes and the expression of solubility of solids. Although it draws on material to be studied in a later chapter, the most compelling justification for Equation 3 is perhaps given by the Nernst equation which demonstrates directly the logarithmic relationship between concentration and electrical energy. As an alternative to empirical justification, Equation 3 can be obtained from the theory of statistical mechanics. The Maxwell-Boltzmann distribution law, which relates concentrations of particles to the exponential of their energy, is clearly kindred to Equation 3. For our purposes this equation is simply considered as God-given.

1.3   Molar Free Energies in Ideal Systems

Consider a species S_i at a concentration X_i in some system. Suppose that we are given the value μ_i^{ref} for the chemical free energy of S_i at some other concentration, X_i^{ref}, in the same system (same P and T). From the principle of additivity of energies, we obtain an expression for the free energy of S_i as a function of X_i:

$$\mu_i = \mu_i^{\text{ref}} + RT \ln \frac{X_i}{X_i^{\text{ref}}} \tag{4}$$

yielding

$$\mu_i = \mu_i^0 + RT \ln X_i \tag{5}$$

where

$$\mu_i^0 = \mu_i^{\text{ref}} - RT \ln X_i^{\text{ref}} \tag{6}$$

This expression is valid only if the free energy of S_i is not influenced by the presence of other species in the system. Indeed, as written in Equation 5 the expression of molar free energy of a species is explicitly dependent on its own concentration (mole fraction X_i) and not on that of any other species, except inasmuch as the total number of moles of the system, and hence $X_i = n_i/\sum n_j$, are affected. This independence of the free energy of a species from the concentrations of all others is the definition of an ideal system—an ideal solution to specialize the discussion to the aqueous phase which is of principal interest to us. It corresponds to a good approximation of the physical reality in either of the two following cases:

1 The system is very dilute, and all individual solute molecules are "ignorant" of each other (i.e., they have no energetic interactions, and their individual free energies are unaffected by each other's presence).

2 The major solutes (those accounting for the bulk of the dissolved species) are considered to be at a fixed concentration, and whatever effects they have on the free energies of another species may be accounted for in the standard value μ_i^0 of the chemical free energy of that species.

Both limiting cases for ideal behavior—infinite dilution or fixed composition of background solutes—are used in practice to express the various thermodynamic parameters. In the first case the standard molar free energy μ_i^0 is the work necessary to produce an infinitesimal concentration of S_i from its constitutive elements in their standard state, all solutes being present at vanishingly low concentrations. In the second, μ_i^0 is the work necessary to produce an infinitesimal concentration of S_i in the presence of a given concentration of other solutes. In both cases the value of the standard free energy μ_i^0 is expressed per mole of species S_i.

Note that due to the logarithmic expression $\ln X_i$, the standard free energy μ_i^0 is the value of μ_i when the concentration of S_i is unity ($X_i = 1$), the so-called "standard state" for S_i. Although convenient, this situation seems somewhat paradoxical since a unit concentration can hardly be considered to be infinitesimal. This is simply due to the implicit inclusion of the $(-RT \ln X_i^{\text{ref}})$

term in the value of μ_i^0 (Equation 6). In effect, to obtain the free energy at a given concentration, X_i, from that obtained (theoretically) at an infinitisimal concentration, we first extrapolate up to unit concentration and then back down to X_i. As seen in the next section, the only problem resulting from such a convention is that the standard free energies are dependent on the chosen concentration scale, each of which corresponds to a different standard state.

It must be recognized that chemical systems behave ideally only over a limited range of concentrations. We shall study in Section 1.7 the important question of nonideal behavior; for now it is worth noting that an ideal thermodynamic description of a real solution is all the better if the chosen reference state is the closer to the actual state.

1.4 Practical Expressions of Molar Free Energies

Recognizing our predilection for moles per liter (M) rather than mole fractions as concentration units, we now establish some convenient formulae for molar free energies, starting with the ideal formula, Equation 5.

Solids

For a species constituting a pure solid phase, the mole fraction is identically unity, and its logarithm is thus identically null:

$$X = 1$$

therefore

$$\mu_i = \mu_i^0 \tag{7}$$

Solutes

For species in the solution phase, the mole fraction is given by

$$X_i = \frac{n_i}{N_w} \tag{8}$$

therefore

$$\mu_i = \mu_i^0 + RT \ln \frac{n_i}{N_w} \tag{9}$$

where N_w is the total number of moles in the solution. We can introduce molar concentrations in Equation 9:

$$(S_i) = \frac{n_i}{V} \tag{10}$$

$$\mu_i = \mu_i^0 - RT \ln \frac{N_w}{V} + RT \ln \frac{n_i}{V} \tag{11}$$

For dilute solutions, N_w/V, the total number of moles per liter of solution is approximately constant and equal to the number of moles of water (55.4 M). The second term of Equation 11 can be thus incorporated in the standard free energy value μ_i^0:

$$\mu_i = \mu_i^0 + RT \ln (S_i) \tag{12}$$

This is the equation that we use in practice for solutes, and it should be remembered that the corresponding standard free energy μ_i^0 is only valid for the chosen molar concentration scale.

Water

The mole fraction of water in dilute aqueous solutions is approximately unity, and we shall consider the molar free energy of water to be constant:

$$\mu_{water} \cong \mu_{water}^0 \tag{13}$$

This approximation, which is satisfactory for the study of chemical phenomena in dilute solutions, does not permit the study of physical phenomena such as the depression in freezing point, osmotic pressure, or other colligative properties.*

Gases

For gases, the mole fraction is equal to the ratio of the partial pressure of the gas to the total pressure of the gas phase:

$$X_i = \frac{n_i}{N_G} = \frac{P_i}{P} \tag{14}$$

$$\mu_i = \mu_i^0 + RT \ln \frac{P_i}{P} \tag{15}$$

where N_G is the total number of moles in the gas phase.

An equation similar to Equation 12 is obtained for gases if the total pressure is taken as constant (e.g., $P = 1$ at):

$$\mu_i = \mu_i^0 + RT \ln (P_i) \tag{16}$$

1.5   Energetics of Chemical Reactions

On the basis of the formulae for the total free energy of a system

$$G = \sum_i n_i \mu_i \tag{1}$$

and of the molar free energy of individual species

$$\mu_i = \mu_i^0 + RT \ln X_i \tag{5}$$

* The colligative properties of a solvent are those that depend on the number of particles, such as molecules or ions, in the solvent.

we can study the energetic consequences of the advancement of a chemical reaction.

First let us consider a simple example, that of a system composed of three species, A, B, and C, in which only one reaction is taking place:

$$2A + 3B = C \tag{17}$$

What is the change dG in the free energy G of the system when the reaction proceeds to form an additional number dn_C moles of C?

$$G = n_A\mu_A + n_B\mu_B + n_C\mu_C \tag{18}$$

$$dG = \underbrace{\left[\frac{dn_A}{dn_C}\mu_A + \frac{dn_B}{dn_C}\mu_B + \frac{dn_C}{dn_C}\mu_C \, dn_C\right]}_{\sum dn_i\mu_i} + \underbrace{\left[n_A\frac{d\mu_A}{dn_C} + n_B\frac{d\mu_B}{dn_C} + n_C\frac{d\mu_C}{dn_C} \, dn_C\right]}_{\sum n_i d\mu_i} \tag{19}$$

Let us show that the second term, $\sum n_i d\mu_i$, is null. Introducing the expression of the molar free energies, and differentiating,*

$$\sum n_i d\mu_i = RT\left[\frac{n_A}{X_A}\frac{dX_A}{dn_C} + \frac{n_B}{X_B}\frac{dX_B}{dn_C} + \frac{n_C}{X_C}\frac{dX_C}{dn_C}\right] dn_C$$

But, by definition of the mole fractions,

$$\frac{n_A}{X_A} = \frac{n_B}{X_B} = \frac{n_C}{X_C} = n_A + n_B + n_C$$

therefore

$$\sum n_i d\mu_i = RT(n_A + n_B + n_C)\frac{d(X_A + X_B + X_C)}{dn_C} \, dn_C$$

and, since $X_A + X_B + X_C = 1$, the last term yields zero as the differential of a constant:

$$\sum n_i d\mu_i = 0. \quad \text{QED} \tag{20}$$

The differential increase in the total energy of the system is thus given solely by the first term, $\sum dn_i\mu_i$; in this case, with the particular stoichiometry of Reaction 17,

$$\frac{dG}{dn_C} = -2\mu_A - 3\mu_B + \mu_C \tag{21}$$

Introducing the expression of the molar free energies (Equation 5),

$$\underbrace{\frac{dG}{dn_C}}_{\Delta G \,=} = \underbrace{(-2\mu_A^0 - 3\mu_B^0 + \mu_C^0)}_{\Delta G^0} + \underbrace{RT\ln\frac{X_c}{X_A^2 \cdot X_B^3}}_{+ \, RT\ln \quad Q} \tag{22}$$

* Note that

$$\frac{d\mu_i^0}{dn_C} = 0; \quad \frac{d\ln X_i}{dn_C} = \frac{1}{X_i}\frac{dX_i}{dn_C}$$

This quantity, which measures the change in the free energy of the system upon progress of the reaction is called the *molar free energy change* of the reaction and is usually denoted (unfelicitously) ΔG.

In general, considering a system made up of any number of species S_i, we can show that Equation 20, which is known as the Gibbs-Duhem relation, is verified for any possible change in the composition of the system. The demonstration is simply a generalization of the derivation in Equations 18 through 20, and in this approach the Gibbs-Duhem relation (Equation 20) is thus taken as a mere mathematical consequence of the functionality of the concentration term in the expression of the molar free energy. The property of μ_i to be a partial differential of $G(\mu_i = \partial G/\partial n_i$; hence the justification of its appellation as a *partial* molar free energy) follows from the Gibbs-Duhem relation and from the fundamental expression of additivity of free energies $G = \sum_i n_i \mu_i$. In classical thermodynamic texts μ_i is usually defined a priori as a partial differential, $\mu_i = \partial G/\partial n_i$, and G as an extensive property of the system (by which it is meant to be a homogeneous function of degree 1 in n_i). Use of Euler's theorem for homogeneous functions then yields $G = \sum_i n_i \mu_i$ and, by differentiation, the Gibbs-Duhem relation, $\sum_i n_i d\mu_i = 0$. The approach taken here appears no less intuitive, and it stresses the need for the molar free energies to verify the Gibbs-Duhem relation in order for the fundamental thermodynamic equations to be strictly applicable. For example, neither the expression of the free energy change of a reaction given by Equation 22 nor the mass law in Equation 33 is strictly correct when the molar scale rather than the mole fraction scale is used and the free energy of species is expressed as in Equations 12 and 16.

To generalize the expression of free energy change, consider now any reaction symbolized by

$$0 = v_1 S_1 + v_2 S_2 + v_3 S_3 + \cdots \tag{23}$$

The v_i's are positive for products, negative for reactants, and null for species not involved in the reaction. We define the degree of advancement ξ of the reaction such that

$$\frac{\partial n_i}{\partial \xi} = v_i \tag{24}$$

The Gibbs-Duhem relation (Equation 20) can then be written:

$$\sum_i n_i \frac{\partial \mu_i}{\partial \xi} = 0$$

and the free energy change for the reaction can be expressed in the general form

$$\Delta G = \frac{\partial G}{\partial \xi} \tag{25}$$

$$\Delta G = \frac{\partial}{\partial \xi} \left(\sum_i n_i \mu_i \right) = \sum_i \frac{\partial n_i}{\partial \xi} \mu_i + \sum_i n_i \frac{\partial \mu_i}{\partial \xi}$$

so that

$$\Delta G = \Delta G^0 + RT \ln Q \tag{26}$$

where

$$\Delta G^0 = \sum_i v_i \mu_i^0 = v_1 \mu_1^0 + v_2 \mu_2^0 + v_3 \mu_3^0 \ldots \tag{27}$$

$$Q = \prod_i X_i^{v_i} = X_1^{v_1} X_2^{v_2} X_3^{v_3} \ldots \tag{28}$$

The free energy change of a reaction is thus made up of two terms: (1) a constant, ΔG^0, called the *standard free energy change* of the reaction which is given by the stoichiometry of the reaction and the standard free energies of the species involved, and (2) a logarithmic term that depends on the composition of the system through the product Q, called the *reaction quotient*.

Reactions that have very large standard free energies, positive or negative, tend to proceed to completion in one direction or the other. Such reactions are very far from equilibrium when products and reactants are present in comparable concentrations, and in this case the first term of Equation 26 is normally much larger than the second. The change in the free energy of the system, as the reaction proceeds, is then approximately constant and equal to the standard free energy change of the reaction (per mole reacted):

$$\Delta G \cong \Delta G^0 = \sum_i v_i \mu_i^0 \tag{29}$$

This relationship is particularly useful for examining the energetics of redox reactions carried out by organisms: photosynthesis, respiration, nitrification, nitrogen fixation, sulfate reduction, and others. Note, however, that it is only an approximate relationship and in many cases the concentration of some reactants such as H^+ can have a large effect on the free energy change of the reaction (see Chapter 7).

1.6   Chemical Equilibrium

If we consider a set of independent chemical reactions describing all the possible independent variations in the composition of a closed system, the minimum of the free energy of the system—and hence its equilibrium state—is obtained mathematically by equating to zero all the differentials of G with respect to the degree of advancement ξ of each reaction, subject to constraints of mass conservation (i.e., mole balances) and positiveness of concentrations:

$$\frac{\partial G}{\partial \xi} = 0 \tag{30}$$

This is the expression of equilibrium for the reaction of advancement ξ: there is a zero gain or loss of energy in the system as the reaction proceeds infinitesimally in either direction. Any other reaction taking place in the system must have a stoichiometry and thus a degree of advancement that are linear combinations of those of the chosen independent reactions. It follows that this reaction's free energy change is also zero at equilibrium. The system is thus at equilibrium when all the possible reactions are at equilibrium, and vice versa:

$$\frac{dG}{\partial \xi} = \Delta G = \sum v_i \mu_i = 0 \quad \text{for all reactions} \tag{31}$$

1.6.1 Mass Law Equation

Introducing the formula for the free energy change of a reaction, Equation 26, into Equation 31, we obtain the equilibrium condition

$$\frac{\partial G}{\partial \xi} = \Delta G = \Delta G^0 + RT \ln Q = 0 \tag{32}$$

This equation is usually written in its exponential form and is known as the mass law equation:

$$Q = \exp\left[-\frac{\Delta G^0}{RT}\right]$$

that is,

$$X_1^{v_1} X_2^{v_2} X_3^{v_3} \ldots = K \tag{33}$$

where

$$K = \exp\left[-\frac{\Delta G^0}{RT}\right] = \exp\left[-\frac{1}{RT}\sum_i v_i \mu_i^0\right] \tag{34}$$

K is called the *equilibrium constant*, and it is directly related to the standard free energy of a reaction (ΔG^0) by the exponential expression in Equation 34, and vice versa:

$$-RT \ln K = \Delta G^0 = \sum v_i \mu_i^0 \tag{35}$$

For convenience, the mass law equation is usually written with molar concentrations and partial pressures rather than mole fractions in the expression of the reaction quotient, that is, utilizing Equations 12 and 16 rather than

Equation 5 for the expressions of the molar free energies. This involves an approximation since the Gibbs-Duhem relation is not strictly verified by Equations 12 and 16, and the expression for the free energy change of a reaction ΔG given by Equation 26 is then not strictly correct. More important, this unit change also involves a change in the standard free energy ΔG^0 which then incorporates some multiple of $RT \ln (55.4)$, depending on the stoichiometry of the reaction. Hence the equilibrium constants that are usually tabulated for the molar scale are valid only for that particular scale.

On the basis of Equations 7, 12, and 16, mass law equations can be written for reactions among solids, gases, and solutes in any combination:

1 Solid-solutes

$$CaCO_3(s) = Ca^{2+} + CO_3^{2-}$$

$$(Ca^{2+})(CO_3^{2-}) = K_{CaCO_3} \quad \text{(solubility constant)}$$

2 Gas-solutes

$$CO_2(g) = CO_2(aq)$$

$$\frac{(CO_2 \cdot aq)}{P_{CO_2}} = K_H \quad \text{(Henry's law constant)}$$

3 Solutes-solvent

$$CO_2(aq) + H_2O = H_2CO_3$$

$$\frac{(H_2CO_3)}{(CO_2 \cdot aq)} = K \quad \text{(hydration constant)}$$

or

$$H_2O = H^+ + OH^-$$

$$(H^+)(OH^-) = K_w \quad \text{(ion product of water)}$$

4 Solutes

$$HAc = H^+ + Ac^-$$

$$\frac{(H^+)(Ac^-)}{(HAc)} = K_a \quad \text{(acidity constant)}$$

or

$$Hg^{2+} + 3Cl^- = HgCl_3^-$$

$$\frac{(HgCl_3^-)}{(Hg^{2+})(Cl^-)^3} = K \quad \text{(complexation constant)}$$

1.6.2   Uniformity of Molar Free Energies at Equilibrium

Consider a species S_i in two different parts A and B of a system at equilibrium. A and B may be two different phases, or they may simply be different locations in the same phase. If S_i is free to transfer from A to B, we can represent that possible variation in the system by the advancement of the reaction

$$S_i^A = S_i^B \qquad (36)$$

Given the values $v_i^A = -1$, and $v_i^B = +1$ of the stoichiometric coefficients in this reaction, we can write the equilibrium condition:

$$\frac{\partial G}{\partial \xi} = \Delta G = \mu_i^B - \mu_i^A = 0$$

or

$$\mu_i^B = \mu_i^A \qquad (37)$$

This result can of course be applied to any part of the system. We have thus demonstrated that at equilibrium the molar free energy of any species must be uniform throughout the system (supposing no absolute barrier to transfer). This necessary condition is useful to study equilibrium among phases. Applied to a solution phase, it results directly in a condition of uniformity of concentrations throughout the solution:

$$\mu_i^0 + RT \ln (S_i^A) = \mu_i^0 + RT \ln (S_i^B)$$

or

$$(S_i^A) = (S_i^B) \qquad (38)$$

We have thus rediscovered the wheel and the thermodynamic necessity of diffusion of sugar in our coffee cup.

1.7   Nonideality Effects

In Equation 5 we postulated that the molar free energy μ_i of a species S_i depends exclusively on the mole fraction of that species. In effect, we have implied that the composition of the rest of the system has no effect on μ_i; this is the very definition of an *ideal* thermodynamic system. In order to study *nonideal effects*— to study the effects of other solutes on the free energy of a given soluble species, for example—we need to escape from thermodynamics per se and examine experimental results or some theory of the mechanisms of interaction among species. Consistent with our intuitive rather than empirical approach, we choose the second.

1.7.1   Debye-Hückel Theory

Chemical species in solution can interact in a variety of ways ranging from covalent bonding, to London-Van der Waals interactions, to long-range electrostatic repulsion and attraction, or even simply to volume exclusion effects when solutes are sufficiently concentrated that they crowd each other. (All these interactions are of course superimposed on the interaction between solute and solvent molecules, the solvation effect, which in aqueous solutions is primarily an electrostatic interaction between polar molecules and the dipole moment of the water molecules.) When these interactions are sufficiently intense, and the determination of when they are is somewhat arbitrary, we simply consider them as resulting in a chemical reaction and the formation of new chemical species. The study of nonideal effects is thus the study of species interactions not accounted for by chemical reactions. In dilute solutions such interactions are mostly due to long-range electrostatic forces among ions: attractions of ions of opposite charge and repulsion of ions of like charge. As a result of these forces the distribution of ions in solution is not uniform. This separation of charges at the molecular scale leads to local variations in the electrical potential of the solution which effectively decrease the total free energy of the system.

This concept is the basis of the theory of nonideal effects in dilute solutions— the Debye-Hückel theory—in which it is assumed that the only interactions among solutes not accounted for by chemical reactions are due to their electrical charge, not their chemical nature. The development of the theory consists in evaluating the relevant electrostatic energy and including it in the expression of the free energy of ions.

From basic electrostatic theory we know that the work necessary to bring a charge c from an electrical potential ψ to a potential ψ' is

$$w = c[\psi' - \psi] \tag{39}$$

If we consider an ion of molar electrical charge, $Z_i F$ (Z_i is the charge number of the ion, for example, $Z = +2$ for Ca^{2+}, $Z = -1$ for Cl^-, and F is the Faraday constant $= 96{,}500$ coulombs equiv^{-1}), the work necessary to bring a mole of this ion from some reference potential ψ^0 to a potential ψ is

$$w = Z_i F(\psi - \psi^0) \tag{40}$$

Based on the principle of additivity of energies and on our convention to account for the total free energy of the system as the sum of the free energies of its constituent molecules, the total energetic effect of the long-range electrostatic interactions can be expressed by adding a term in the molar free energy of each ion:

$$\mu_i = \mu_i^0 + RT \ln (S_i) + Z_i F \psi \tag{41}$$

We are choosing here the molar scale, and the term $Z_i F \psi^0$ is naturally included in μ_i^0. The problem is then to evaluate the potential ψ applicable to each ion. The original derivation of Debye and Hückel, which involves a mixture of classical thermodynamics and electrostatics, is straightforward but tedious, so we shall only give the results. The derivation involves some assumptions (mobile ions behaving as point charges in a continuous medium of uniform dielectric constant) and the definition of parameters:

1   The *ionic strength* of the solution I, given by

$$I = \tfrac{1}{2} \sum_i Z_i^2 (S_i) \tag{42}$$

TABLE 2.1

Values of the Parameter a (10^{-10} m) for Some Ions

$a = 3$	$a = 4$	$a = 5$	$a = 6$	$a = 8$	$a = 9$
OH^-	HCO_3^-	CO_3^{2-}			
F^-, Cl^-, Sr^-, I^-	$H_2PO_4^-, HPO_4^{2-}$				
CN^-	PO_4^{3-}				
NO_2^-, NO_3^-					
ClO_4^-					
HS^-	SO_4^{2-}	S^{2-}			
	HSO_3^-	$S_2O_4^{2-}$			
	$S_2O_3^{2-}$	SO_3^{2-}			
$HCOO^-$	CH_3COO^-	COO^{2-}			
$H_2citrate^-$	$glycine^-$	$Hcitrate^{2-}$			
	$H_2glycine^+$	$citrate^{3-}$			
K^+	Na^+	Sr^{2+}	Ca^{2+}	Mg^{2+}	H^+
Ag^+		Ba^{2+}	Cu^{2+}		Al^{3+}
NH_4^+	$CdCl^+$	Cd^{2+}	Zn^{2+}		Fe^{3+}
		Hg^{2+}	Sn^{2+}		Cr^{3+}
		Pb^{2+}	Mn^{2+}		
			Fe^{2+}		
			Ni^{2+}		
			Co^{2+}		

Source: Adapted from J. Kielland.[5]

where (S_i) is the molar concentration of the species S_i of charge number Z_i. I has dimensions of moles per liter.*

2　An adjustable size parameter, a, introduced in the Debye-Hückel derivation as the distance of closest approach between the center of adjacent ions and corresponding roughly to the radius of the hydrated ion. (a is usually expressed in the old Angstrom units: $1 \text{ Å} = 10^{-10} \text{ } m$. Table 2.1 gives the value of a for various ions.)

If we choose the origin of potentials to be that applicable at infinite dilution (and include it in the expression of μ_i^0), the theoretical value of ψ is given by

$$\psi = -A \frac{RT}{F} Z_i \frac{I^{1/2}}{1 + BaI^{1/2}} \qquad (43)$$

* There appears to be a great deal of confusion regarding the dimensions of I. In many classical textbooks I is considered implicitly or explicitly as a dimensionless number following the tradition of Lewis and Randall (1921) who originated the concept of ionic strength and did not explicitly state the dimensions of I.[4] This tradition has led to the common practice of omitting the dimensions for the parameter A in Equation 43 and defining the parameter B with dimensions of inverse length.

in which the constants A and B depend on the absolute temperature T and the dielectric constant of the system. In water at 25°C

$$A \cong 1.17 \; mol^{-1/2} \; l^{1/2}$$

$$B \cong 0.33 \times 10^8 \; mol^{-1/2} \; l^{1/6}$$

If a is expressed in angstroms and I in moles per liter, $B = 0.33$. Then

$$\mu_i = \mu_i^0 + RT \ln (S_i) - ARTZ_i^2 \frac{I^{1/2}}{1 + BaI^{1/2}} \tag{44}$$

Note that the electrostatic interactions among ions result in a decrease of their molar free energies. We can calculate the relative magnitude of the non-ideal term, considering, for example, a doubly charged ion of millimolar concentration in a medium of ionic strength $I = 10^{-2}$ M. The concentration term in the expression of μ_i is

$$RT \ln (10^{-2}) = -4.6RT$$

while the nonideal term is

$$-1.17 \frac{4 \times 0.1}{1 + 0.1} RT = -0.43RT \quad \text{(taking } a = 3\text{Å)}$$

Under these conditions the contribution of the nonideal term to the molar free energy is about 10% of that of the concentration term. Note also that according to this theory there is no nonideal effect for uncharged species, including all solids and gases.

1.7.2 Effect of Nonideality on Equilibrium

We have derived the condition of equilibrium of a chemical reaction, the mass law equation, ignoring the presence of the nonideal term in the expression of the molar free energies. In order to include this term, we now define the *activity coefficient* γ_i of the species S_i by the equation

$$RT \ln \gamma_i = Z_i F \psi \tag{45}$$

When the potential ψ is given by its theoretical value according to the Debye-Hückel theory,

$$\ln \gamma_i = -AZ_i^2 \frac{I^{1/2}}{1 + BaI^{1/2}} \tag{46}$$

and

$$\mu_i = \mu_i^0 + RT \ln (S_i) + RT \ln \gamma_i \tag{47}$$

For uncharged solutes which behave ideally, the value of the activity coefficient γ_i is simply unity.

The problem of expressing the mass law equation can then be treated in one of two ways:

1. The last two terms can be combined to define the *activity* $\{S_i\}$ of the species S_i

$$\{S_i\} = \gamma_i(S_i) \tag{48}$$

The mass law equation for a general reaction

$$0 = v_1 S_1 + v_2 S_2 + v_3 S_3 \ldots$$

is then written

$$\gamma_1^{v_1}(S_1)^{v_1}\gamma_2^{v_2}(S_2)^{v_2} \ldots = K = \exp\left[-\frac{1}{RT} \sum_i v_i \mu_i^0 \right]$$

therefore

$$\{S_1\}^{v_1}\{S_2\}^{v_2}\{S_3\}^{v_3} \ldots = K \tag{49}$$

2. Alternatively, the first and the last term can be combined to define the *concentration equilibrium constant* K_c

$$\ln K_c = -\frac{1}{RT} \sum_i v_i \mu_i^0 - \sum_i v_i \ln \gamma_i \tag{50}$$

therefore

$$K_c = K\gamma_1^{-v_1}\gamma_2^{-v_2}\gamma_3^{-v_3} \ldots \tag{51}$$

In this case the mass law equation is written

$$(S_1)^{v_1}(S_2)^{v_2}(S_3)^{v_3} \ldots = K_c \tag{52}$$

In both cases the convenient form of the mass law equation is retained. In the first case this is achieved by replacing each species concentration by an idealized quantity, the species' activity. This activity is smaller than the concentration, and its difference with the concentration is a measure of the decreased reactivity of the ion due to the stabilizing effect of its electrostatic interactions with other ions. In the second case, the standard free energy change of each reaction is decreased, while the ions themselves are considered to behave ideally. This is strictly identical to considering a reference state that includes the effects of the major solutes (those contributing most of the ionic strength of the system) on the free energy of the ions: $RT \ln \gamma_i$ is simply included in μ_i^0.* This second approach is a great deal more convenient from a numerical point of view, and we shall use it preferentially.

---

* The *reference state* of a solute is strictly defined as that for which the activity coefficient is considered to be unity.

On the basis of Equation 51 we can evaluate the concentration equilibrium constant of a reaction at a given ionic strength I from a constant defined for infinite dilution (i.e., zero ionic strength) and a knowledge of the activity coefficients at the ionic strength I. If the constant is originally given for a finite ionic strength I_0, the formula becomes

$$K_c = K_0 \left[\frac{\gamma_1^0}{\gamma_1}\right]^{\nu_1} \left[\frac{\gamma_2^0}{\gamma_2}\right]^{\nu_2} \ldots \tag{53}$$

Because the activity (rather than the concentration) of the hydrogen ion is usually measured, *mixed acidity constants* are often used for acid-base reactions. For example, the mixed acidity constant for the dissociation reaction

$$HA = H^+ + A^- \tag{54}$$

is defined as

$$K_a' = \frac{\{H^+\}(A^-)}{(HA)} \tag{55}$$

The activity coefficient of H^+ is then eliminated from the formula for the calculation of the mixed acidity constant at any ionic strength.

Tabulations of equilibrium constants (e.g., Sillén and Martell,[6] Martell and Smith, and Smith and Martell[7]) normally contain information on the pertinent ionic strength condition as well as temperature and pressure. Often the values of the constants are extrapolated to zero ionic strength, and these "infinite dilution constants" are those given in the appropriate tables at the beginning of the following chapters.

1.7.3 "Nonideal" Nonideal Effects

Although aesthetically pleasing, the Debye-Hückel theory involves a certain number of approximations and does not account for all interactions among solutes. Molecules interact through means other than long-range electrostatic forces, and the activity coefficients of ions do not exactly follow the "ideal" nonideal formula of Debye and Hückel, while the activity coefficients of neutral molecules are observed to deviate from unity.

Empirical Formulae for Activity Coefficients at High Ionic Strength. Although activity coefficients of individual ions cannot be measured, the mean activity coefficients of binary electrolytes can be obtained experimentally by various methods such as solubility measurements or electrochemical determinations of free energies. The mean activity coefficient $\gamma_\pm$ is related to the single ion activity coefficients, γ_+ and γ_-, by the formula

$$(Z_+ + Z_-) \log \gamma_\pm = Z_- \log \gamma_+ + Z_+ \log \gamma_-$$

where Z_+ and Z_- are the charge numbers of the cation and the anion, respectively. As can be seen in Figure 2.1, the Debye-Hückel formula does not predict accurately the activity coefficients of simple electrolytes at high ionic strength; even at an ionic strength of 0.1 M, the activity coefficient of such an important salt as NaCl is measurably underpredicted by the formula.

This is because at high electrolyte concentrations, the crowding of the hydrated solutes begins to affect their free energies (the point charge approximation becomes invalid), and the chemical nature of the species, not just their electrical charge, plays a role in their mutual interactions. In order to extend the applicability of the Debye-Hückel formula to higher ionic strength systems, a variety of empirical and semiempirical expressions have been proposed. One

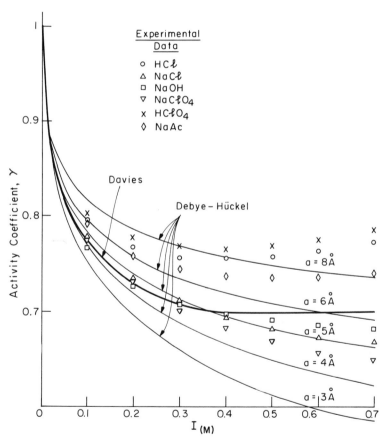

Figure 2.1 Activity coefficients of 1:1 electrolytes as a function of ionic strength. The experimental data represent the *mean* activity coefficients of the respective electrolytes (taken from Harned and Owen, 1958[3]). The lines correspond to the predictions of the Debye-Hückel (Equation 46) and the Davies formulae (Equation 56) for singly charged ions. The original value $b = 0.2$ has been used for the empirical parameter of the Davies formula; Davies himself later suggested a slightly higher value: $b = 0.3$.

of the simplest and most widely used expressions is that of Davies who eliminated the size parameter a and added a linear term to the Debye-Hückel formula:

$$\ln \gamma_i = -AZ_i^2 \left(\frac{I^{1/2}}{1 + I^{1/2}} - bI \right) \tag{56}$$

By optimizing for measured activity coefficients below $I = 0.1\ M$, the empirical parameter b was estimated by Davies to be in the range 0.2 to 0.3. The principal advantage of Davies' formula is to provide, as exhibited by the data, a quasi-constant value of the activity coefficients in the range $I = 0.3$ to 0.7 M. In aquatic systems the ionic strength rarely exceeds 0.7 M, and the inaccuracies introduced by using the Davies equation are usually smaller than the sources of errors and uncertainties. However, when the objective is to analyze accurately the interactions of seawater ions or to study the chemistry of concentrated brines, a more sophisticated approach becomes necessary, one that accounts for specific as well as unspecific ion interactions and predicts the large increase in activity coefficients at high ionic strengths.

The most commonly used solution to this problem is the Bronsted-Guggenheim[8,9] specific ion interaction model and its extensions by Scatchard, Mayer,[10,11,12] and others. The general formula for this model consists of a virial expansion of the form:

$$\ln \gamma_i = \ln \gamma_{DH} + \sum_j B_{ij}(S_j) + \sum_j \sum_k C_{ijk}(S_j)(S_k) + \cdots \tag{57}$$

The first term is simply the Debye-Hückel activity coefficient. The *second virial coefficients* (B_{ij}) account for specific interactions among pairs of ions, the *third virial coefficients* (C_{ijk}) for specific interactions among three ions, and so on. [Note that in this formula (S_j) represents the *molar* concentrations of S_j.] Like γ_{DH}, the higher-order virial coefficients are functions of the ionic strength, and successful empirical formulations for these coefficients are at the core of the thermodynamic description of concentrated electrolytes.[13] Up to ionic strengths of about 4 M, good agreement with experimental data can be obtained using an expansion that stops with the second virial coefficients. The higher-order terms become important only in the most concentrated systems.

Activity Coefficient of Neutral Molecules.   Neutral molecules also are found to behave nonideally in solution. This is most often observed as a *salting out* effect such as the decrease in the solubility of sucrose in the presence of sodium chloride. This effective increase in the activity coefficient of sucrose ($\gamma_{sucrose} > 1$) is mostly due to the interaction between sucrose and the sodium and chloride ions through modification of the dielectric constant of water. The added sucrose lowers the dielectric constant of water, thus increases the free energy of the charged ions and, in effect, contributes to the total free energy of the system beyond its "own" free energy: $dG > \mu_{sucrose}\ dn_{sucrose}$. This effect is taken into account by assigning an activity coefficient greater than 1 to sucrose. In practice, the activity coefficients of neutral molecules are negligibly different from 1 at ionic strengths less than 0.1 M (log $\gamma \cong 0.1\ I$), and we shall neglect such non-ideality effects.*

Note also that our whole discussion of nonideal behavior has been limited to the aqueous phase. At high pressures—a condition of little interest to us—gases also deviate from ideal behavior; they no longer verify Boyle-Mariotte's law. This is accounted for by defining *fugacity coefficients* and *fugacities*, which are to partial pressures what activity coefficients and activities are to concentrations.

1.7.4 Nonideal Effects as Ideal Processes

It is important to realize that the basic distinction between ideal and nonideal interactions among molecules is somewhat arbitrary. As noted earlier, we choose to consider some interactions as "ideal" and call those chemical re-actions. Other interactions we consider "nonideal" and relegate them to the role of correction factors on ideal effects. In some situations either of these ap-proaches can be—and is—taken, resulting in different descriptive language and apparently different mathematical formulations to describe the same chemical reality.

For example, the attractions between two ions of opposite charges in sea-water can be described by using a negative second virial coefficient in the "nonideal" specific ion interaction formula, Equation 57. Alternatively, such attraction can also be considered to produce an independent chemical species—an ion pair—whose formation is governed by the *"ideal"* mass law equation, Equation 33. From a thermodynamic point of view both approaches are equally valid, and choosing among them is thus governed strictly by practical considerations. Note that, if the interactions are attractive and not too large, the second-order virial expansion of the activity coefficient is mathematically very similar to the expression derived from the formation of ion pair complexes: compare

$$\ln\left[\frac{\gamma_{DH}(S_i)_T}{\{S_i\}}\right] = -\sum B_{ij}(S_j)$$

with

$$\ln\left[\frac{\gamma_{DH}(S_i)_T}{\{S_i\}}\right] = \ln\left[1 + \sum K_{ij}(S_j)\right]$$

In general, all types of *attractive* interactions among ions can be equally well accounted for by the formation of particular chemical species (the ideal solution approach) or by specific modifications of activity coefficients (the nonideality

---

* One should not confuse the mean activity coefficient of electrolytes with the activity coefficients of neutral molecules. The former represents a mean thermodynamic quantity for both the cation and the anion in a fully dissociated electrolyte. The latter describes the nonideal behavior of the single species in solution.

correction approach). However, in the case of concentrated solutions where repulsive forces among ions of like charge become important, the large resulting increase in the activity (i.e., the reactivity) of the ions *must be* accounted for by increasing the activity coefficients.

1.8 Effects of Pressure and Temperature on Equilibrium

Having based our whole thermodynamic development on the premise of fixed pressure and temperature, we are not in the best of positions to discuss the effects that changes in pressure and particularly temperature may have on equilibrium conditions in chemical systems. While we have so far been able to study the advancement of chemical reactions by relying solely on the notion of molar free energy, we must now belatedly introduce other fundamental thermodynamic quantities such as *molar volume, entropy,* and *enthalpy*. In this process it will become finally apparent that the free energy function G, on which we have based our whole discussion, is distinct from the more intuitive internal energy function, denoted E in classical thermodynamic texts. For example, an increase in pressure at constant volume necessitates no mechanical energy and corresponds to no change in the internal energy of a system. It corresponds, however, to a free energy change.

1.8.1   Effect of Pressure

Consider that a chemical system of volume V is subjected to an increase in pressure dP (fixed temperature; fixed composition). The increase in the free energy of the system is written

$$dG = VdP \tag{58}$$

thus defining the volume as the partial differential of the free energy with respect to the pressure. (See Section 1.9.2 for an intuitive justification of this result.) In order to account for this change in the total free energy of the system as a sum of changes in the molar free energies of individual molecules, dG must also be given by

$$dG = \sum_i n_i \frac{\partial \mu_i}{\partial P} dP \tag{59}$$

To evaluate the pressure differential of the molar free energies, it is then necessary to break down the total volume of the system into molar volumes:

$$'V = \sum_i n_i \overline{V}_i \tag{60}$$

Each (partial) molar volume $\overline{V}_i$ is defined as the increase in the total volume of the system upon the addition of an infinitesimal number of moles of the species S_i:

$$\overline{V}_i = \frac{\partial V}{\partial n_i} \tag{61}$$

For solids, $\bar{V}_i$ is simply the ratio of the molecular weight M_i of S_i to the density ρ:

$$\bar{V}_i = \frac{M_i}{\rho}$$

For gases, $\bar{V}_i$ is the ratio of the total gas volume V_g to the total number of gas molecules n_g, and thus it is the same for all the gaseous species in the same phase:

$$\bar{V}_i = \frac{V_g}{n_g} = \frac{RT}{P}$$

For solutes, $\bar{V}_i$ does not correspond to such a simple notion. In fact the partial molar volumes of aqueous solutes are often negative since addition of a salt usually results in a decrease in the total volume of the solution. $\bar{V}_i$ for a solute is also dependent, in principle, upon the composition of the solution.

Substituting Equation 60 in 58, comparing with Equation 59, and recognizing that the identity must be verified for any composition of the system, we find

$$\frac{\partial \mu_i}{\partial P} = \bar{V}_i = \frac{\partial V}{\partial n_i} \tag{62}$$

With the pressure dependence of the molar free energy being entirely accounted for in the standard molar free energy*

$$\frac{\partial \mu_i}{\partial P} = \frac{\partial}{\partial P} (\mu_i^0 + RT \ln X_i) = \frac{\partial \mu_i^0}{\partial P}$$

the final expression is

$$\frac{\partial \mu_i^0}{\partial P} = \bar{V}_i \tag{63}$$

On the basis of Equation 63 we can calculate the effect of pressure on equilibrium constants by differentiating Equation 35:

$$\frac{\partial \ln K}{\partial P} = -\frac{1}{RT} \frac{\partial}{\partial P} (\sum_i v_i \mu_i^0)$$

therefore

$$\frac{\partial \ln K}{\partial P} = -\frac{\Delta V^0}{RT} \tag{64}$$

where

$$\Delta V^0 = \sum_i v_i \bar{V}_i \tag{65}$$

The term ΔV^0 represents the change in volume of the system due to the advancement of the reaction. Equation 64 has a qualitatively pleasing aspect: a reaction

---

* As is the case for ionic strength, the effects of pressure on molar free energy can also be accounted for with an appropriate activity coefficient.

that results in an expansion of the system is favored (pushed to the right) by a decrease in pressure, and vice versa. This is of course no more than a particular application of Le Chatelier's principle. If ΔV^0 can be considered to be constant over the range of pressure considered—often not a bad approximation—the value of the equilibrium constant at a pressure P can be calculated from that at a pressure P_0:

$$\ln \frac{K_P}{K_{P_0}} = -\frac{\Delta V^0}{RT} (P - P_0) \tag{66}$$

For example, following Stumm and Morgan,[14] let us consider the reaction for calcium carbonate dissolution:

$$CaCO_3(s) = Ca^{2+} + CO_3^{2-}$$

with the data

$$\overline{V}_{CaCO_3} = +36.9 \text{ cm}^3 \text{ mol}^{-1}$$

$$\overline{V}_{Ca} = -17.7 \text{ cm}^3 \text{ mol}^{-1}$$

$$\overline{V}_{CO_3} = -3.7 \text{ cm}^3 \text{ mol}^{-1}$$

This reaction is chosen as an example because it has a rather large negative volume change:

$$\Delta V^0 = -3.7 - 17.7 - 36.9 = -58.3 \text{ cm}^3 \text{ mol}^{-1}$$

$$-\frac{\Delta V^0}{RT} = \frac{58.3 \text{ cm}^3 \text{ mol}^{-1}}{8.31 \times 298 \text{ J mol}^{-1}} = 2.35 \times 10^{-2} \text{ cm}^3 \text{ J}^{-1} \cong 2.38 \times 10^{-3} \text{ at}^{-1}$$

Then for every 1 atmosphere increase in pressure, the solubility product is increased by a factor

$$\frac{K_P}{K_{P_0}} = \exp(2.38 \times 10^{-3}) = 1.0024 \tag{67}$$

Thus despite the relatively large negative volume change of this reaction, the effect of pressure on equilibrium is rather small. In aquatic chemistry such pressure effects can generally be neglected, except in the deep ocean. In that case, the partial molar volumes cannot be considered constant as in Equation 66. They are markedly affected by pressure and ionic strength; for example, the increase in calcite solubility in deep oceanic waters is less than predicted by equation (67).

1.8.2 Effect of Temperature

The effect of temperature on equilibrium (for a system with fixed pressure and invariant composition) can be studied in a way similar to that of pressure by defining the entropy S of the system as the partial differential of the free energy with respect to temperature,*

$$dG = -SdT \tag{68}$$

and defining in the same way the partial molar entropies $\bar{S}_i$ by

$$\bar{S}_i = \frac{\partial S}{\partial n_i} \tag{69}$$

with

$$S = \sum n_i \bar{S}_i \tag{70}$$

so that

$$\frac{\partial \mu_i}{\partial T} = -\bar{S}_i \tag{71}$$

Equation 71 is similar to Equation 62. However, the notion of entropy is customarily one that is thought to be somewhat obtuse and is introduced in classical thermodynamic texts through the study of engine efficiency—the Carnot cycle, a subject not particularly close to our present preoccupations. The similarity with the study of pressure effects in fact stops with Equation 71, as the presence of the absolute temperature in the concentration dependent term of the molar free energy $[RT \ln(X_i)]$ greatly complicates the rest of the derivation.

Without stretching our intuition any further, let us then proceed more directly and accept that the standard free energy μ_i^0 of a species can be expressed as a linear function of temperature. Using the absolute temperature scale and keeping our notation coherent with that normally used in thermodynamics (inelegant as it may seem here to denote *linear coefficients* by H_i and S_i), we then write

$$\mu_i^0 = H_i^0 - T\bar{S}_i^0 \tag{72}$$

For the purpose of this simplified derivation, we take $\bar{H}_i^0$ to be constant over the range of temperatures considered.

The expression of the molar free energy thus becomes

$$\mu_i = \mu_i^0 + RT \ln X_i$$

$$\mu_i = \bar{H}_i^0 - T\bar{S}_i^0 + RT \ln X_i$$

Dividing by T and differentiating with respect to T,

$$\frac{\partial(\mu_i/T)}{\partial T} = \frac{\partial(\mu_i^0/T)}{\partial T} = \frac{\partial(\bar{H}_i^0/T)}{\partial T} = -\frac{\bar{H}_i^0}{T^2} \tag{73}$$

* As introduced in Equations 58 and 68, the variations in pressure and temperature of a closed system must necessarily satisfy the generalized Gibbs-Duhem formula:

$$SdT - VdP + \sum n_i \, d\mu_i = 0$$

The effect of temperature on equilibrium constants is obtained by differentiating Equation 35:

$$\frac{\partial \ln K}{\partial T} = -\frac{1}{R}\frac{\partial}{\partial T}\left(\sum_i \frac{v_i \mu_i^0}{T}\right)$$

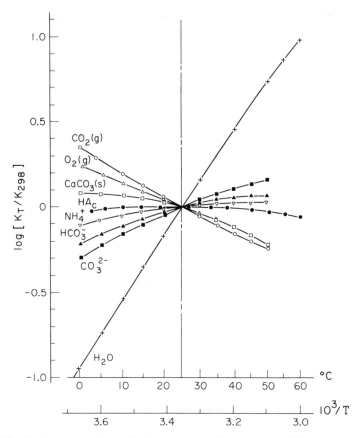

Figure 2.2 Variations of a few equilibrium constants with temperature. The reactions, the equilibrium constants at 25°C, and the corresponding data sources are as follows:

$CO_2(g):CO_2(g) = CO_2(aq)$; $\log K_{298} = -1.464$ (Harned and Davis, 1943[15]).

$O_2(g):O_2(g) = O_2(aq)$; $\log K_{298} = -3.066$ (Carpenter, 1966[16]).

$CaCO_3(s):CaCO_3(s) = Ca^{2+} + CO_3^{2-}$; $\log K_{298} = -8.475$ (Jacobson and Langmuir, 1974[17]).

HAc: $HAc = H^+ + Ac^-$; $\log K_{298} = -4.756$ (Harned and Ehlers, 1933[18]).

$NH_4^+:NH_4^+ = H^+ + NH_3(aq)$; $\log K_{298} = -9.249$ (*CRC Handbook of Chemistry and Physics*, 62nd ed., 1981[19]).

$HCO_3^-:CO_2 + H_2O = H^+ + HCO_3^-$; $\log K_{298} = -6.351$ (Harned and Davis, 1943[15]).

$CO_3^{2-}:HCO_3^- = H^+ + CO_3^{2-}$; $\log K_{298} = -10.330$ (Harned and Scholes, 1941[20]).

$H_2O:H_2O = H^+ + OH^-$; $\log K_{298} = -13.9965$ (*CRC Handbook of Chemistry and Physics*, 62nd ed., 1981[19]).

which results in

$$\frac{\partial \ln K}{\partial T} = + \frac{\Delta H^0}{RT^2} \tag{74}$$

where

$$\Delta H^0 = \sum_i v_i \overline{H}_i^0 \tag{75}$$

The relationship between equilibrium constants at two temperatures T_0 and T is obtained by integrating Equation 74:

$$\ln \left(\frac{K_T}{K_{T_0}}\right) = -\frac{\Delta H^0}{R}\left(\frac{1}{T} - \frac{1}{T_0}\right) \tag{76}$$

This formula, which is known as the van't Hoff equation, holds rather well over the range of temperatures considered in aquatic chemistry. It is particularly useful because ΔH^0 is in fact a quantity measurable a priori: it is the heat absorbed by the chemical reaction as it proceeds to the right at constant pressure and is called the *change in enthalpy* of the reaction. Although we have little basis to understand this fundamental nature of $\overline{H}_i^0$ as introduced in Equation 72, it remains that Equation 74 is also intuitively pleasing if we have Le Chatelier's intuition: an exothermic reaction ($\Delta H^0 < 0$) is favored (K increases) by a decrease in temperature, and vice versa.

Figure 2.2 illustrates the magnitude of the temperature effect on various equilibrium constants. The effect is rather large, with some equilibrium constants changing by a factor of 10 over the 35°C span typically encountered in natural waters. Unlike pressure corrections, temperature corrections of equilibrium constants cannot in general be neglected in aquatic chemistry. Although in this text we shall assume for simplicity a constant temperature of 25°C, it is important to make appropriate temperature corrections when studying a particular water body at another temperature. Note that some of the lines on Figure 2.2 exhibit noticeable curvature and thus do not obey Equation 76. This can be accounted for in more complete formulae by considering the change in the heat capacity of the system due to the advancement of the reaction.

1.9 Concentration Gradients in Equilibrium Systems

We have seen in Section 1.6.2 that one of the consequences of thermodynamic equilibrium in a solution phase is the uniform concentration of species throughout the phase. This result is not strictly correct, but the very small or very large distances over which equilibrium concentration gradients exist in aquatic chemistry usually obscure or preclude their observation.

Temperature gradients are not possible in equilibrium systems, but pressure or electrical potential gradients are quite common. Neither atmospheric nor hydrostatic pressure gradients require a continuous input of energy or result in perpetual motion, nor does the decline in electrical potential away from a charged surface. The result of uniform concentrations in Section 1.6.2 was obtained by ignoring such gradients, an omission that we wish to repair here.

1.9.1   Extended Expression of Molar Free Energy

To study the effects of pressure and electrical potential gradients on chemical equilibrium, we follow our now well-established methodology and extend the expression of molar free energies (Equation 5), to include the contribution of each species to the corresponding mechanical or electrical energy of the system [respectively, $\overline{V}_i(P - P_0)$ as seen in Section 1.8.1, and $Z_iF(\psi - \psi_0)$ as seen in Section 1.7.1]. Other types of energies could be included if we wished to consider other processes such as surface tension. In the study of tall water columns and their pressure gradients we *have* to include gravitational energy. Since the molar contribution of a species to the total mass m of a system is simply its molecular weight $M_i(m = \sum n_iM_i)$, the molar gravitational energy is given by

$$M_ig(z - z_0)$$

where g is the gravity of the earth (taken here as constant for justifiable simplicity) and $(z - z_0)$ is some elevation above a reference elevation z_0.

The extended expression of molar free energy is thus written

$$\mu_i = \mu_i^0 + RT \ln X_i + \overline{V}_i(P - P_0) + Z_iF(\psi - \psi_0) + M_ig(z - z_0) \quad (77)$$

Like the simpler expression, Equation 5, this equation satisfies the Gibbs-Duhem relation. The condition of uniformity of the molar free energy in an equilibrium system is thus applicable

$$\mu_i = \text{constant throughout the system} \quad (37)$$

as is the equation

$$\sum_i v_i\mu_i = 0 \quad (31)$$

if the species are involved in a chemical reaction with stoichiometry v_i.

Equation 77 provides a particularly convenient way to study the effects of pressure or electrical potential gradients on species concentrations (or vice versa, as in the case of osmotic pressure). We shall apply this equation in Chapter 8 to the study of the distribution of ions near an electrically charged surface. For now, as an illustration of the methodology, let us use it to obtain the equilibrium distribution of chemical species in a tall water column, omitting the electrical term. For convenience we replace the mole fraction by the activity of the species:

$$\mu_i = \mu_i^0 + RT \ln \{S_i\} + \overline{V}_i(P - P_0) + M_ig(z - z_0) \quad (78)$$

where

$$\{S_i\} = 1 \text{ for solids}$$

$$= (S_i) \text{ for ideal solutes}$$

$$= P_i \text{ for ideal gases}$$

□ □ □

1.9.2   Distribution of Species in a Tall Water Column

Water.   Consider first the distribution of the solvent (the water is taken as incompressible):

$$\{water\} = 1; \quad M_w = \rho_w \bar{V}_w$$

therefore

$$\mu_w = \mu_w^0 + \bar{V}_w(P - P_0) + \rho_w \bar{V}_w g(z - z_0)$$

Taking the reference pressure to be that of the atmosphere ($P_0 = P_{at}$) at the surface of the water, which is itself taken as the origin of elevations, $z_0 = 0$, we derive the hydrostatic pressure formula from the condition of uniformity of molar free energy:

$$\mu_w = constant = \mu_w^0$$

therefore

$$P = P_{at} - \rho_w gz \tag{79}$$

(Remember that z is negative at depth.)

Conversely, on the basis of the hydrostatic pressure formula—which is merely an Archimedean statement of mechanical equilibrium—the form of the pressure term in the free energy expression can be justified. By describing the equilibrium between two adjacent layers of an incompressible fluid as both a balance of mechanical forces and an equality of free energies, one can derive Equation 58.

Solids.   An equally familiar result is obtained when the equation is applied to a solid species:

$$\{S_i\} = 1; \quad M_i = \rho_i \bar{V}_i \quad [P_0 = P_{at}; z_0 = 0]$$

therefore

$$\mu_i = \mu_i^0 + \bar{V}_i(P - P_{at}) + \rho_i \bar{V}_i gz$$

Introducing Equation 79,

$$\mu_i = \mu_i^0 + \bar{V}_i(\rho_i - \rho_w)gz$$

The molar free energy of the solid cannot in general be uniform in the water column:

1   $\rho_i > \rho_w$, the solid is denser than water and sinks (at equilibrium μ_i must be minimum and thus z has the maximum negative value possible).

2   $\rho_i < \rho_w$, the solid is lighter than water and floats (at equilibrium μ_i must be minimum and thus z has its maximum value $z = 0$).

Solutes.   Let us finally apply Equation 78 to a solute S_i and obtain a less commonly known result:

$$\mu_i = \mu_i^0 + RT \ln(S_i) + \bar{V}_i(P - P_{at}) + M_i gz$$

Introducing Equation 79,

$$\mu_i = \mu_i^0 + RT \ln(S_i) + (M_i - \rho_w \overline{V}_i)gz$$

The condition of uniformity of the molar free energy then yields

$$\mu_i = \text{constant} = \mu_i^0 + RT \ln(S_i)_0$$

where $(S_i)_0$ is the concentration of the solute at the surface; therefore

$$RT \ln \frac{(S_i)}{(S_i)_0} = -[M_i - \rho_w \overline{V}_i]gz$$

or

$$(S_i) = (S_i)_0 \exp - \left(\frac{M_i - \rho_w \overline{V}_i}{RT} gz \right) \tag{80}$$

The distribution of solutes at equilibrium in a tall water column depends on the size and magnitude of the quantity $(M_i - \rho_w \overline{V}_i)$. The concentrations of the many species that have negative partial molar volumes in water (e.g., Ca^{2+} and CO_3^{2-}, see Section 1.81) increase exponentially with depth ($z < 0$).

1.9.3   Chemical Equilibrium in a Tall Water Column

The fundamental condition of equilibrium for a chemical reaction

$$\sum v_i \mu_i = 0 \tag{30}$$

applied to Equation 78 yields

$$\{S_1\}^{v_1}\{S_2\}^{v_2}\{S_3\}^{v_3} \ldots = K_z \tag{81}$$

where

$$-RT \ln[K_z] = \sum v_i \mu_i^0 + \sum v_i \overline{V}_i (P - P_{at}) + \sum v_i M_i gz$$

Introducing

$$-RT \ln K_0 = \sum v_i \mu_i^0 \tag{35}$$

$$\sum v_i \overline{V}_i = \Delta V^0 \tag{65}$$

$$P = P_{at} - \rho_w gz \tag{79}$$

$$\sum v_i M_i = 0 \quad \text{(conservation of mass)}$$

therefore

$$K_z = K_0 \left[\exp \left(\frac{\rho_w gz}{RT} \Delta V^0 \right) \right] \quad \text{(compare with 66)} \tag{82}$$

In reactions involving only solutes, the concentration variations due to depth in fact compensate for the effect of pressure on the equilibrium constant, and the reaction is at equilibrium throughout the water column.

For a reaction involving a solid, the situation is different. At equilibrium the solid must either float at the top or rest at the bottom. Consider, for example, a tall water column containing a saturating concentration of $CaCO_3$. In this case the solid sinks. To calculate its equilibrium, we take the origin of elevation and pressures at the bottom of the water column, i.e., at $z = 0$, $P = P_{max}$:

$$(Ca^{2+})_z = (Ca^{2+})_0 \exp\left[\frac{-M_{Ca} + \rho_w \overline{V}_{Ca}}{RT} gz\right]$$

With $\rho_w = 1$ and z in kilometers,

$$(Ca^{2+})_z = (Ca^{2+})_0 \exp(-0.236z)$$

$$(CO_3^{2-})_z = (CO_3^{2-})_0 \exp\left[\frac{-M_{CO_3} + \rho_w \overline{V}_{CO_3}}{RT} gz\right]$$

$$(CO_3^{2-})_z = (CO_3^{2-})_0 \exp(-0.260z)$$

$$K_z = K_0 \exp\left[\frac{\rho_w gz}{RT} (\overline{V}_{Ca} + \overline{V}_{CO_3} - \overline{V}_{CaCO_3})\right]$$

$$K_z = K_0 \exp(-0.238z)$$

Consider the ratio of the concentration product to the solubility product at each depth:

$$\frac{(Ca^{2+})_z(CO_3^{2-})_z}{K_z} = \exp\frac{gz}{RT} [\rho_w \overline{V}_{CaCO_3} - M_{Ca} - M_{CO_3}]$$

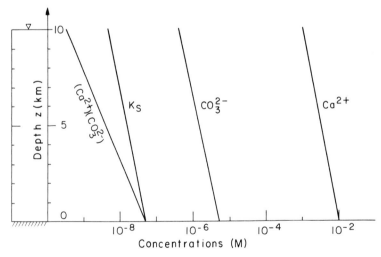

Figure 2.3 Theoretical equilibrium distribution of calcium and carbonate ions in a tall water column saturated with $CaCO_3(s)$. In the absence of mixing, the solid calcium carbonate rests on the bottom where the ion product $(Ca^{2+})(CO_3^{2-})$ is equal to the solubility product K_s. At shallower depths the decrease in concentrations of the individual ions due to the effects of pressure on free energies more than makes up for the decrease in the solubility product K_s. The solution is thus undersaturated compared to $CaCO_3(s)$ at all depths except the bottom.

Introducing $M_{Ca} + M_{CO_3} = M_{CaCO_3} = \overline{V}_{CaCO_3}\rho_{CaCO_3}$,

$$\frac{(Ca^{2+})_z(CO_3^{2-})_z}{K_z} = \exp\frac{gz}{RT}\,\overline{V}_{CaCO_3}(\rho_w - \rho_{CaCO_3}) \tag{83}$$

Since $\rho_w < \rho_{CaCO_3}$, this ratio is less than one at all depths except at the bottom (see Figure 2.3). The water is thus in equilibrium with $CaCO_3$ at the bottom and undersaturated throughout the water column despite the decrease in solubility of the solid at lower pressures.

In order to obtain a cycle of precipitation at the surface and of dissolution at depth as is observed in the oceans, a mixing process must be invoked in addition to the effect of pressure on the solubility product of $CaCO_3$.

□ □ □

2. ELEMENTS OF KINETICS

In both our discussion of mass conservation and our sketch of chemical thermodynamics, we have considered chemical reactions as simple stoichiometric relations among species in a system. Our symbolism for a chemical reaction shows that some species can react together to form some products in a given proportion, and thermodynamics tells us to what extent the reaction is energetically favorable. Neither says anything regarding *how* the reaction actually takes place (the path of the reaction), what are the kinetics of the reaction, or even if in fact it can proceed at all. At ordinary pressures and temperatures some reactions such as the formation of nitrous oxides from nitrogen and oxygen gas, which are thermodynamically quite favorable, do not proceed in the absence of biological activity—and mercifully so, for the integrity of our oxygen supply. In order to understand the chemistry of our environment, we must then ultimately understand the kinetics, and hence the mechanisms, of the reactions involved.

The kinetics of chemical reactions in natural waters are complicated by a multitude of factors that are carefully eliminated from the beakers of the chemical kineticist. In addition to being subjected to complex transport processes and supporting biological activity that promotes very specific reactions, aquatic systems usually contain a multitude of trace compounds that may catalyze or inhibit the reactions of interest. Surface waters also contain short-lived but very active radicals produced by solar ultraviolet radiation. As a result aquatic chemical kinetics is an extremely complicated and largely undeveloped discipline which is still awaiting a systematic treatment.[21] Here we merely define some terminology and present a few principles of chemical kinetics to serve as a foundation for our discussions of kinetic processes in aquatic systems presented in various later chapters, as they appear topical.

2.1 Elementary Processes

The formation of a complex species such as the trichloromercuric ion $HgCl_3^-$ from its constituents, the mercuric and the chloride ions, does not proceed in the one step symbolized by the reaction:

$$Hg^{2+} + 3Cl^- = HgCl_3^- \tag{84}$$

The probability of simultaneous encounter and reaction among four separate ions is obviously very low, and it is reasonable to imagine that the reaction takes place as a stepwise formation of complexes with increasing chloride coordination:

$$Hg^{2+} + Cl^- = HgCl^+ \tag{85}$$

$$HgCl^+ + Cl^- = HgCl_2^0 \tag{86}$$

$$HgCl_2^0 + Cl^- = HgCl_3^- \tag{87}$$

The three stepwise reactions so written may appear simple enough to represent a mechanism, and, depending on what we exactly mean by that, they actually do. However, the simplified symbolism that we are using here is misleading: the free mercuric ion Hg^{2+} is not really free, it is in fact coordinated to six water molecules: $Hg^{2+} = Hg(H_2O)_6^{2+}$. The complexation with chloride ion is thus an exchange of a water molecule for a chloride ion, and this process is thought to take place by formation of an intermediary *ion pair* (or, more correctly, an *outer sphere complex*) before expulsion of the water molecule; for example,

$$Hg(H_2O)_5Cl^+ + Cl^- = Hg(H_2O)_5Cl^+ \cdot Cl^-$$

$$Hg(H_2O)_5Cl^+ \cdot Cl^- = Hg(H_2O)_4Cl_2^0 + H_2O$$

Even if we lump these two steps together and consider the symbolism of Reaction 86 as representing an "elementary" reaction step, we should recall that an essential additional step for the overall reaction to proceed is that the reacting ions diffuse to each other.

From a mechanistic point of view, chemical reactions may thus comprise several elementary steps, including the formation of intermediary, possibly transient, species and the necessary transport of reactants to each other. Elementary chemical reactions may involve one, two, or (rarely) three individual molecules (*unimolecular*, *bimolecular*, or *termolecular* reactions), they may be reversible or irreversible, and they may take place in series or in parallel (*consecutive* or *concurrent*) reactions.

2.2 Rate Laws

The mathematical description of the kinetics of chemical reactions—the differential equations of the time rate of change of concentrations as a function of concentrations—are known as rate laws. Elementary chemical reactions have very simple rate laws:

1 In unimolecular reactions

$$AB \xrightarrow{k} A + B \tag{88}$$

$$\frac{d(A)}{dt} = \frac{d(B)}{dt} = -\frac{d(AB)}{dt} = k(AB) \tag{89}$$

2 In bimolecular reactions

$$A + B \xrightarrow{k'} AB \tag{90}$$

$$\frac{d(AB)}{dt} = -\frac{d(A)}{dt} = -\frac{d(B)}{dt} = k'(A)(B) \tag{91}$$

Just from the probabilistic nature of the reactions we expect the concentration dependency of these rate laws. The kinetics of the reversible reaction

$$A + B \underset{k_b}{\overset{k_f}{\rightleftarrows}} AB \tag{92}$$

are then described by the rate expression

$$\frac{d(AB)}{dt} = -\frac{d(A)}{dt} = -\frac{d(B)}{dt} = k_f(A)(B) - k_b(AB) \tag{93}$$

If the reaction is at equilibrium, all the time differentials are null:

$$\frac{(AB)}{(A)(B)} = \frac{k_f}{k_b} = K \tag{94}$$

This important relationship equating the equilibrium constant to the ratio of the rate constants is known as the law of *microscopic reversibility*.[1,22] It applies to all reactions that are single-step processes, including many important reactions in aquatic systems such as protonation of weak acids, hydrolysis and complexation of metal ions, and many electron transfer reactions. The law of microscopic reversibility also applies obviously to reactions consisting of a series of reversible elementary steps (consecutive opposing reactions) in which there is but a single rate-limiting step, all others being considered at (pseudo) equilibrium. In general, overall reactions involving several elementary steps may have quite complicated rate laws that bear no relation to the stoichiometry of the overall reactions. In particular, the rate laws of reactions that have irreversible or concurrent steps do not consist simply of two polynomial expressions of opposite signs.

Reactions Involving an Irreversible Step. Consider, for example, a typical enzymatic reaction: a substrate S combines reversibly with an enzyme E to form a complex ES which then dissociates irreversibly into a product P and the regenerated enzyme E:

$$S + E \underset{k_b}{\overset{k_f}{\rightleftarrows}} ES \qquad (95)$$

$$ES \overset{k}{\rightarrow} P + E \qquad (96)$$

Choosing the initial conditions,

$$(ES)_0 = 0$$

$$(E)_0 = E_0$$

The rate laws for the two reactions give the differential equation for ES:

$$\frac{d(ES)}{dt} = +k_f(S)(E) - k_b(ES) - k(ES)$$

Introducing the mole balance $(ES) + (E) = E_0$, and assuming that the substrate concentration (S) is relatively large (thus constant), we can integrate this equation:

$$(ES) = \frac{k_f E_0(S)}{k_b + k + k_f(S)} \left[1 - \exp(-k_b - k - k_f(S))t\right]$$

The rate law of formation of the product P can then be written

$$\frac{d(P)}{dt} = k(ES) = \frac{kE_0(S)}{(k_b + k)/k_f + (S)} \left[1 - \exp(-k_b - k - k_f(S))t\right]$$

After sufficient time the exponential term becomes negligible, leading to a steady-state solution in the form of the Michaelis-Menten or Monod expression:

$$\left[\frac{d(P)}{dt}\right]_{\infty} = \frac{V_{max}(S)}{K + (S)} \qquad (97)$$

where

$$V_{max} = kE_0$$

$$K = \frac{k_b + k}{k_f}$$

Such hyperbolic expression is widely applicable—and even more widely applied—to a number of enzymatic and transport processes.

Concurrent Reactions. As an example of kinetic process involving concurrent reactions, consider the formation of bicarbonate ion from carbon dioxide in solution which is of great importance in aquatic chemistry (see Chapter 4):

$$CO_2(aq) + H_2O = HCO_3^- + H^+$$

This reaction is thought to consist of two parallel elementary reactions:

$$CO_2 + H_2O \underset{k_b}{\overset{k_f}{\rightleftharpoons}} HCO_3^- + H^+ \tag{98}$$

$$CO_2 + OH^- \underset{k_b'}{\overset{k_f'}{\rightleftharpoons}} HCO_3^- \tag{99}$$

(The first of these two reactions is actually composed of two steps in series: a "slow" hydration of CO_2 to carbonic acid H_2CO_3 followed by a fast dissociation of the carbonic acid.)

The total rate law is written

$$\frac{d(HCO_3^-)}{dt} = \frac{-d(CO_2)}{dt} = k_f(CO_2) - k_b(HCO_3^-)(H^+)$$

$$+ k_f'(CO_2)(OH^-) - k_b'(HCO_3^-)$$

resulting in

$$\frac{d(HCO_3^-)}{dt} = [k_f + k_f'(OH^-)](CO_2) - [k_b' + k_b(H^+)](HCO_3^-) \tag{100}$$

This form of the rate law is observed experimentally[23] (and can be used to justify the foregoing elementary steps) with constants

$$k_f = 3 \times 10^{-2}\ s^{-1}; \qquad k_b = 7 \times 10^4\ M^{-1}\ s^{-1};$$
$$k_f' = 8.5 \times 10^3\ M^{-1}\ s^{-1}; \quad k_b' = 2 \times 10^{-4}\ s^{-1} \tag{101}$$

2.3 Order of Reaction

A reaction such as Reaction 88, $AB \rightarrow A + B$, is said to be first order. Reaction 90, $A + B \rightarrow AB$, is said to be second order; it is also said to be first order in A and first order in B. A reaction with a rate law of the form

$$\frac{d(\text{product})}{dt} = kA^\lambda B^\mu C^\nu$$

is said to be of $(\lambda + \mu + \nu)$th order, of λth order in A, and so on. Cases of zeroth-order reactions—reactions with fixed rates—are encountered in natural waters, for example, in the dissolution of some solids.

Consider Reaction 90, $A + B \rightarrow AB$. If reactant B is in great excess of A, the concentration of B remains effectively constant. The dependency of the rate law on B can then be included in the rate constant $[k'' = k'(B)]$, and the reaction is called a *pseudo-first-order reaction* (in A). This case is more the rule than the exception in natural waters. An extreme case is that of reactions involving water as a reactant (e.g., Reaction 98); the water concentration is always effectively constant, so the order of the reaction with respect to water never appears in the rate expression. A reaction that follows a hyperbolic rate law, such as Equation 97, is first order at low-substrate concentrations and zeroth order at high-substrate concentrations.

2.4   Transition State Theory

Because of the relative ease with which thermodynamic analysis can be made compared to kinetic analysis, there is much interest in theories that may provide a link between kinetics and thermodynamics. One may seek an intuitive justification for the existence of such a link by extension of our earlier feather metaphor. Consider a feather lying on the floor of a classroom with an open window. If there is a sufficient draft, the feather eventually will find its way through the window down onto a more stable, lower energy, state at the street pavement. Clearly, the kinetics of this process are independent of the exact floor where the room is situated but very much dependent on the height of the window above the classroom floor. Were we to repeat this experiment many times and raise the windowsill, we would certainly find some correlation between the height of the windowsill and the time necessary for the feather to get down onto the street. The height of the energy barrier to be overcome before reaching a lower energy state governs the overall kinetics of the process. This is the concept behind most theories relating kinetics and thermodynamics.

Consider, for example, the all important bimolecular reaction

$$A + B \underset{}{\overset{k_f}{\rightleftarrows}} AB \tag{92}$$

and assume that the forward reaction is in fact composed of two steps, the reversible formation of an intermediary high energy compound $AB^{\#}$ followed by an energetically favorable reaction of $AB^{\#}$ into the final product AB:

$$A + B = AB^{\#}; \quad K^{\#} \tag{102}$$

$$AB^{\#} \overset{k}{\rightarrow} AB \tag{103}$$

By assuming (pseudo) equilibrium between the *activated state* $AB^{\#}$ and the reactants A and B, the rate law can be written

$$-\frac{d(A)}{dt} = -\frac{d(B)}{dt} = \frac{d(AB)}{dt} = k(AB^{\#}) = kK^{\#}(A)(B)$$

which leads to

$$k_f = kK^{\#} \tag{104}$$

Considering that the constant k may be the same for various reactions—better yet, that it is universal—we can then link the kinetics of the overall reaction to the thermodynamics of the activated compound. Suppose, for example, that in a first approximation k is independent of temperature, pressure, or ionic strength. The results that have been established for the effect of these parameters on equilibrium constants are then directly applicable to rate constants.

Consider, in particular, the effect of temperature on k_f. In accordance with our earlier notation (see Equation 72), let us write the standard free energy change of the activated compound as a linear function of temperature and define $\Delta G^{\#}$, $\Delta H^{\#}$, and $\Delta S^{\#}$ so that

$$RT \ln K^{\#} = -\Delta G^{\#} = -(\Delta H^{\#} - T \Delta S^{\#})$$

and

$$k_f = kK^{\#} = k \exp\left(\frac{-\Delta G^{\#}}{RT}\right) \qquad (105)$$

The temperature dependent term can then be explicitly isolated:

$$k_f = \left[k \exp\left(\frac{\Delta S^{\#}}{R}\right) \right]\left[\exp\left(\frac{-\Delta H^{\#}}{RT}\right) \right] \qquad (106)$$

thus we have

$$k_f = \lambda e^{-E_a/RT} \qquad (107)$$

where

$$\lambda = k \exp\left(\frac{\Delta S^{\#}}{R}\right) \quad \text{and} \quad E_a = \Delta H^{\#}$$

This last equation is in fact the Arrhenius equation, which describes the effect of temperature on chemical kinetics according to an energy of activation. E_a. E_a ranges from 10 to 100 kJ mol^{-1}, the typical value of 50 kJ mol^{-1} (12 kcal mol^{-1}) yielding the well-known rule of thumb that reaction rates double for every 10°C increase in temperature. In the framework of transition state theory the Arrhenius equation is formally similar to the van't Hoff equation. [Note that E_a is not strictly equal to the enthalpy of the (endothermic) formation of the activated state. E_a also contains a term from k which is proportional to the absolute temperature, not independent of it ($E_a = \Delta H^{\#} + RT$). However, such proportionality corresponds to only a 3% rate increase for every 10°C at room temperature, a small contribution to the total temperature effect.]

2.5 Linear Free Energy Relations (LFER)

In many instances it is observed that the rate and equilibrium constants for a series of related elementary reactions are highly correlated. For example, Figure 2.4 shows how dissociation rates of various nickel-ligand complexes are almost inversely proportional to the corresponding equilibrium constants.[21] According to the law of microscopic reversibility, such proportionality is consistent with a quasi-uniform rate constant for the corresponding reverse reactions, that is, the formation of the various complexes:

$$\frac{k_b'}{k_b} = \frac{1/K'}{1/K} = \frac{k_b'/k_f'}{k_b/k_f}$$

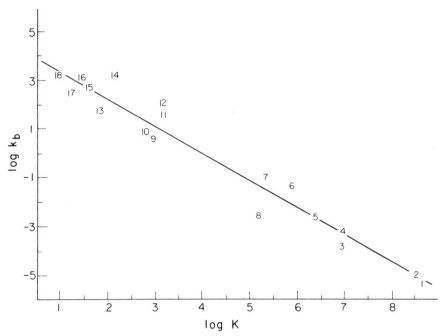

Figure 2.4 Linear free energy relationship for nickel complexes. The ordinate is the log of the dissociation constant of the various complexes (sec^{-1}), and the abscissa is the log of the stability constant (M^{-1}). 1 = histidine, 2 = phenanthroline, 3 = dipyridyl, 4 = salicylate, 5 = sulfosalicylate, 6 = proline, 7 = alanine, 8 = oxalate, 9 = imidazole, 10 = ammonia, 11 = ethylmalonate, 12 = HEDTA, 13 = pyridine, 14 = phthalate, 15 = lactate, 16 = acetate, 17 = thiocyanate, 18 = sulfate. Adapted from Hoffmann. 1981. [21]

therefore

$$k'_f = k_f$$

From a mechanistic point of view this is justified by considering that the rate of formation of each complex is controlled primarily by the rate of loss of a co-ordinated water molecule from the nickel-ligand outer sphere complex (see Chapter 6). This process can be shown to be only moderately dependent on the nature of the ligand.

According to the transition state theory a linear relation between rate and equilibrium constants is expected if the differences in the energies of activation are proportional to the differences in the free energies of formation of the products (hence the name LFER):

$$E_a^1 - E_a^2 = \alpha(\Delta G_1^0 - \Delta G_2^0) \tag{108}$$

If the entropy terms are the same (a condition that is thus necessary for a reaction series to exhibit a LFER), Equation 106 yields

$$\ln k_f^1 - \ln k_f^2 = -\frac{1}{RT}(\Delta H_1^{\#} - \Delta H_2^{\#}) = \frac{1}{RT}(E_a^1 - E_a^2)$$

Introducing Equation 108,

$$\ln k_f^1 - \ln k_f^2 = -\frac{\alpha}{RT}(\Delta G_1^0 - \Delta G_2^0)$$

therefore

$$\ln k_f^1 - \ln k_f^2 = -\alpha(\ln K_1 - \ln K_2) \tag{109}$$

Equation 109 is the general form of a LFER. The theoretical conditions under which such proportionality of differences in activation energies and in free energies of formation is applicable to a reaction series have been extensively studied by kineticists.[21] (The coefficient of proportionality α need not be unity.) There is then, in the absence of more specific information, a sound theoretical basis for utilizing empirically derived LFER's to estimate kinetic parameters.

2.6   Partial and Pseudoequilibrium

Although it would seem ideal to have complete kinetic information for any chemical system of interest (i.e., rate laws for all the reactions), it is usually impossible and often unnecessary. Many chemical reactions are fast, as compared to the time scales of interest to aquatic chemists, and can effectively be considered at equilibrium. Because of the great mathematical complexity that results from coupling the rate laws of even a few reactions, it is quite undesirable to include more than the absolutely necessary kinetic data to describe chemical systems, even in the rare cases where such data are all available.

Although precise rate laws are usually not available, we have semiquantitative (order-of-magnitude) knowledge of the rates of many of the chemical processes taking place in natural waters, from microseconds for ionization reactions to millennia for some geological processes in sediments. Equilibrium "models" are then made to include this semiquantitative knowledge of kinetics by considering all reactions with characteristic times shorter than a chosen time scale to reach equilibrium and all with longer characteristic times not to take place at all. This approximation of *partial* equilibrium is all the better if the chemical reactions considered have widely different time scales.

The choice of the time scale of interest is dependent on mixing characteristics of the system and on the questions being asked of the chemical model. For example, we shall study models of the carbonate system in which the gas exchange reaction with carbon dioxide in the atmosphere is considered either to reach equilibrium or not to take place at all, depending on whether we are interested in hourly or seasonal behavior of a surface water. Situations of course occur where the time scale of interest does not permit the "freezing" of the kinetics of some (usually few) important reactions that happen to proceed over that very time scale. A kinetic description of these reactions then becomes a necessity, and it is usually superimposed on a *pseudoequilibrium* model for all the faster reactions. The conditions under which such a pseudoequilibrium approximation is applicable are discussed in the next three subsections. The major point to be emphasized here is that equilibrium models for natural waters are *partial* equilibrium models and that the choice of reactions that are included or excluded permits the capture of some kinetic information in the models.

□ □ □

In order to define mathematically the conditions under which chemical reactions may be considered approximately at equilibrium, it is necessary to examine the set of differential equations that describes the time evolution of chemical systems. Only in very simple situations can explicit closed-form general solutions be obtained, and we shall thus restrict our analysis to three elementary examples in order to illustrate the applicability of equilibrium and pseudo-equilibrium conditions.

2.6.1   One Reversible First-Order Reaction in a Closed System

Consider, first, the simple case of a closed system containing two species, A and B, reacting according to the reversible first-order or pseudo-first-order reaction:

$$A \underset{k_b}{\overset{k_f}{\rightleftharpoons}} B$$

with initial conditions A_0 and B_0. The differential equations are written:

$$\frac{dA}{dt} = -k_f A + k_b B \tag{110}$$

$$\frac{dB}{dt} = +k_f A - k_b B \tag{111}$$

Adding these two equations together and integrating leads to the familiar mole balance equation:

$$A + B = A_0 + B_0 = T \tag{112}$$

Combining Equations 112 and 110,

$$\frac{dA}{dt} = -(k_f + k_b)A + k_b T$$

Integrating then gives the solution for A:

$$A = A_\infty + (A_0 - A_\infty)e^{-\lambda t} \tag{113}$$

where

$$\lambda = k_f + k_b \tag{114}$$

$$A_\infty = \frac{k_b}{k_f + k_b} T \tag{115}$$

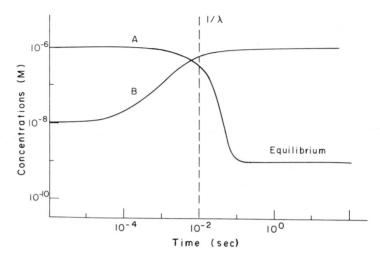

Figure 2.5 Example of the kinetics of a first-order reaction in a closed system. The conditions are those described in Section 2.6.1, with the values $A_0 = 10^{-6}$ M, $B_0 = 10^{-8}$ M, $k_f = 10^2$ sec^{-1}, $k_b = 10^{-1}$ sec^{-1}. Equilibrium between A and B is achieved for times in large excess of the characteristic time $1/\lambda = 10^{-2}$ sec.

A symmetrical solution is obtained for B:

$$B = B_\infty + (B_0 - B_\infty)e^{-\lambda t} \tag{116}$$

where

$$B_\infty = \frac{k_f}{k_f + k_b} T \tag{117}$$

A_∞ and B_∞ are the equilibrium concentrations of A and B.

This solution to the kinetic problem thus introduces a characteristic time $1/\lambda = 1/(k_f + k_b)$. As illustrated in Figure 2.5, for $t \ll 1/\lambda$ the reaction has made no tangible progress, and the initial conditions prevail in the system. For $t \gg 1/\lambda$ the reaction is complete, and the system can be considered at equilibrium. How fast the concentration of a reactant is within, say, 5% of its equilibrium value depends on its initial concentration:

$$\frac{|A_0 - A_\infty|}{A_\infty} e^{-\lambda t} < 0.05$$

therefore

$$t > -\frac{1}{\lambda} \ln\left(0.05 \frac{A_\infty}{A_0 - A_\infty} \right)$$

and

$$t > \frac{3}{\lambda} - \frac{1}{\lambda} \ln \frac{A_\infty}{A_0 - A_\infty}$$

If the initial value of A is within 5% of its equilibrium value, this condition is satisfied at all (positive) times. Note that the time necessary to approach the equilibrium concentration is different for the two reactants. Due to initial conditions, one of the reactants may be close to its equilibrium concentration while the other is not.

2.6.2 One Reversible First-Order Reaction in an Open System

Introducing a modest complication to our previous example, let us now consider a flow rate Q through the well-mixed system (volume $= V$, hydraulic rate $r = Q/V$, influent concentrations A_i and B_i, and effluent concentrations A and B). The governing equations are now

$$\frac{dA}{dt} = rA_i - (k_f + r)A + k_b B \tag{118}$$

$$\frac{dB}{dt} = rB_i - (k_b + r)B + k_f A \tag{119}$$

For simplicity, let us consider the initial conditions: $A_0 = B_0 = 0$. (Note that this choice is not indifferent; as noted earlier, initial conditions have a bearing on how fast equilibrium is reached.) Let us solve first for the total concentration $T = A + B$. Adding Equations 118 and 119 and defining $T_i = A_i + B_i$:

$$\frac{dT}{dt} = rT_i - rT$$

therefore

$$T = T_i(1 - e^{-rt}) \tag{120}$$

As expected, the total concentration is unaffected by the chemical reaction, and its progress toward its stable value T_i (steady state) depends only on the hydraulic rate r.

Consider now Equation 118. Obtaining from it an expression for B and, by differentiation, an expression for dB/dt, and introducing these in Equation 119, we obtain a second-order differential equation for A:

$$\frac{d^2A}{dt^2} + (\lambda + r)\frac{dA}{dt} + r\lambda A = r\lambda A_\infty \tag{121}$$

where

$$\lambda = k_f + k_b + r \tag{122}$$

and

$$A_\infty = \frac{k_b(A_i + B_i) + rA_i}{\lambda} \tag{123}$$

The solution to this equation is written

$$A = A_\infty + \frac{rA_i - \lambda A_\infty}{\lambda - r}e^{-rt} - \frac{r(A_i - A_\infty)}{\lambda - r}e^{-\lambda t} \tag{124}$$

Symmetrically for B,

$$B = B_\infty + \frac{rB_i - \lambda B_\infty}{\lambda - r} e^{-rt} - \frac{r(B_i - B_\infty)}{\lambda - r} e^{-\lambda t} \tag{125}$$

where

$$B_\infty = \frac{k_f(A_i + B_i) + rB_i}{\lambda} \tag{126}$$

To analyze the conditions under which A and B are close to being at equilibrium with each other, let us first consider the final steady-state concentrations A_∞ and B_∞. The ratio

$$\frac{A_\infty}{B_\infty} = \frac{k_b(A_i + B_i) + rA_i}{k_f(A_i + B_i) + rB_i} \tag{127}$$

is close to the equilibrium ratio k_b/k_f in either of two cases:

1

$$\frac{A_i}{B_i} = \frac{k_b}{k_f}$$

In this case $A_\infty = A_i$ and $B_\infty = B_i$.

2

$$r \ll \frac{A_i + B_i}{A_i} k_b$$

and

$$r \ll \frac{A_i + B_i}{B_i} k_f$$

In this case at least one of the reaction rates, k_f or k_b, is much greater than the hydraulic rate r:

$$r \ll k_b \quad \text{or} \quad r \ll k_f$$

and thus

$$r \ll k_f + k_b + r = \lambda \tag{128}$$

Note that Conditions **2** are always met if both reaction rates are much greater than r, that is, $r \ll k_b$ and $r \ll k_f$.

Let us show that these Conditions **1** and **2** which make the steady-state composition close to chemical equilibrium are also those that pertain to pseudo-equilibrium much before the steady state is achieved.

Consider Equations 124 and 125. If Condition **1** is verified, the last terms (in $e^{-\lambda t}$) drop, and the equations simplify to

$$A = A_i(1 - e^{-rt})$$
$$B = B_i(1 - e^{-rt})$$

so that

$$\frac{A}{B} = \frac{A_i}{B_i} = \frac{A_\infty}{B_\infty}$$

This is the trivial case where the reactants are effectively preequilibrated in the inflow and no reaction ever takes place as the concentrations build up in the system.

To examine Conditions **2**, let us express A_∞ and λ in the first two terms of Equation 124:

$$A = \frac{k_b(A_i + B_i) + rA_i}{k_f + k_b + r} - \frac{k_b(A_i + B_i)}{k_f + k_b}\,e^{-rt} - [\ldots]e^{-\lambda t}$$

and similarly for B

$$B = \frac{k_f(A_i + B_i) + rB_i}{k_f + k_b + r} - \frac{k_f(A_i + B_i)}{k_f + k_b}\,e^{-rt} - [\ldots]e^{-\lambda t}$$

For any time much in excess of the characteristic reaction time $1/\lambda$, the last terms in $e^{-\lambda t}$ can be neglected. Conditions **2** (including Equation 128 implicitly) are then necessary and sufficient conditions for A and B to take the form

$$A = A_\infty(1 - e^{-rt})$$

$$B = B_\infty(1 - e^{-rt})$$

where

$$A_\infty = \frac{k_b(A_i + B_i)}{k_f + k_b}$$

$$B_\infty = \frac{k_f(A_i + B_i)}{k_f + k_b}$$

The ratio A/B then remains at the equilibrium value k_b/k_f for all times in excess of the characteristic reaction time ($t \gg 1/\lambda$). As illustrated in Figure 2.6, such times that are sufficient for chemical equilibrium to be achieved are reached much before the steady state:

$$\frac{1}{\lambda} \ll t \ll \frac{1}{r}$$

This condition under which a chemical reaction remains at equilibrium while the concentration of reactants and products change over time is properly called pseudoequilibrium. In this example, if Conditions **2** are met, the two very different characteristic times $1/r$ (hydraulic residence time) and $1/\lambda$ (reaction time) define three time scales:

$t \ll \dfrac{1}{\lambda}$	The concentrations of A and B slowly build up in the system due to the inflow. The chemical reaction does not proceed.
$\dfrac{1}{\lambda} \ll t \ll \dfrac{1}{r}$	A state of pseudoequilibrium is reached; $A/B = k_b/k_f$, but A and B change with time
$t \gg \dfrac{1}{r}$	The final steady state is obtained:

$$A = A_\infty,\; B = B_\infty,\quad \text{and}\quad A_\infty/B_\infty = k_b/k_f.$$

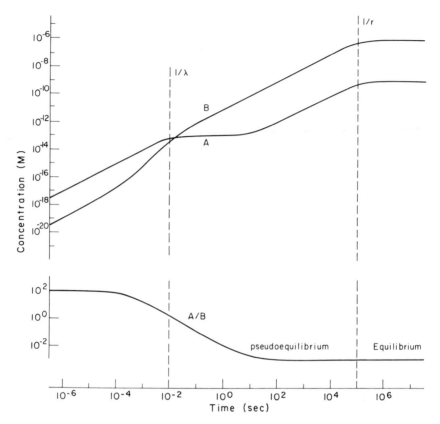

Figure 2.6 Example of the kinetics of a first-order reaction in a well-stirred flow-through system. The conditions are those described in Section 2.6.2 with the values $A_0 = B_0 = 0$, $A_i = 10^{-6}$ M, $B_i = 10^{-8}$ M, $r = 10^{-5}$ sec^{-1}, $k_f = 10^2$ sec^{-1}, $k_b = 10^{-1}$ sec^{-1}. Because the hydraulic residence time (10^5 sec) is much longer than the characteristic reaction time (10^{-2} sec), A and B reach a state of chemical pseudoequilibrium, much before attaining their final steady state concentrations.

2.6.3   Two Competing First-Order Reactions in a Closed System

Let us examine now a closed system containing a species A that reacts reversibly to form two products, B and C, according to the first-order, or pseudo-first-order, reactions:

$$A \underset{b_1}{\overset{f_1}{\rightleftarrows}} B$$
$$A \underset{b_2}{\overset{f_2}{\rightleftarrows}} C$$

The governing differential equations are written

$$\frac{dA}{dt} = -(f_1 + f_2)A + b_1B + b_2C \qquad (129)$$

$$\frac{dB}{dt} = f_1 A - b_1 B \tag{130}$$

$$\frac{dC}{dt} = f_2 A - b_2 C \tag{131}$$

The initial conditions are $A = A_0, B = B_0, C = C_0$. Adding the three equations together leads to the mole balance equation:

$$A + B + C = A_0 + B_0 + C_0 = T \tag{132}$$

Let us treat this problem in a general way. Focusing first on the concentration of A, we obtain a second-order differential equation by substituting C from Equation 132 into Equation 129, solving for B, then substituting B and its differential into Equation 130:

$$\frac{d^2 A}{dt} + \alpha \frac{dA}{dt} + \beta A = \beta A_\infty \tag{133}$$

where

$$\alpha = f_1 + f_2 + b_1 + b_2 \tag{134}$$

$$\beta = b_1 b_2 + b_1 f_2 + b_2 f_1 \tag{135}$$

$$A_\infty = \frac{b_1 b_2 T}{\beta} \tag{136}$$

The characteristic rates λ_1 and λ_2 corresponding to this differential equation are the solutions of the quadratic

$$\lambda^2 - \alpha\lambda + \beta = 0 \tag{137}$$

Under the conditions of this problem ($f_1, f_2, b_1, b_2 > 0$), λ_1 and λ_2 can be shown to be both real and positive, guaranteeing a stable and nonoscillating solution of the form

$$A = K_1 e^{-\lambda_1 t} + K_2 e^{-\lambda_2 t} + A_\infty \tag{138}$$

The constants K_1 and K_2 are determined by the initial conditions A_0, B_0, and C_0. The corresponding solutions for B and C are found to be

$$B = \frac{f_1 K_1}{b_1 - \lambda_1} e^{-\lambda_1 t} + \frac{f_1 K_2}{b_1 - \lambda_2} e^{-\lambda_2 t} + B_\infty \tag{139}$$

$$C = \frac{f_2 K_1}{b_2 - \lambda_1} e^{-\lambda_1 t} + \frac{f_2 K_2}{b_2 - \lambda_2} e^{-\lambda_2 t} + C_\infty \tag{140}$$

where

$$B_\infty = \frac{f_1}{b_1} A_\infty = \frac{f_1 b_2 T}{\beta} \tag{141}$$

$$C_\infty = \frac{f_2}{b_2} A_\infty = \frac{f_2 b_1 T}{\beta} \tag{142}$$

This system is thus characterized by two reaction times: $1/\lambda_1$ and $1/\lambda_2$. For times much shorter than these reaction times, initial conditions prevail in the system; for much longer times, equilibrium is achieved. The most interesting case is in between: Are there intermediate times where one reaction has reached (pseudo) equilibrium while the other one is still proceeding? If $1/\lambda_2$ is the longer of the two characteristics times, a necessary and sufficient condition is given by

$$\lambda_2 \ll b_1 \tag{143}$$

since for $t \gg 1/\lambda_1$ the equations for A and B become, in this case,

$$A = K_2 e^{-\lambda_2 t} + A_\infty$$

$$B = \frac{f_1 K_2}{b_1} e^{-\lambda_2 t} + B_\infty = \frac{f_1}{b_1} A$$

(Symmetrically, C will be at pseudoequilibrium with A if $\lambda_2 \ll b_2$ and $t \gg 1/\lambda_2$.) Calculating λ_2 from Equation 137 then gives the condition for pseudo-equilibrium:

$$b_1 \gg \tfrac{1}{2}[\alpha - (\alpha^2 - 4\beta)^{1/2}]$$

Since $b_1 < \alpha$, the two terms on the right must almost cancel each other for the inequality to be verified, and thus

$$\beta \ll \alpha$$

The square root can then be **approximated**:

$$b_1 \gg \frac{1}{2}\left[\alpha - \left(\alpha - \frac{2\beta}{\alpha}\right)\right]$$

therefore

$$b_1 \gg \frac{\beta}{\alpha} = \frac{b_1 b_2 + b_1 f_2 + b_2 f_1}{b_1 + b_2 + f_1 + f_2}$$

Necessary and sufficient conditions for this inequality to be verified (and thus for pseudoequilibrium of the first reaction to be obtainable) are

$$b_1 \gg b_2 \quad \text{and} \quad b_1 \gg f_2. \tag{144}$$

The characteristic reaction rates are then given by

$$\lambda_1 = b_1 + f_1 \tag{145}$$

$$\lambda_2 = \frac{b_1 b_2 + b_1 f_2 + b_2 f_1}{b_1 + f_1} \tag{146}$$

For any time $t \gg 1/\lambda_1$ the reaction between A and B is then at pseudoequilibrium. Since the two characteristic reaction rates are widely different ($1/\lambda_1 \ll 1/\lambda_2$) this pseudoequilibrium can be reached much ahead of the final equilibrium, as illustrated in Figure 2.7.

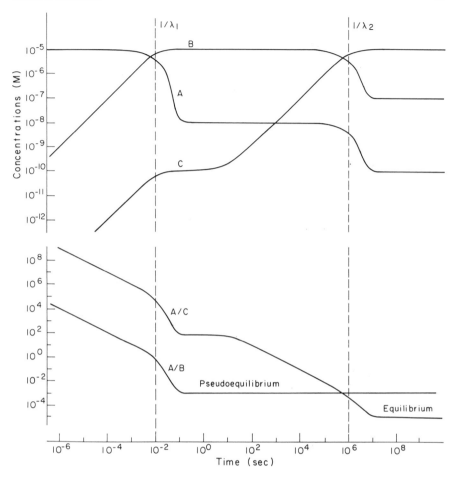

Figure 2.7 Example of simultaneous kinetics for two competing first-order reactions in a closed system. The conditions are those described in Section 2.6.3 with the values:

$$A_0 = 10^{-5} \ M, \ B_0 = C_0 = 0$$
$$A \rightleftarrows B, f_1 = 10^2 \ \text{sec}^{-1}, \ b_1 = 10^{-1} \ \text{sec}^{-1}$$
$$A \rightleftarrows C, f_2 = 10^{-3} \ \text{sec}^{-1}, \ b_2 = 10^{-8} \ \text{sec}^{-1}$$

Because the first reaction ($A \rightleftarrows B$) is much faster than the second reaction ($A \rightleftarrows C$), A and B reach a state of pseudoequilibrium while C is still being formed. The initial kinetics of the formation of B are not affected by the slow formation of C. By contrast, the kinetics of the eventual formation of C are very much controlled by the pseudoequilibrium of A with the more stable product B. Eventually, the slow formation of C, which is even more stable than B, dominates the equilibrium composition of the system.

Under these conditions the characteristic reaction time for the first reaction $(1/\lambda_1)$ is what it would be in the absence of the second reaction, as can be seen by comparing Equations 145 and 114. On the contrary, the characteristic reaction time for the second reaction $(1/\lambda_2)$ is generally much affected by the reaction that has achieved pseudoequilibrium. However, if either reaction is energetically unfavorable,

$$b_1 \gg f_1 \quad \text{or} \quad b_2 \gg f_2$$

yields

$$\lambda_2 = b_2 + f_2$$

The characteristic reaction time of the second reaction is then also unaffected by the first reaction.

As an application of this result, consider now A to be an intermediary in the reaction from B to C and the second reaction to be irreversible $(b_2 = 0)$. This situation is similar to that described for the transition state theory if one of the initial reactants is in large excess and the formation of the activated compound (A) is pseudo first order. Under what conditions can B and A be at pseudo-equilibrium and the rate of formation of C be given by

$$\lambda_2 = \frac{b_1}{f_1} f_2$$

as accepted for the transition state theory? Introducing $b_2 = 0$ in the pseudo-equilibrium conditions, Equations 144, and in the expression for λ_2, Equation 146, we obtain

$$b_1 \gg f_2 \quad \text{and} \quad t \gg \frac{1}{\lambda_1} = \frac{1}{b_1 + f_1}$$

$$\frac{b_1 f_2}{b_1 + f_1} = \frac{b_1}{f_1} f_2 \rightarrow f_1 \gg b_1$$

Thus the formation of the activated state must be energetically unfavorable (a tautology), and the rate of formation of the product from the activated state must be slower than the rate of formation of the activated state.

The three preceding examples illustrate some of the general features of equilibrium and pseudoequilibrium conditions in chemical systems. They also illustrate how complex these conditions can be, even in very simple cases. Although generalizations are difficult, it appears that, unless initial conditions are redhibitory (see Section 2.6.1), a reaction can be considered at equilibrium when the time is in large excess of the characteristic times of both the forward and reverse reactions (e.g., $t \gg 1/k_f$ and $1/k_b$ for a first-order reaction). Though sufficient, this condition is far from necessary, as illustrated in the preceding examples.

The results of the second example (Section 2.6.2) can be generalized to situations where mixing (transport) and chemical reactions are occurring simultaneously. A condition of local pseudoequilibrium is achieved whenever the forward and backward rate constants are much larger than the mixing rate. Note that if the concentrations of reactants and products are very different from each other, local pseudoequilibrium may be achieved even if one of the rate constants is relatively small.

□ □ □

REFERENCES

1   G. N. Lewis and M. Randall, *Thermodynamics*, 2nd ed.; revised by K. S. Pitzer and L. Brewer, McGraw-Hill, New York, 1961.
2   H. S. Harned and B. B. Owen, *The Physical Chemistry of Electrolytic Solutions*, 3rd ed., Van Nostrand-Reinhold, New York, 1958.
3   R. A. Robinson and R. H. Stokes, *Electrolyte Solutions*, 2nd ed., Butterworths, London, 1959.
4   G. N. Lewis and M. Randall, *J. Am. Chem. Soc.*, **43**, 1112–1154 (1921).
5   J. Kielland, *J. Am. Chem. Soc.*, **59**, 1675 (1937); as cited by J. N. Butler, *Ionic Equilibrium: A Mathematical Approach*, Addison-Wesley, North Reading, Ma., 1964, p. 434.
6   L. G. Sillen and A. E. Martell, *Stability Constants of Metal Ion Complexes*, Special Publication 17, Chemical Society, London, 1964; Supplement 1, Special Publication 25, Chemical Society, London, 1971.
7   A. E. Martell and R. M. Smith, *Critical Stability Constants*, vol. 1, *Amino Acids*, Plenum, New York, 1974; R. M. Smith and A. E. Martell, vol. 2, *Amines*, Plenum, New York, 1975; A. E. Martell and R. M. Smith, vol. 3, *Other Organic Ligands*, Plenum, New York, 1977; R. M. Smith and A. E. Martell, vol. 4, *Inorganic Ligands*, Plenum, New York, 1976.
8   J. N. Brønsted, *J. Am. Chem. Soc.*, **44**, 877–898 (1922).
9   E. A. Guggenheim, *Philos Mag.*, **19**, 588–643 (1935).
10   G. Scatchard, *Chem. Rev.*, **19**, 309–327 (1936).
11   J. E. Mayer, *J. Chem. Phys.*, **18**, 1426–1436 (1950).
12   G. Scatchard, *J. Am. Chem. Soc.*, **90**, 3124–3127 (1968).
13   K. S. Pitzer, *J. Phys. Chem.*, **77**, 268–277 (1973).
14   W. Stumm and J. J. Morgan, *Aquatic Chemistry*, 2nd ed., Wiley, New York, 1981, pp. 76–78.
15   H. S. Harned and R. Davis, *J. Am. Chem. Soc.*, **65**, 2030–2037 (1943).
16   J. H. Carpenter, *Limnol. Oceanogr.*, **11**, 264–277 (1966).
17   R. L. Jacobson and D. Langmuir, *Geochim. Cosmochim. Acta*, **38**, 301–318 (1974).
18   H. S. Harned and R. W. Ehlers, *J. Am. Chem. Soc.*, **55**, 652–656 (1933).
19   *CRC Handbook of Chemistry and Physics*, 62nd ed., 1981.
20   H. S. Harned and S. R. Choles, *J. Am. Chem. Soc.*, **63**, 1706–1709 (1941).
21   M. R. Hoffmann, *Env. Sci. Technol.*, **15**, 345–353 (1981).
22   J. O. Edwards, *Inorganic Reaction Mechanisms*, Benjamin, Elmsford, N.Y., 1965.
23   W. Stumm and J. J. Morgan, *Aquatic Chemistry*, 2nd ed., Wiley, New York, 1981, pp. 88, 211.

PROBLEMS

2.1   Consider the hydrated ionic radii for Na^+ (0.36 nm) and Cl^- (0.33 nm). What is the approximate distance *between* the hydrated ions in seawater $[(NaCl)_T) \cong 0.7\ M]$? At what $(NaCl)_T$ concentration do you expect the activities of the ions to be affected by crowding?

2.2   Discuss from a thermodynamic point of view the question of co-existence of two solid phases of the same composition, such as $CaCO_3$ (calcite) and $CaCO_3$ (aragonite).

2.3   Consider a compound X (acid, base, or salt) that dissociates according to the reaction

$$X = Y + Z$$

What should the standard energy change of the reaction be at the minimum for the dissociation to be considered complete ($>99.9\%$) for all possible compositions in dilute solutions (concentrations $\leq 0.5\ M$)?

2.4   Consider a $10^{-3}\ M$ solution of acetic acid; the dissociation reaction $HAc = H^+ + Ac^-$; and the standard free energies $\mu^0_{H^+} = 0$, $\mu^0_{HAc} = -95.5\ kcal\ mol^{-1}$, $\mu^0_{Ac^-} = -89.0\ kcal\ mol^{-1}$.

   a. What is the equilibrium constant for the reaction on the molar scale? On the mole fraction scale?
   b. Neglecting all other reactions, make a plot of the free energy of the system (G) and of the free energy change of the reaction (ΔG), as the dissociation proceeds from $(Ac^-) = 0$ to $(Ac^-) = 10^{-3}\ M$.
   c. Calculate the equilibrium composition of the solution.
   d. In the context of the Gibbs-Duhem relationship, how constant (quantitatively) is the mass law quotient expressed on the molar scale?

2.5   Using the Davies equation, calculate the ionic strength effect.

   a. On the solubility of $BaSO_4$ ($pK_s = 9.7$), as $NaCl$ is added to pure water up to $0.5\ M$ [only species Na^+, Cl^-, Ba^{2+}, SO_4^{2-}, $BaSO_4(s)$].
   b. On the pH ($= -\log\{H^+\}$), as $NaOH$ is added to pure water up to $1\ M$ (only species H^+, OH^-, Na^+).
   c. On the pH, as HCl is added to pure water up to $1\ M$ (only species H^+, OH^-, Cl^-).

2.6   A $10^{-3}\ M$ solution of acetic acid (see Problem 2.4) is transported from the warm surface of a river ($T = 30°C$, $P = 1$ at, $I = 10^{-2}\ M$) to the bottom of the ocean ($T = 4°C$, $P = 400$ at, $I = 0.7\ M$). Given $\bar{H}^0_{H^+} = 0$, $\bar{H}^0_{HAc} = 116.4\ kcal\ mol^{-1}$, $\bar{H}^0_{Ac^-} = 116.8\ kcal\ mol^{-1}$, $\bar{V}_{H^+} = 0$, $\bar{V}_{HAc} = 50.7\ cm^3\ mol^{-1}$, $\bar{V}_{Ac^-} = 41.5\ cm^3\ mol^{-1}$, calculate the total change in the acidity constant.

2.7   Consider an intracellular fluid at a potential ψ with respect to the extracellular medium ($\psi_{out} = 0$).

   a. Determine the ratio of concentrations inside to outside for a freely diffusing ion of charge number Z.
   b. Define the equation for equilibrium (outside, inside, and across membrane) for a weak acid $AH = A^- + H^+$; K_a.

2.8   Two species, X and Y, are added to a system at initial concentrations $(X)_0$ and $(Y)_0$. A product Z is formed according to the bimolecular reversible reaction:

$$X + Y \underset{k_b}{\overset{k_f}{\rightleftharpoons}} Z$$

Supposing that $(X)_0 \ll (Y)_0$ and $k_b \ll k_f(Y)_0$:

a. Find the (approximate) equilibrium $(t = \infty)$ composition of the system.

b. Find the (approximate) composition of the system at all times.

CHAPTER THREE

ORGANIZATION AND SOLUTION OF CHEMICAL EQUILIBRIUM PROBLEMS

The equilibrium state of a closed chemical system is defined by its free energy minimum within the constraints of mass conservation. In Chapter 1 we saw how a proper choice of components leads to a set of mole balance equations that expresses fully the mass conservation condition. In Chapter 2 we showed how, for a complete set of independent chemical reactions, the corresponding set of mass law equations is equivalent to the conditions of free energy minimum. These two sets of equations, mole balances and mass laws, define a well-posed mathematical problem that can be shown to have one, and only one, solution—the set of species concentrations corresponding to the equilibrium composition. In this chapter we focus our attention on such a chemical equilibrium problem, examining how best to organize and most simply solve it.

Posing a chemical equilibrium problem in proper mathematical terms is a matter of ensuring that all the necessary analytical and thermodynamic information is translated into a complete set of algebraic equations. We shall first show how to organize a problem in the form of a "tableau," already introduced in Chapter 1 as a convenient way to express the stoichiometric relations between species and components. Once the mathematical problem is properly posed, its solution would appear to require only some numerical manipulations of little chemical interest. It is indispensable, however, that one be able to perform these manipulations which can often be confusing and unwieldy. Fortunately, the mathematics of chemical equilibrium present some remarkable particularities that make them amenable to convenient techniques such as drastic approximations, graphical solutions, and iterative schemes. These will be presented in a series of examples illustrating the different types of equilibrium problems encountered in aquatic chemistry. Though our objective here is only to solve the problems conveniently, it should be remembered that the mathematical symbolism is a representation of the chemical reality and that the numerical particularities must correspond to actual chemical behavior. There is thus some chemistry to be learned by solving chemical equilibrium problems; conversely, chemical intuition can greatly simplify one's way through the numerical maze of mole balance and mass law equations.

Much of the material in the following chapters dealing with aquatic chemical processes can be understood in depth only if one has sufficient mastery of the chemical equilibrium problem. This chapter is thus detailed and slow-footed, and some of the examples may even appear to be overexplained, to ensure complete comprehension of the methodology.

1. ORGANIZATION AND TABLEAUX

Using the same example as in Chapter 1, let us consider a classical equilibrium problem specified as follows:

1 A recipe for how the system is made up (Recipe #1):

$$(H_2S)_T = 10^{-3} \ M$$

$$(Na_2SO_4)_T = 10^{-2} \ M$$

2 A list of species in the system:

$$(H_2O), H^+, OH^-, Na^+, SO_4^{2-}, H_2S, HS^-, S^{2-}$$

3 A list of independent reactions with their equilibrium constants:

$$H_2O = H^+ + OH^-; \quad K_w = 10^{-14.0} \tag{1}$$

$$H_2S = HS^- + H^+; \quad K_1 = 10^{-7.0} \tag{2}$$

$$HS^- = S^{2-} + H^+; \quad K_2 = 10^{-13.9} \tag{3}$$

For simplicity we assume here that these equilibrium constants are valid for the ionic strength of the system (approximately, $I = 3 \times 10^{-2} \ M$). In practice the choice of appropriate equilibrium constants is of course critical.

In order to write a complete set of equations—mole balances and mass laws—for the equilibrium composition of this system, it is convenient to organize the information from (1), (2), and (3) in a tableau, as we did in Chapter 1. Taking (H_2O), H^+, S^{2-}, Na^+, and SO_4^{2-} as components yields Tableau 3.1.

In this tableau a column of equilibrium constants has been added. The constants are written for the mass laws corresponding to the formation of the species from the components. They are readily obtained from those given in

TABLEAU 3.1

Species/Components	H^+	S^{2-}	Na^+	SO_4^{2-}	$\log K$
H^+	1				
OH^-	−1				−14.0
Na^+			1		
SO_4^{2-}				1	
H_2S	2	1			+20.9
HS^-	1	1			+13.9
S^{2-}		1			
Recipe					*TOTX*
H_2S	2	1			$10^{-3}\,M$
Na_2SO_4			2	1	$10^{-2}\,M$

Problem Specification **3** by considering the reactions corresponding to each line of the tableau. The constant for the hydroxyl ion OH^- is quite straightforward:

$$(OH^-) = K(H^+)^{-1} \leftrightarrow H_2O = H^+ + OH^-$$

therefore

$$\log K = \log K_w = -14.0$$

Calculation of the sulfide constants requires some manipulation:

$$(H_2S) = K(H^+)^2(S^{2-}) \leftrightarrow 2H^+ + S^{2-} = H_2S$$

This reaction is obtained by adding Reactions 2 and 3

$$
\begin{array}{lll}
H_2S = HS^- + H^+; & \log K_1 = -7.0 \\
HS^- = S^{2-} + H^+; & \log K_2 = -13.9 \\
\hline
H_2S = S^{2-} + 2H^+; & \log K = -20.9
\end{array}
$$

and changing the direction of the reaction and thus the sign of $\log K$. Note that the logarithms of the equilibrium constants are subjected to the same linear operations as the corresponding reactions. Similarly for HS^-

$$(HS^-) = K(H^+)(S^{2-}) \leftrightarrow H^+ + S^{2-} = HS^-$$

therefore

$$\log K = -\log K_2 = 13.9$$

The exponents in these mass law expressions are precisely the coefficients of the corresponding lines of the tableau. A constant of unity is implied for the expressions of the species that are also components.

Tableau 3.1 in this form contains all the necessary information for the equilibrium problem, and a complete set of equations is readily written:

Mole Balances (from each column)

$$TOTH = (H^+) - (OH^-) + 2(H_2S) + (HS^-) = 2(H_2S)_T = 2 \times 10^{-3} \, M \quad (4)$$

$$TOTS = (H_2S) + (HS^-) + (S^{2-}) = (H_2S)_T = 10^{-3} \, M \quad (5)$$

$$TOTNa = (Na^+) = 2(Na_2SO_4)_T = 2 \times 10^{-2} \, M \quad (6)$$

$$TOTSO_4 = (SO_4^{2-}) = (Na_2SO_4)_T = 10^{-2} \, M \quad (7)$$

Mass Laws (from each line)

$$(OH^-) = 10^{-14}(H^+)^{-1} \quad (8)$$

$$(H_2S) = 10^{20.9}(H^+)^2(S^{2-}) \quad (9)$$

$$(HS^-) = 10^{13.9}(H^+)(S^{2-}) \quad (10)$$

There are seven unknowns in this equilibrium problem (the concentrations of each of the seven species) and seven equations (four mole balances and three mass laws). Were we to consider (H_2O) as an additional unknown, we would simply have to consider the corresponding column and mole balance equation; the result would be, approximately, $(H_2O) = 55.4 \, M$.

2. THE MATHEMATICAL PROBLEM

Since the species concentrations are given explicitly as a function of the components by the mass law equations, it is always possible to reduce the number of equations and the number of unknowns by substituting the mass laws into the mole balances. One is then left with a system of nonlinear equations equal to the number of components and with the concentrations of the components themselves as the only unknowns. Although such substitution and reduction in the number of equations is tempting, and is in fact practical for computer algorithms, it should be firmly resisted. Instead we want to show how one can avoid manipulating equations and solving high-degree polynomials.

The lazy man's method is trial and error; his constant hope (hypothesis) is that the solution of the problem is trivial. If we consider the four mole balance equations, two are indeed trivial and need no further consideration:

$$(Na^+) = 2 \times 10^{-2} \, M$$

$$(SO_4^{2-}) = 10^{-2} \, M$$

Sodium sulfate is merely a background electrolyte in this system. Since the sodium and the sulfate ions are not reacting with any other species, their influence on the equilibrium composition of the system is restricted to nonideal effects. Once the contributions of sodium and sulfate to the ionic strength of the system have been accounted for, and the applicable equilibrium constants have been chosen, such background ions can simply be ignored in the solution of the problem.

For the other two mole balances given by Equations 4 and 5, it would be very convenient if all the terms but one could be neglected from the sum; that is, if the concentration of one species were much larger than that of the others and effectively equal to the total concentration of the corresponding component. This is usually the case. (Note that if $TOTX = 0$, then two terms of opposite signs have to be equal.)

Suppose that we choose arbitrarily one species and hypothesize that its concentration is much larger than that of the other species appearing in the same mole balance. There are three possible outcomes to such an arbitrary choice: (A) we are right, (B) we are wrong because some other concentration is in fact much larger, and (C) we are wrong because some other concentration(s) is (are) of comparable magnitude. We wish to show that in Case A the problem is readily solved; in Case B the inappropriateness of the choice is immediately apparent, leading to a better alternative choice; and in Case C the problem is solved by a simple iteration procedure.

Case A.   Giving ourselves the benefit of a good chemical or mathematical intuition, let us assume: $(H_2S) \gg$ all other concentrations. The $TOTH$ equation then yields

$$(H_2S) = 10^{-3} \ M*$$

We are then faced with a difficulty as the $TOTS$ equation is identically verified. The problem is poorly behaved numerically since both $TOTH$ and $TOTS$ simplify to the same result: $H_2S = 10^{-3} \ M$. To resolve this difficulty, we have to eliminate (H_2S) between these two equations (as we did in Chapter 1 with H_2O):

$$TOTH - 2TOTS = (H^+) - (OH^-) - (HS^-) - 2(S^{2-}) = 0$$

Note that this equation would have been obtained directly as the $TOTH$ equation if we had chosen H_2S as a component instead of S^{2-}. In general, and we shall make this a rule, those species that we suspect to be the most abundant should be chosen as components.

If our general hypothesis that there are great differences among concentrations holds true, this last equation has only three possible solutions:

1. $(H^+) = (OH^-)$.
Using $(H_2S) = 10^{-3}$ and substituting into the mass laws, we obtain

$$(H^+) = 10^{-7}; \quad (S^{2-}) = 10^{-9.9}; \quad (HS^-) = 10^{-3}$$

This is impossible since (HS^-) is computed to be of the same magnitude as (H_2S), contradicting our initial assumption, $(H_2S) \gg$ all other concentrations.

2. $(H^+) = (HS^-)$.

---

* Units (M) will often be omitted from the numerical expression of the concentrations.

This equation together with $(H_2S) = 10^{-3}$ can be introduced into the mass laws to yield

$$(S^{2-}) = 10^{-13.9}; \quad (H^+) = 10^{-5} = (HS^-); \quad (OH^-) = 10^{-9}$$

This result is reasonable.

3. $(H^+) = 2(S^{2-})$.

Substitution into the mass laws as above, yields

$$(H^+) = 10^{-7.97} = (S^{2-}); \quad (HS^-) = 10^{-2.04}$$

This is again impossible as (HS^-) would be much too large.

Choice **2** provides the only solution to the equilibrium problem. Verification by substituting the calculated concentrations back into the original equations is straightforward—and necessary, as one may lose track of some important term during the simplification procedure.

Case B. Let us consider some less felicitous choices for our initial guess of the dominant term in the *TOTH* equation:

1. $(OH^-) \gg$ all other concentrations.

This is clearly impossible since (OH^-) has a negative sign in the *TOTH* equation.

2. $(HS^-) \gg$ all other concentrations.

Therefore

$$(HS^-) = 2 \times 10^{-3}$$

Substituting in the *TOTS* expression, we obtain

$$(H_2S) + (S^{2-}) = -10^{-3}$$

which is impossible since concentrations cannot be negative.

3. $(H^+) \gg$ all other concentrations.

Therefore

$$(H^+) = 2 \times 10^{-3}$$

Substituting into the mass law equations yields

$$(H_2S) = 10^{15.5}(S^{2-})$$

$$(HS^-) = 10^{11.2}(S^{2-})$$

From which we conclude

$$(H_2S) \gg (HS^-) \gg (S^{2-})$$

The *TOTS* equation then simplifies to

$$(H_2S) = 10^{-3}$$

which contradicts our hypothesis $(H^+) \gg (H_2S)$.

Although this last guess is wrong, at least it satisfies the requirement that a solution of a weak acid actually be acidic $[(H^+) > (OH^-)]$. It also leads to a relatively mild contradiction. From this initial bad guess the correct guess $[(H_2S) = 10^{-3} \, M]$ and the solution of the problem are easily obtained.

Case C. To illustrate this case, let us consider the same system as in the previous example (same species, same reactions) but with a different recipe (Recipe #2):

$$(H_2S)_T = 10^{-3} \, M$$

$$(NaOH)_T = 5 \times 10^{-4} \, M$$

We have replaced the background electrolyte with a small concentration of strong base. Having learned from experience, we choose H^+, H_2S, and Na^+ as components as demonstrated in Tableau 3.2. Thus

Mole Balances

$$TOTH = (H^+) - (OH^-) - (HS^-) - 2(S^{2-}) = -(NaOH)_T = -5 \times 10^{-4} \quad (11)$$

$$TOTH_2S = (H_2S) + (HS^-) + (S^{2-}) = (H_2S)_T = 10^{-3} \quad (12)$$

$$TOTNa = (Na^+) = (NaOH)_T = 5 \times 10^{-4} \quad (13)$$

Mass Laws

$$(OH^-) = 10^{-14}(H^+)^{-1} \quad (14)$$

$$(HS^-) = 10^{-7}(H_2S)(H^+)^{-1} \quad (15)$$

$$(S^{2-}) = 10^{-20.9}(H_2S)(H^+)^{-2} \quad (16)$$

In the absence of great inspiration, let us resort to our previous initial guess: $(H_2S) \gg$ all other concentrations.

TABLEAU 3.2

Species/Components	H^+	H_2S	Na^+	$\log K$
H^+	1			
OH^-	-1			-14.0
Na^+			1	
H_2S		1		
HS^-	-1	1		-7.0
S^{2-}	-2	1		-20.9
Recipe				$TOTX$
H_2S		1		$10^{-3} \, M$
NaOH	-1		1	$5 \times 10^{-4} \, M$

From the $TOTH_2S$ equation, we then obtain $(H_2S) = 10^{-3}$. It is immediately apparent that this guess is wrong since, according to the $TOTH$ equation, some other concentration, (OH^-), (HS^-), or (S^{2-}), must be of the order of 5×10^{-4}. We will proceed, however, in the usual manner and look for a solution to the $TOTH$ equation, assuming one of the terms to be much larger than the others:

1. $(OH^-) = 5 \times 10^{-4} = 10^{-3.3}*$.
Therefore

$$(H^+) = 10^{-10.7}; \quad (S^{2-}) = 10^{-2.5}; \quad (HS^-) = 10^{0.7}$$

These concentrations are much too high to satisfy the $TOTH_2S$ equation.
2. $(HS^-) = 5 \times 10^{-4} = 10^{-3.3}$.
Therefore

$$(H^+) = 10^{-6.7}; \quad (S^{2-}) = 10^{-10.5}; \quad (OH^-) = 10^{-7.3}$$

This looks like it is close to a solution; let us substitute back into the mole balance equations:

$$TOTH = 10^{-6.7} - 10^{-7.3} - 10^{-3.3} - 2 \times 10^{-10.5} = -10^{-3.3}$$
$$TOTH_2S = 10^{-3} + 10^{-3.3} + 10^{-10.5} = 10^{-2.8} \neq 10^{-3}$$

The mass laws are identically verified. Since the $TOTH$ equation is well verified for the assumed value of (HS^-), let us iterate on the value of (H_2S) in order to obtain a better agreement with $TOTH_2S$. Given that the pH is approximately neutral, $(S^{2-}) = 10^{-20.9}(H_2S)(H^+)^{-2}$ will be small, and thus

$$(H_2S) = TOTH_2S - (HS^-) = 10^{-3} - 10^{-3.3} = 10^{-3.3}$$

and from the mass laws

$$(H^+) = 10^{-7}; \quad (OH^-) = 10^{-7}; \quad (S^{2-}) = 10^{-9.9}$$

This is the sought-for solution, as can be verified by substituting into the original equations.

Note how little algebra has been performed to obtain this answer. It is typical of the mathematics of chemical equilibrium problems that concentrations of species are orders of magnitude different from each other (due to the multiplicative nature of mass laws) and that only a few terms are actually important in each mole balance equation. Even when initial approximations are not strictly valid, it is usually easier to iterate to a more precise solution than to try solving a polynomial of high degree. Intelligent reasoning and guessing make iterating relatively simple, rarely requiring more than a couple of attempts before success is achieved.

---

* It is useful to remember the convenient equalities $5 = 10^{0.7}$ and $2 = 10^{0.3}$.

3. THE GRAPHICAL METHOD

Another way to take advantage of the peculiar mathematics involved in chemical equilibrium problems is to utilize logarithmic graphs, typically a diagram of log concentration versus pH. Consider the equations that include the concentrations of sulfide species in our example:

Mole Balance Equation

$$TOTS = (H_2S) + (HS^-) + (S^{2-}) = (H_2S)_T = 10^{-3} \tag{17}$$

Mass Law Equations

$$(H_2S) = 10^{+7.0}(HS^-)(H^+) \tag{18}$$

$$(HS^-) = 10^{13.9}(S^{2-})(H^+) \tag{19}$$

(written here for Reactions 2 and 3).

By themselves these three equations completely define the concentrations of the sulfide species as a function of (H^+) and are applicable, whatever concentrations of other acids and bases are also in the system (e.g., they are the same equations for Recipe #1 and Recipe #2). For a given total concentration, $(H_2S)_T$, a unique graph of sulfide species concentrations versus hydrogen ion concentration can thus be developed. This graph has a particularly convenient shape on a log versus log scale, and we refer to it as a "log C-pH" diagram.

3.1   Log C-pH Diagram for Sulfide

To develop the graph of species concentrations as a function of pH, shown in Figure 3.1, we use the logarithmic forms of the governing equations and make all appropriate approximations. First, the concentration plots for H^+ and OH^- are obtained directly from the definition of pH and the ion product of water (Reaction 1):

$$\log(H^+) = -pH \quad \text{(a straight line of slope } -1)$$

$$\log(OH^-) = -14 + pH \quad \text{(a straight line of slope } +1)$$

The sulfide species are slightly more difficult to graph because their concentrations change dramatically near the relevant pK_a's (negative logs of acidity constants). For example, consider the mass laws for the various sulfide species as the pH is varied from low to high values:

Case A : pH < 7.   Given the mass laws, Equations 18 and 19, and $(H^+) \gg 10^{-7}$, we may conclude that H_2S is the dominant species at this pH:

$$(H_2S) = 10^7(H^+)(HS^-) \rightarrow (H_2S) \gg (HS^-)$$

$$(HS^-) = 10^{13.9}(H^+)(S^{2-}) \rightarrow (HS^-) \gg (S^{2-})$$

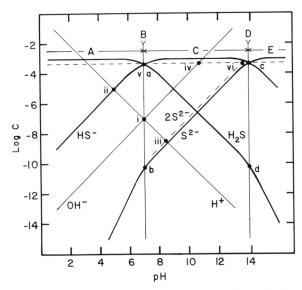

Figure 3.1 Log C-pH diagram for the sulfide system; graphical solutions for Recipe #1 (point ii) and Recipe #2 (point v).

On the basis of the mole balance of Equation 17 the concentration of H_2S is thus practically equal to the total sulfide concentration:

$$\log(H_2S) = \log TOTS = -3 \quad \text{(a straight horizontal line)}$$

Substituting this result into the logarithmic forms of the mass laws, we then obtain the graphs for (HS^-) and (S^{2-}):

$$\log(HS^-) = \log(H_2S) - 7 - \log(H^+)$$

$$= -10 + pH \quad \text{(a straight line of slope } +1\text{)}$$

$$\log(S^{2-}) = \log(HS^-) - 13.9 - \log(H^+)$$

$$= -23.9 + 2\,pH \quad \text{(a straight line of slope } +2\text{)}$$

Case B: pH = 7. Substituting $(H^+) = 10^{-7}$ into the mass laws gives us

$$(H_2S) = (HS^-)$$

$$(S^{2-}) = 10^{-6.9}(HS^-) \ll (HS^-)$$

Neglecting (S^{2-}) in the sulfide mole balance given by Equation 17 yields the following:

$$(H_2S) = (HS^-) = 0.5 \times 10^{-3} = 10^{-3.3}$$

therefore

$$\log(H_2S) = \log(HS^-) = -3.3 \quad \text{(point a)}$$

$$\log(S^{2-}) = -6.9 - 3.3 = -10.2 \quad \text{(point b)}$$

Case C: $7 < pH < 13.9$. Given $10^{-7} \gg (H^+) \gg 10^{-13.9}$, and following the same reasoning as in Case A, we find

$$(H_2S) = 10^7(H^+)(HS^-) \rightarrow (H_2S) \ll (HS^-)$$

$$(HS^-) = 10^{13.9}(H^+)(S^{2-}) \rightarrow (HS^-) \gg (S^{2-})$$

HS^- is thus the dominant sulfide species, and, according to the mole balance of Equation 17, its concentration is approximately equal to the total sulfide concentration:

$$\log(HS^-) = \log TOTS$$

$$= -3 \quad \text{(horizontal line)}$$

therefore

$$\log(H_2S) = 7 + \log(H^+) + \log(HS^-)$$

$$= 4 - pH \quad \text{(straight line of slope } -1)$$

$$\log(S^{2-}) = \log(HS^-) - 13.9 - \log(H^+)$$

$$= -16.9 + pH \quad \text{(straight line on slope } +1)$$

Case D: $pH = 13.9$.

$$(H^+) = 10^{-13.9}$$

$$(H_2S) = 10^{-6.9}(HS^-) \ll (HS^-)$$

$$(HS^-) = (S^{2-})$$

Neglecting (H_2S) in the sulfide mole balance yields

$$(HS^-) = (S^{2-}) = 0.5 \times 10^{-3} = 10^{-3.3}$$

therefore

$$\log(HS^-) = \log(S^{2-}) = -3.3 \quad \text{(point c)}$$

$$\log(H_2S) = -10.2 \quad \text{(point d)}$$

Case E: $pH > 13.9$.

$$(H^+) \ll 10^{-13.9}$$

$$(H_2S) = 10^7(HS^-)(H^+) \rightarrow (H_2S) \ll (HS^-)$$

$$(HS^-) = 10^{13.9}(H^+)(S^{2-}) \rightarrow (HS^-) \ll (S^{2-})$$

Hence S^{2-} is the dominant sulfide species:

$$\log(S^{2-}) = \log TOTS = -3 \quad \text{(horizontal line)}$$

and therefore

$$\log(HS^-) = 13.9 + \log(H^+) + \log(S^{2-})$$
$$= 10.9 - pH \quad \text{(straight line of slope } -1)$$
$$\log(H_2S) = 7 + \log(H^+) + \log(HS^-)$$
$$= 17.9 - 2\,pH \quad \text{(straight line of slope } -2)$$

Interpolation of the various lines in the neighborhood of points a, b, c, and d yields the diagram of Figure 3.1. The characteristics of such a diagram are uniform among all log C-pH graphs for weak acids and bases:

1 Various species are dominant in pH ranges delineated by the pK_a's, and their concentrations are constant (horizontal lines) and equal to the total weak acid-weak base concentration within these pH ranges.
2 The graphs of the nondominant species are given by straight lines of slopes $+2, +1, -1, -2$, and so on, depending on how many protons they must gain or lose to yield the dominant species in each pH range.
3 At each pK_a the concentrations of the species that dominate on each side of the pK_a are both equal to one-half the total weak acid-weak base concentration.

Note that the graphs for (H^+) and (OH^-) are independent of the particular system considered. Note also that the graphs for the weak acid-weak base species $[(H_2S),(HS^-),(S^{2-})]$ are only a function of the total concentration (given fixed equilibrium constants) and that they translate up and down as a block according to the particular total concentration in the system (S_T in our example).

3.2 The Graphical Method for Recipe #1

Let us go back to our example as defined by Recipe #1: $(H_2S)_T = 10^{-3}\,M$, $(Na_2SO_4)_T = 10^{-2}\,M$. As before, the convenient TOTH equation is obtained by using H_2S as a component [along with (H_2O), H^+, Na^+, and SO_4^{2-}]:

$$TOTH = (H^+) - (OH^-) - (HS^-) - 2(S^{2-}) = 0 \qquad (20)$$

According to our general hypothesis of great differences among concentrations, the solution to this equation is given by one of the following conditions:

(i) $(H^+) = (OH^-) \gg (HS^-), (S^{2-})$
(ii) $(H^+) = (HS^-) \gg (OH^-), (S^{2-})$
(iii) $(H^+) = 2(S^{2-}) \gg (OH^-), (HS^-)$

The points corresponding to the Equalities (i), (ii), and (iii) are marked on the graph (see Figure 3.1). The only point where the inequalities are verified is (ii), and a pH of 5 is read on the graph.

Note. Even if the graph is so rough that it permits only a very vague estimation of pH (say, ± 1 unit), it still should be immediately clear that the only possible solution is provided by Equality (ii), the intersection of the (H^+) and (HS^-) lines. From there on the problem is readily solved analytically.

Note. To obtain a graphical solution, it is critical that the proper $TOTH$ equation be utilized, one that corresponds to the major species as components (in this case H_2S). For example, according to the assumption of great differences among concentrations, the $TOTH$ equation corresponding to S^{2-} as a component

$$TOTH = (H^+) - (OH^-) + 2(H_2S) + (HS^-) = 2(H_2S)_T = 2 \times 10^{-3} \ M \quad (21)$$

would yield six possible solutions:

$(H^+) = 2 \times 10^{-3} = 10^{-2.7} \gg$ all other concentrations. At pH $= 2.7$,

$(H_2S) = 10^{-3}$ (see graph), and (H_2S) is thus approximately of the same magnitude as (H^+) which contradicts the hypothesis.

$(H_2S) = 10^{-3} \gg$ all others. This yields any pH in the range of 3.0 to 6.5.

$(HS^-) = 2 \times 10^{-3} = 10^{-2.7} \gg$ all others, which is impossible, since $TOTS$ is only 10^{-3}.

$(H^+) = (OH^-) \gg$ all others (including $10^{-2.7}$). Obviously, it cannot be that $10^{-7} \gg 10^{-2.7}$.

$2(H_2S) = (OH^-) \gg 10^{-2.7}$. Looking at the graph, one can see that this is also impossible.

$(HS^-) = (OH^-) \gg 10^{-2.7}$. Again this is impossible.

The correct hypothesis is the second one, but it is inconclusive because we have chosen the wrong components (and thus the wrong $TOTH$ equation) for the given system.

3.3   The Graphical Method for Recipe #2

Consider now our example with Recipe #2: $(H_2S)_T = 10^{-3} \ M$; $(NaOH)_T = 5 \times 10^{-4} \ M$. The corresponding $TOTH$ equation with H_2S as a component is written:

$$TOTH = (H^+) - (OH^-) - (HS^-) - 2(S^{2-}) = -0.5 \times 10^{-3} = -10^{-3.3} \quad (22)$$

To obtain a graphical solution, we draw the $\log C = -3.3$ line on the graph and look for a point where

(iv) $(OH^-) = 10^{-3.3} \gg (H^+), (HS^-), (S^{2-})$

(v) $(HS^-) = 10^{-3.3} \gg (H^+), (OH^-), (S^{2-})$

(vi) $2(S^{2-}) = 10^{-3.3} \gg (H^+), (OH^-), (HS^-)$

The points corresponding to the Equalities (iv), (v), and (vi) are shown in Figure 3.1. The only point where the inequalities are verified is (v), and a pH of 7 is read on the graph.

Note. As is now amply clear for the graph in Figure 3.1, the pH is a critical variable in all problems involving weak acid-weak base systems, all problems of interest in natural waters. Some approximate knowledge of the equilibrium pH value permits an a priori choice of the most abundant species as components and leads directly to many simplifications in the mole balance equations. In general, it is practical to consider pH [or (H^+)] as the principal unknown and to start the calculation with a rough guess of its value.

4. SOME PRACTICAL CONSIDERATIONS

In various places in this chapter we have pointed out some particularly convenient ways to pose and solve chemical equilibrium problems. Before studying a series of examples, let us summarize here some of these considerations as a practical methodology. Some of the "rules" set out here for solids and gases are made in anticipation of things to come.

1. As much as is possible (depending on intuition for chemistry or mathematics or on knowledge of an actual system being modelled) attempt to obtain some range of values for the critical parameters; foremost among these is pH.

2. Sketch out useful graphs (mainly log C-pH diagrams).

3. Choose components for the system in the following order:

a (H_2O) implicitly.

b (H^+).

c Species with fixed activities, that is solids or gases at fixed partial pressure.

d "Major" (most abundant) soluble species to round out the set.

This set of components will be called the *principal components*. It is not always possible to determine a priori the principal components. As the trial and error procedure for solving an equilibrium problem progresses, so does the choice of components.

4. Set up the necessary tableau.

5. Write all the equations and solve them by trial and error. To do this, make extensive use of the graphs whenever it is practical and always start with the general hypothesis that concentrations are orders of magnitude different from each other. If the system of equations appears numerically ill-behaved, it is probable that the set of components should be modified.

Eventually, once the necessary skills for solving equilibrium problems have been developed, graphs and tableaux can be pictured mentally and need not actually be put on paper. Their methodological value, however, is never lost.

Let us now examine a series of examples and illustrate some typical difficulties.

5. EXAMPLE 1: WEAK ACID AND STRONG BASE WITH BACKGROUND ELECTROLYTE

Recipe   0.1 M NaCl
                  $9 \times 10^{-3}\ M$ NaOH
                  $10^{-2}\ M$ HAc

Species (H_2O), H^+, OH^-, Na^+, Cl^-, Ac^-, HAc

Reactions $H_2O = H^+ + OH^-$; $pK_w^0 = 14.00$
                  $HAc = H^+ + Ac^-$; $pK_a^0 = 4.76$

5.1   Ionic Strength Corrections

Correction of equilibrium constants for $I = 0.1\ M$, using the Davies equation, yields

$$K_w = \frac{K_w^0}{\gamma_H \times \gamma_{OH}}$$

therefore

$$pK_w = pK_w^0 + \log \gamma_H + \log \gamma_{OH}$$

$$pK_w = 13.78$$

$$K_a = K_a^0 \frac{\gamma_{HAc}}{\gamma_H \times \gamma_{Ac}}$$

therefore

$$pK_a = pK_a^0 + \log \gamma_H + \log \gamma_{Ac} - \log \gamma_{HAc}$$

$$pK_a = 4.54$$

In other examples of this section we shall dispense with these nonideality corrections and use approximate constants (e.g., $pK_w = 14.0$).

5.2   Choice of Components

By convention, we always choose H_2O (implicitly) and H^+ (explicitly) as components. Other obvious choices in this case are Na^+, Cl^-, and HAc or Ac^-. To decide whether HAc or Ac^- is the principal component (i.e., the major acetate species), we need to know whether the equilibrium pH is below or above the $pK_a = 4.54$. Although we expect the small excess of acid over base to be reflected in a slightly acidic pH (below 7), it is unlikely that this pH will be below the pK_a. In the absence of better information, *a good rule of thumb is to choose the component(s) that yield the smallest absolute numerical value for the right-hand side of the TOTH equation.* In this case

$$\text{Components } (H_2O), H^+, Na^+, Cl^-, HAc \rightarrow TOTH = -(NaOH)_T$$
$$= -9 \times 10^{-3}$$

$$\text{Components } (H_2O), H^+, Na^+, Cl^-, Ac^- \rightarrow TOTH = -(NaOH)_T$$
$$+ (HAc)_T = 10^{-3}$$

Ac^- is thus the principal component to be chosen.

5.3   Tableau and Equations

With the choice of (H_2O), H^+, Na^+, Cl^-, and Ac^- as principal components, Tableau 3.3 is obtained.

TABLEAU 3.3

	H^+	Na^+	Cl^-	Ac^-	$\log K$
H^+	1				
OH^-	-1				-13.78
Na^+		1			
Cl^-			1		
Ac^-				1	
HAc	1			1	$+4.54$
NaCl		1	1		$10^{-1}\ M$
NaOH	-1	1			$9 \times 10^{-3}\ M$
HAc	1			1	$10^{-2}\ M$

The mole balances are written for each column of the tableau:

$$TOTH = (H^+) - (OH^-) + (HAc) = -(NaOH)_T + (HAc)_T = 10^{-3} \quad (23)$$

$$TOTNa = (Na^+) = (NaCl)_T + (NaOH)_T = 0.109 \quad (24)$$

$$TOTCl = (Cl^-) = (NaCl)_T = 0.1 \quad (25)$$

$$TOTAc = (Ac^-) + (HAc) = (HAc)_T = 10^{-2} \quad (26)$$

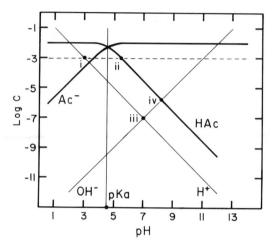

Figure 3.2 Log C-pH diagram for acetate; graphical solution of Example 1 (point ii).

The mass laws for the noncomponent species are obtained from the corresponding lines in Tableau 3.3:

$$(OH^-) = K_1(H^+)^{-1} \tag{27}$$

$$(HAc) = K_2(H^+)(Ac^-) \tag{28}$$

The constants K_1 and K_2 are K_w and K_a^{-1}, respectively ($10^{-13.78}$ and $10^{+4.54}$).

5.4   Log C-pH Diagram

To make the solution scheme more tangible, let us develop the appropriate log C-pH diagram. According to Equations 26 and 28, the concentration plots for Ac⁻ and HAc are obtained as before:

$$pH < 4.54; \quad (HAc) = 10^{-2}\ M; \quad (Ac^-) = 10^{-6.54}(H^+)^{-1}$$

$$pH > 4.54; \quad (Ac^-) = 10^{-2}\ M; \quad (HAc) = 10^{+2.54}(H^+)$$

$$pH = 4.54; \quad (Ac^-) = (HAc) = 0.5 \times 10^{-2}\ M = 10^{-2.3}\ M$$

Interpolation gives the rest of the plots for Figure 3.2. The diagram has the standard form, showing the change in dominance of the two acetate species at the pK_a.

5.5   Solution

Consider the system of Equations 23 to 28. In this system of six equations with six unknowns, two are trivial:

$$(Na^+) = 0.109\ M \tag{24}$$

$$(Cl^-) = 0.1\ M \tag{25}$$

Instead of starting to manipulate the remaining four equations, consider our general hypothesis of great differences among concentrations. The $TOTH$ equation should simplify to one of the following four possibilities:

(i) $(H^+) = 10^{-3} M \gg (OH^-), (HAc)$

(ii) $(HAc) = 10^{-3} M \gg (H^+), (OH^-)$

(iii) $(H^+) = (OH^-) \gg (HAc), 10^{-3}$

(iv) $(HAc) = (OH^-) \gg (H^+), 10^{-3}$

Each of the four equalities is shown on Figure 3.2. Because of the corresponding inequalities, Equality (i) is impossible $[(H^+) < (HAc)]$, and so are Equalities (iii) $[(H^+) \ll 10^{-3}]$ and (iv) $[(HAc) \ll 10^{-3}]$. Thus, Equality (ii) is the only one for which the corresponding inequalities come close to being verified. A pH in the range of 5 to 6 is then read on the graph. A more exact solution can be obtained algebraically, using Equality (ii) in the mole balance of Equation 26:

$$(Ac^-) = 10^{-2} - (HAc) = 9 \times 10^{-3} = 10^{-2.05}$$

Introducing the values of (HAc) and (Ac^-) in the mass law given by Equation 28 yields

$$10^{-3} = 10^{4.54} \times 10^{-2.05}(H^+)$$

therefore

$$(H^+) = 10^{-5.49}$$

This result can be verified by substitution into Equations 23 and 26:

$$10^{-5.49} - 10^{-8.29} + 10^{-3} = 10^{-3} \quad \text{(from Equation 23)}$$

$$10^{-2.05} + 10^{-3} = 10^{-2} \quad \text{(from Equation 26)}$$

$$0.991 \times 10^{-2} = 10^{-2}$$

0.9% is an acceptable error, but a more exact solution is obtained by carrying more significant digits: $(Ac^-) = 10^{-2.046}$.

To be precise, the pH is calculated as the negative log of the hydrogen ion *activity*:

$$pH = -\log(H^+) - \log \gamma_H$$

therefore

$$pH = 5.49 + 0.11 = 5.60$$

In future examples we shall not bother with the activity correction and take $pH = -\log(H^+)$.

Note that, if we had mistakenly chosen HAc as our principal component originally, we would have become quickly aware that this choice was wrong. The corresponding $TOTH$ equation would have led to $(Ac^-) = 9 \times 10^{-3}$ and thus $(HAc) = 10^{-3} < (Ac^-)$, which is contrary to the hypothesis that HAc is the principal component.

6. EXAMPLE 2: DIPROTIC WEAK ACID (WEAK BASE) AND STRONG ACID

Recipe 10^{-1} M KHS
                 10^{-2} M K_2S
                 10^{-2} M HCl

Species (H_2O), H^+, OH^-, K^+, Cl^-, H_2S, HS^-, S^{2-} (no gas phase).

Reactions $H_2O = H^+ + OH^-$; $pK_w = 14.0$
             $H_2S = H^+ + HS^-$; $pK_1 = 7.0$
             $HS^- = H^+ + S^{2-}$; $pK_2 = 13.9$

Choice of Components

Straightforward choices for components are (H_2O), H^+, K^+, and Cl^-. For the sulfide species there is some uncertainty as to which one may be dominant since the system is not drastically basic or acidic. Our rule of thumb—to choose the component(s) that yields the smallest absolute numerical value for the right-hand side of the $TOTH$ equation—makes HS^- the correct choice since it yields $TOTH = 0$ (Tableau 3.4).

TABLEAU 3.4

	H^+	K^+	Cl^-	HS^-	$\log K$
H^+	1				
OH^-	-1				-14.0
K^+		1			
Cl^-			1		
H_2S	1			1	$+7.0$
HS^-				1	
S^{2-}	-1			1	-13.9
KHS		1		1	10^{-1} M
K_2S	-1	2		1	10^{-2} M
HCl	1		1		10^{-2} M

Mole Balances

$$TOTH = (H^+) - (OH^-) + (H_2S) - (S^{2-}) = -(K_2S)_T + (HCl)_T = 0 \quad (29)$$

$$TOTK = (K^+) = (KHS)_T + 2(K_2S) = 1.2 \times 10^{-1} \ M \quad (30)$$

$$TOTCl = (Cl^-) = (HCl)_T = 10^{-2} \ M \quad (31)$$

$$TOTHS = (H_2S) + (HS^-) + (S^{2-}) = (KHS)_T + (K_2S)_T = 1.1 \times 10^{-1} \ M \quad (32)$$

Mass Laws

$$(OH^-) = 10^{-14}(H^+)^{-1} \tag{33}$$

$$(H_2S) = 10^{+7}(H^+)(HS^-) \tag{34}$$

$$(S^{2-}) = 10^{-13.9}(H^+)^{-1}(HS^-) \tag{35}$$

Solution Equations 30 and 31 are trivial and do not merit our further attention:

$$(K^+) = 1.2 \times 10^{-1} \; M$$

$$(Cl^-) = 10^{-2} \; M$$

To help solve the other equations, consider the log C-pH diagram corresponding to this system. In the domain where HS^- is the dominant sulfide species (defined by $7 < pH < 13.9$, according to mass laws given by Equations 34 and 35), the sulfide species concentrations are approximated by

$$(HS^-) = TOTHS = 1.1 \times 10^{-1} = 10^{-0.96}$$

$$(H_2S) = 10^7(H^+)10^{-0.96} = 10^{6.04}(H^+)$$

$$(S^{2-}) = 10^{-13.9}(H^+)^{-1}10^{-0.96} = 10^{-14.86}(H^+)^{-1}$$

The rest of the diagram can be obtained by carrying through the other approximations: $(H_2S) = 10^{-0.96}$ for $pH < 7$ and $(S^{2-}) = 10^{-0.96}$ for $pH > 13.9$. See Figure 3.3.

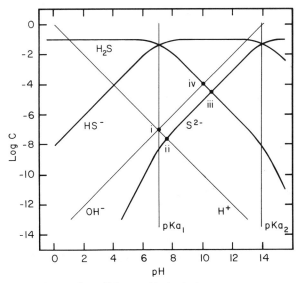

Figure 3.3 Log C-pH diagram for sulfide; graphical solution of Example 2 (in the neighborhood of points iii and iv).

According to the hypothesis of large inequalities among concentrations, the solution to Equation 29 will be given by one of four possibilities:

(i) $(H^+) = (OH^-) \gg (H_2S), (S^{2-})$

(ii) $(H^+) = (S^{2-}) \gg (H_2S), (OH^-)$

(iii) $(H_2S) = (S^{2-}) \gg (H^+), (OH^-)$

(iv) $(OH^-) = (H_2S) \gg (H^+), (S^{2-})$

It can be seen graphically, or by simple reasoning, that none of these conditions can be fulfilled exactly: Conditions (i) and (ii) are impossible, while (iii) and (iv) yield approximately

$$(H_2S) \cong (S^{2-}) \cong (OH^-) = 10^{-4}\ M \gg (H^+) \qquad (36)$$

The solution must thus be obtained for

$$-(OH^-) + (H_2S) - (S^{2-}) = 0$$

$$-10^{-14}(H^+)^{-1} + 10^{6.04}(H^+) - 10^{-14.86}(H^+)^{-1} = 0$$

and

$$(H^+)^2 = \frac{10^{-14.86} + 10^{-14}}{10^{6.04}} = 10^{-19.98}$$

$$(H^+) = 10^{-9.99}\ M$$

Then

$$(OH^-) = 10^{-4.01}\ M$$

$$(S^{2-}) = 10^{-4.87}\ M$$

$$(H_2S) = 10^{-3.95}\ M$$

Verification by substituting into Equations 29 and 32 yields

$$10^{-9.99} - 10^{-4.01} + 10^{-3.95} - 10^{-4.87} = 0 \quad \text{(Equation 29)}$$

$$9.88 \times 10^{-7} = 0$$

$$10^{-0.96} + 10^{-4.87} + 10^{-3.95} = 1.1 \times 10^{-1} \quad \text{(Equation 32)}$$

$$0.1098 = 0.11$$

Note that the verification of Equation 29 is good not just because 9.88×10^{-7} is a small number but because it is a small number compared to 10^{-4}. This example with three concentrations, (H_2S), (S^{2-}), and (OH^-) approximately equal at equilibrium is about as complicated as equilibrium problems can get. Very rarely will the final equation be more complicated than a quadratic.

7. EXAMPLE 3: PRECIPITATION OF A HYDROXIDE SOLID

Recipe $10^{-4} M$ $FeCl_3$
                    $10^{-4} M$ $NaOH$

Species (H_2O), H^+, OH^-, Na^+, Cl^-, Fe^{3+}, $FeOH^{2+}$, $Fe(OH)_2^+$
                    $Fe(OH)_3(s)$, $Fe(OH)_4^-$ (one solid phase)

Reactions $H_2O = H^+ + OH^-$; $pK_w = 14.0$
                    $FeOH^{2+} = Fe^{3+} + OH^-$; $pK_1 = 11.8$
                    $Fe(OH)_2^+ = Fe^{3+} + 2OH^-$; $pK_2 = 22.3$
                    $Fe(OH)_4^- = Fe^{3+} + 4OH^-$; $pK_3 = 34.4$
                    $Fe(OH)_3(s) = Fe^{3+} + 3OH^-$; $pK_s = 38.8$

Choice of Components

If the solution is indeed saturated with respect to $Fe(OH)_3(s)$ (a hypothesis that we shall verify later), then addition of more $Fe(OH)_3(s)$ to the system will leave the aqueous solution unchanged. In effect, if we are interested primarily in the composition of the aqueous phase, and we are, then the amount of solid is arbitrary, and $(Fe(OH)_3 \cdot s)$ is not a proper variable to describe the system. Note also that the activity of the solid is fixed at unity and is unrelated to the "concentration" of the solid in the aqueous phase. For these reasons we shall eliminate the solid "concentration" from all mole balance equations but one. This is achieved directly in the manner that we have used previously to eliminate the water concentration: $Fe(OH)_3(s)$ will be chosen as a component. The choice of principal components is then straightforward: (H_2O), H^+, Na^+, Cl^-, $Fe(OH)_3(s)$, as listed in Tableau 3.5.

TABLEAU 3.5

	H^+	Na^+	Cl^-	$Fe(OH)_3$	log K
H^+	1				
OH^-	-1				-14.0
Na^+		1			
Cl^-			1		
Fe^{3+}	3			1	$+3.2$
$FeOH^{2+}$	2			1	$+1.0$
$Fe(OH)_2^+$	1			1	-2.5
$Fe(OH)_3(s)$				1	
$Fe(OH)_4^-$	-1			1	-18.4
$FeCl_3$	3		3	1	$10^{-4} M$
$NaOH$	-1	1			$10^{-4} M$

Mole Balances

$$TOTH = (H^+) - (OH^-) + 3(Fe^{3+}) + 2(FeOH^+) + Fe(OH)_2^+$$

$$- (Fe(OH)_4^-) = 3(FeCl_3)_T - (NaOH)_T = 2 \times 10^{-4} \qquad (37)$$

$$TOTNa = (Na^+) = (NaOH)_T = 10^{-4} \qquad (38)$$

$$TOTCl = (Cl^-) = 3(FeCl_3)_T = 3 \times 10^{-4} \qquad (39)$$

$$TOTFe(OH)_3 = (Fe^{3+}) + (FeOH^{2+}) + (Fe(OH)_2^+) + (Fe(OH)_3 \cdot s)$$

$$+ (Fe(OH)_4^-) = (FeCl_3)_T = 10^{-4} \qquad (40)$$

Mass Laws

$$(OH^-) = 10^{-14}(H^+)^{-1} \qquad (41)$$

$$(Fe^{3+}) = 10^{3.2}(H^+)^3 \qquad (42)$$

$$(FeOH^{2+}) = 10^{1.0}(H^+)^2 \qquad (43)$$

$$(Fe(OH)_2^+) = 10^{-2.5}(H^+) \qquad (44)$$

$$(Fe(OH)_4^-) = 10^{-18.4}(H^+)^{-1} \qquad (45)$$

Note how conveniently the unit activity of the solid phase is included in the various mass law equations.

The last mole balance given by Equation 40 is not necessary to obtain the equilibrium composition of the aqueous phase and can be solved after the others to obtain the solid concentration $(Fe(OH)_3 \cdot s)$.

Solution   Equations 38 and 39 are trivial:

$$(Na^+) = 10^{-4} \ M$$

$$(Cl^-) = 3 \times 10^{-4} \ M$$

The hydrolysis reactions of iron define four pH regions where each of the four hydrolysis species dominate. This can be seen most simply on a log C-pH diagram where the mass laws (Equations 41 to 45) are plotted as straight lines of various slopes (see Figure 3.4).

According to the hypothesis of large inequalities among concentrations, the solution to the $TOTH$ equation is most probably given by one of four possibilities:

(i) $(H^+) = 2 \times 10^{-4} \gg$ all other concentrations
(ii) $(Fe^{3+}) = 0.66 \times 10^{-4} \gg$ all other concentrations
(iii) $(FeOH^{2+}) = 10^{-4} \gg$ all other concentrations
(iv) $(Fe(OH)_2^+) = 2 \times 10^{-4} \gg$ all other concentrations

Other possibilities could be considered in which two terms of opposite signs in the $TOTH$ equation would be equal and $\gg 2 \times 10^{-4}$. However, the intersections

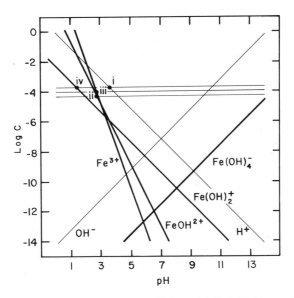

Figure 3.4 Log C-pH diagram for Fe(III) at equilibrium with $Fe(OH)_3(s)$; graphical solution of Example 3 (point i).

of the (OH^-) and $(Fe(OH)_4^-)$ graphs with the others in Figure 3.4 all yield concentrations $\ll 2 \times 10^{-4}$.

It is readily seen on Figure 3.4 that Equality (i) is the correct choice:

$$(H^+) = 2 \times 10^{-4} = 10^{-3.7}$$

therefore

$$pH = 3.7$$

Thus, according to the mass laws,

$$(OH^-) = 10^{-10.3} \ M$$

$$(Fe^{3+}) = 10^{-7.9} \ M$$

$$(FeOH^{2+}) = 10^{-6.4} \ M$$

$$(Fe(OH)_2^+) = 10^{-6.2} \ M$$

$$(Fe(OH)_4^-) = 10^{-14.7} \ M$$

From Equation 40 we can calculate (if we must) the "concentration" of the solid phase, that is, the number of moles of solid per liter of solution:

$$(Fe(OH)_3 \cdot s) = 10^{-4} - 10^{-7.9} - 10^{-6.4} - 10^{-6.2} - 10^{-14.7}$$

$$= 9.89 \times 10^{-5} \ M \quad \text{(moles per liter of solution)}$$

The solid is thus only 1% dissolved.

It is not always possible to guess correctly whether a given solid phase should or should not be present at equilibrium. When there are few such solids, a trial and error procedure will yield the correct answer. For example, in this third problem the composition of the aqueous phase can be solved entirely with the guess that $Fe(OH)_3(s)$ precipitates. At the end of the calculation, the concentration of the solid is calculated as the difference between total iron in the system and the sum of all soluble iron species (Equation 40). If the calculated "concentration" of the solid is positive $[(Fe(OH)_3 \cdot s) > 0]$, this solution is correct. If it is negative, there is simply not enough iron in this system to reach saturation, and the problem should be entirely re-solved without considering the solid phase—and hence choosing another Fe component. Conversely, the problem can be solved initially by not considering the solid phase. When the composition of the aqueous phase is obtained, the solubility product of ferric hydroxide should be checked. If K_s is exceeded, the problem should be re-solved considering the solid to be present; if it is not, the correct solution has been obtained. In complex situations where there are many possible solids involving common components, this trial and error procedure can become quite difficult, and short of trying all possibilities (which can number in the thousands), there is in fact no easy way to obtain the correct set of solids.

8. EXAMPLE 4: WEAK DIPROTIC ACID IN EQUILIBRIUM WITH ITS GAS PHASE

Recipe 10^{-3} M NaOH and $CO_2(g)$ at a fixed partial pressure of $10^{-3.5}$ at

Species (H_2O), H^+, OH^-, Na^+, $CO_2(g)$, $H_2CO_3^*$, HCO_3^-, CO_3^{2-}

Reactions
$$H_2O = H^+ + OH^-; \qquad pK_w = 14.0$$
$$CO_2(g) + H_2O = H_2CO_3^*; \qquad pK_H = 1.5$$
$$H_2CO_3^* = H^+ + HCO_3^-; \qquad pK = 6.3$$
$$HCO_3^- = H^+ + CO_3^{2-}; \qquad pK = 10.3$$

Note. $H_2CO_3^*$ is a conventional notation representing both $CO_2(aq)$ and H_2CO_3 in solution. The concentration of $CO_2(aq)$ is in fact almost three orders of magnitude greater than that of H_2CO_3.

Choice of Components

In the previous example, we chose the solid $Fe(OH)_3$ as one of the components because its activity was fixed ($=1$) and unrelated to its "concentration." We now have a similar situation for the gas phase which has an arbitrary and unknown volume but a fixed partial pressure of CO_2. The activity of $CO_2(g)$ is fixed by $P_{CO_2} = 10^{-3.5}$ at, but its concentration (in moles per liter of solution) is unknown and effectively indifferent. Again we choose CO_2 as a component in

TABLEAU 3.6

	H^+	Na^+	CO_2	$\log K$
H^+	1			
OH^-	-1			-14.0
Na^+		1		
$CO_2(g)$			1	
$H_2CO_3^*$			1	-1.5
HCO_3^-	-1		1	-7.8
CO_3^{2-}	-2		1	-18.1
$NaOH$	-1	1		$10^{-3}\ M$
CO_2			1	?

order to "isolate" its concentration in one equation only. The resulting component set is then (H_2O), H^+, Na^+, CO_2, as listed in Tableau 3.6.

Mole Balances

$$TOTH = (H^+) - (OH^-) - (HCO_3^-) - 2(CO_3^{2-})$$

$$= -(NaOH)_T = -10^{-3}\ M \tag{46}$$

$$TOTNa = (Na^+) = (NaOH)_T = 10^{-3} \tag{47}$$

$$TOTCO_2 = (CO_2 \cdot g) + (H_2CO_3^*) + (HCO_3^-) + (CO_3^{2-}) = (CO_2)_T = ? \tag{48}$$

Mass Laws

$$(OH^-) = 10^{-14}(H^+)^{-1} \tag{49}$$

$$(H_2CO_3^*) = 10^{-1.5}P_{CO_2} = 10^{-5} \tag{50}$$

$$(HCO_3^-) = 10^{-7.8}(H^+)^{-1}P_{CO_2} = 10^{-11.3}(H^+)^{-1} \tag{51}$$

$$(CO_3^{2-}) = 10^{-18.1}(H^+)^{-2}P_{CO_2} = 10^{-21.6}(H^+)^{-2} \tag{52}$$

By substituting the concentrations of noncomponent species, as given by their mass law expressions into the mole balance equations, the system is reduced to two equations, Equations 46 and 47, and two unknowns, which are the concentrations of the components H^+ and Na^+. The third mole balance given by Equation 48 would permit subsequent calculation of the "concentration" of $CO_2(g)$ if the total amount of CO_2 in the system were known. As it is, the problem is constrained, not by knowledge of the analytical (i.e., total) concentration of carbonate but by imposition of the activity of a component, $CO_2(g)$. A similar situation would be encountered if we imposed the presence of a particular solid phase without specifying the analytical concentration of the corresponding component, or if we considered the pH known without specifying the concentration of some strong acid or base in the system.

Solution   Equations 47 and 50 are trivial:

$$(Na^+) = 10^{-3} \, M$$

$$(H_2CO_3^*) = 10^{-5} \, M$$

The acid-base reactions of the carbonate system define three pH regions where each of the three aqueous carbonate species dominates. This can be seen on a log C-pH diagram (Figure 3.5) where the mass laws (Equations 49 to 52) are plotted.

According to the hypothesis of large inequalities among concentrations, the solution of the $TOTH$ equation is most likely given by one of three possibilities:

(i) $(OH^-) = 10^{-3} \gg (H^+), (HCO_3^-), (CO_3^{2-})$

(ii) $(HCO_3^-) = 10^{-3} \gg (H^+), (OH^-), (CO_3^{2-})$

(iii) $(CO_3^{2-}) = 0.5 \times 10^{-3} \gg (H^+), (OH^-), (HCO_3^-)$

Again, other possibilities are given by two terms of opposite signs being equal and much larger than 10^{-3} in the $TOTH$ equation; however, the intersections of the (H^+) graph with the others in Figure 3.5 yield concentrations $\ll 10^{-3}$.

By trial and error it is found graphically or numerically that Equality (ii) is the correct solution:

$$(HCO_3^-) = 10^{-3.0} \, M$$

$$10^{-11.3}(H^+)^{-1} = 10^{-3.0} \quad \text{(from Equation 51)}$$

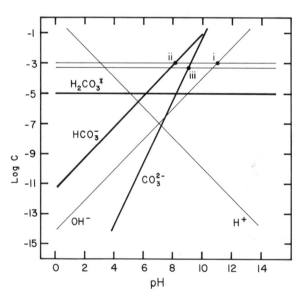

Figure 3.5   Log C-pH diagram for carbonate at equilibrium with $CO_2(g)$; graphical solution of Example 4 (point ii).

therefore

$$(H^+) = 10^{-8.3} \, M$$

$$pH = 8.3$$

All other species are then obtained:

$$(OH^-) = 10^{-5.7} \, M \quad \text{(from Equation 49)}$$

$$(CO_3^{2-}) = 10^{-5.0} \, M \quad \text{(from Equation 52)}$$

Verification of the mole balance given by Equation 46 shows only a 2% error in the $TOTH$ equation; iteration to a more precise solution is not warranted as the equilibrium constants themselves are not known that precisely. This problem is particularly easy to solve because the principal components are known a priori and the hypothesis of large inequalities among concentrations holds well.

9. EXAMPLE 5: USE OF IONIZATION FRACTION PARAMETERS

Consider again the all-important carbonate system, but this time with no gas phase equilibrium; that is, the aqueous phase is considered completely isolated from the gas phase. Given a total inorganic carbon concentration $(CO_2)_T = C_T$, we wish to calculate the pH of the solution for any strong base addition, $(NaOH)_T = B_T$. In other words, we wish to calculate the alkalimetric titration curve for any given concentration, C_T, of carbonic acid.

Recipe
$$(CO_2)_T = C_T$$
$$(NaOH)_T = B_T$$

Species
$$(H_2O), \; H^+, \; OH^-, \; Na^+, \; H_2CO_3^*, \; HCO_3^-, \; CO_3^{2-}$$

Reactions
$$H_2O = H^+ + OH^-; \qquad pK_w = 14.0$$
$$H_2CO_3^* = H^+ + HCO_3^-; \quad pK_{a1} = 6.3$$
$$HCO_3^- = H^+ + CO_3^{2-}; \quad pK_{a2} = 10.3$$

Choice of Components

By its very nature this problem spans a very wide pH range, and each of the three carbonate species is a principal component in its own region of dominance. Our choice of component among the carbonate species is thus indifferent, and we arbitrarily choose $H_2CO_3^*$ (Tableau 3.7).

Mole Balances

$$TOTH = (H^+) - (OH^-) - (HCO_3^-) - 2(CO_3^{2-}) = -B_T \qquad (53)$$

$$TOTNa = (Na^+) = B_T \qquad (54)$$

$$TOTCO_2 = (H_2CO_3^*) + (HCO_3^-) + (CO_3^{2-}) = C_T \qquad (55)$$

TABLEAU 3.7

	H^+	Na^+	$H_2CO_3^*$	$\log K$
H^+	1			
OH^-	-1			-14.0
Na^+		1		
$H_2CO_3^*$			1	
HCO_3^-	-1		1	-6.3
CO_3^{2-}	-2		1	-16.6
$NaOH$	-1	1		B_T
CO_2			1	C_T

Mass Laws

$$(OH^-) = 10^{-14}(H^+) \tag{56}$$

$$(HCO_3^-) = 10^{-6.3}(H^+)^{-1}(H_2CO_3^*) \tag{57}$$

$$(CO_3^{2-}) = 10^{-16.6}(H^+)^{-2}(H_2CO_3^*) \tag{58}$$

Since we wish to derive a *general* formula relating pH to B_T, we cannot make a priori valid approximations, and our previous methodology is of little help. The key to this problem is to exploit the universal functionality that relates the concentrations of individual weak acid species to their pK_a's, their total concentration, and the pH, as is exhibited in the universal shape of log C-pH diagrams.

By substitution of Equations 57 and 58 into 55, an explicit formula for $(H_2CO_3^*)$ is obtained:

$$(H_2CO_3^*) = \alpha_0 C_T \tag{59}$$

where

$$\alpha_0 = (1 + 10^{-6.3}(H^+)^{-1} + 10^{-16.6}(H^+)^{-2})^{-1} \tag{60}$$

Substituting Equation 60 back into Equations 57 and 58 gives us similar formulae for (HCO_3^-) and (CO_3^{2-}):

$$(HCO_3^-) = \alpha_1 C_T \tag{61}$$

$$(CO_3^{2-}) = \alpha_2 C_T \tag{62}$$

where

$$\alpha_1 = (10^{6.3}(H^+) + 1 + 10^{-10.3}(H^+)^{-1})^{-1} \tag{63}$$

$$\alpha_2 = (10^{16.6}(H^+)^2 + 10^{10.3}(H^+) + 1)^{-1} \tag{64}$$

The *ionization fractions* α_0, α_1, and α_2 represent the fractions of the total carbonate present in each of the particular carbonate species at any pH. By

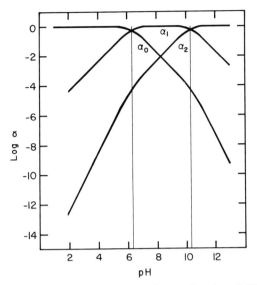

Figure 3.6 Ionization fractions of the carbonate species as a function of pH: $\alpha_0 = (H_2CO_3^*)/C_T$; $\alpha_1 = (HCO_3^-)/C_T$; $\alpha_2 = (CO_3^{2-})/C_T$.

necessity, the sum of the three ionization fractions must always equal unity, as can be verified by adding Equations 60, 63, and 64:

$$\alpha_0 + \alpha_1 + \alpha_2 = 1 \qquad (65)$$

A plot of the log of the ionization fractions as a function of pH, according to Equations 60, 63, and 64, exhibits the precise characteristics of a log C-pH graph for weak acids with a normalization to 0 [$=\log(1)$] on the vertical axis (Figure 3.6). The ionization fraction formulation is thus general for all weak acids and bases; the expressions of each of the α parameters for other acids is obtained by substituting the appropriate acidity constants in Equations 60, 63, and 64.

The solution to our problem is now derived by introducing Equations 56, 61, and 62 into the *TOTH* equation (Equation 53):

$$(H^+) - 10^{-14}(H^+)^{-1} - \alpha_1 C_T - 2\alpha_2 C_T = -B_T \qquad (66)$$

Equation 66 is an implicit function of (H^+) which can theoretically be solved for any given concentrations C_T and B_T. Note, however, that, while we have obtained a compact formulation for the sought-for alkalimetric titration curve, the problem of calculating the pH has really not been resolved. Equation 66 is a third-degree polynomial in (H^+), and the easy solutions that come from making the appropriate simplifications in each pH range have been all but obscured. What Equation 66 is in fact convenient for is the inverse problem of calculating B_T for a given pH and C_T. This is indeed the most straightforward way to obtain the graph of a titration curve: for a series of hydrogen ion concentrations, (H^+), both the ionization fractions and the left-hand side of Equation 66 are readily

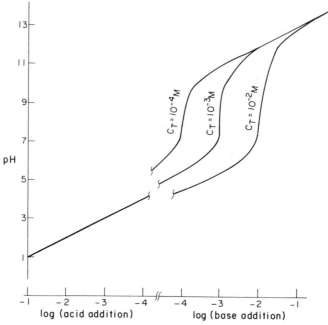

Figure 3.7 Acid-base titration of a CO_2 solution, for various concentrations of the total carbon concentration. Equation 66 is the general formula for such titration curves.

calculated. The results of such calculations for $C_T = 10^{-4}$, 10^{-3}, and 10^{-2} M are shown in Figure 3.7.

10.   COMPUTER CALCULATIONS FOR COMPLEX SYSTEMS

The chemical systems that are representative of natural waters are a good deal more complex than those of the preceding examples. The solutions to the corresponding equilibrium problems can be quite tedious to obtain by hand and sometimes difficult. As a result a number of computer algorithms have been developed to solve a wide spectrum of chemical equilibrium problems, from the very particular and simple (e.g., acid-base titrations) to the most general. One major family of general computer programs designed for the study of aquatic systems, REDEQL and MINEQL,[1,2,3] is based on an algorithm that organizes the chemical equilibrium problem precisely in the manner that we have presented here. A tableau is set up including all the stoichiometric and thermo-dynamic information necessary for solving the problem. The major difference in the solution technique is that a general numerical method rather than a series of approximations is used to solve the set of mole balance and mass law equations. We now give a succinct description of MINEQL as an archetype of such computer algorithms.

A simplified flow chart of the program is shown in Figure 3.8. For each calculation the user chooses a set of components out of a list that includes most

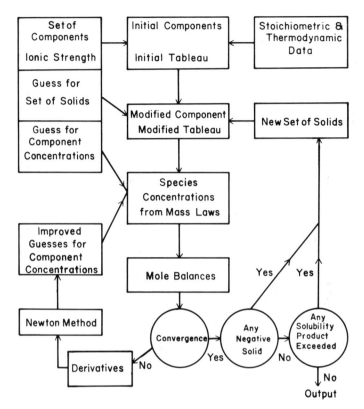

Figure 3.8 Flow chart of a general computer algorithm for solving chemical equilibrium problems.[1,2,3]

common solutes in natural waters. A first tableau is then defined according to the stored stoichiometric and thermodynamic information. This set of components and the corresponding tableau are subsequently modified according to an initial (user-provided) guess for the solids likely to be present in the system, choosing solids as components.

The solution routine then takes over. It consists of a slightly modified Newton method applied to the mole balance equations in which the concentrations of the noncomponent species have been substituted by their mass law expressions. The master variables that are iterated upon are then the concentrations of the soluble components. General algorithms for computing mass laws, mole balances, and the derivatives of the mole balances with respect to the concentrations of the soluble components are included in this routine.

Upon convergence, according to a preset criterion, the solid concentrations are checked for positiveness, and the solubility products are compared to the corresponding ion concentration products. (Some solids can be allowed to be supersaturated, while gases and some user's chosen solids are not allowed to dissolve.) For computational reasons the modification of the set of solids is performed one at a time according to a hierarchical rule: the most "negatively concentrated" solid is dissolved at each iteration until no undersaturated solid is found. Supersaturated solids are then precipitated sequentially according to how much the solubility products are exceeded. When the correct set of solids has been obtained, the problem has been solved. Most chemical equilibrium problems pertinent to natural waters are numerically well behaved, and convergence is typically obtained rapidly, even with very poor initial guesses.

One of the more interesting results of performing equilibrium calculations on very complex chemical systems is that the solution is often obvious, a posteriori!

REFERENCES

1 F. M. M. Morel and J. J. Morgan, *Env. Sci. Technol.*, **6**, 58 (1972).
2 R. E. McDuff and F. M. M. Morel, Technical Report EQ-7302, Keck Laboratory for Environmental Engineering Sciences, California Institute of Technology, Pasadena, 1973.
3 J. C. Westall, J. L. Zachary, and F. M. M. Morel, Technical Note 18, R. M. Parsons Laboratory for Water Resources and Hydrodynamics, Massachusetts Institute of Technology, Cambridge, 1976.

PROBLEMS

3.1   Consider the reactions

$$CO_2(g) + H_2O = H_2CO_3^*; \qquad pK_H = 1.5$$

$$H_2CO_3^* = H^+ + HCO_3^-; \quad pK_{a1} = 6.3$$

$$HCO_3^- = H^+ + CO_3^{2-}; \quad pK_{a2} = 10.3$$

$$CaCO_3(s) = Ca^{2+} + CO_3^{2-}; \quad pK_s = 8.3$$

Choose the principal components, and write the right-hand side (numerical value) of the corresponding mole balances for each of the following systems:

a. Recipe $(NaHCO_3)_T = 10^{-3}\ M$, $(HCl)_T = (NaOH)_T = 10^{-2}\ M$, no gas or solid phase.

b. Recipe $(NaHCO_3)_T = (HCl)_T = 10^{-3}\ M$, $(NaOH)_T = 10^{-2}\ M$, no gas or solid phase.

c. Recipe $(NaHCO_3)_T = (NaOH)_T = 10^{-3}\ M$, $(HCl)_T = 10^{-2}\ M$, no gas or solid phase.

d. Recipe $(Na_2CO_3)_T = 10^{-3}\ M$, $(HCl)_T = (NaOH)_T = 10^{-2}\ M$, no gas or solid phase.

e. Recipe $(NaHCO_3)_T = 10^{-3}\ M$, $P_{CO_2} = 10^{-3.5}$ at, no solid phase.

f. Recipe $(NaHCO_3)_T = 10^{-3}\ M$, $(Ca(OH)_2)_T = 10^{-2}\ M$, $CaCO_3(s)$ precipitates, no gas phase.

g. Recipe $(NaHCO_3)_T = 10^{-3}\ M$, $(Ca(OH)_2)_T = 10^{-2}\ M$, $P_{CO_2} = 10^{-3.5}$ at, $CaCO_3(s)$ precipitates.

3.2 Sketch log C-pH diagrams, and find approximate graphical solutions for the following equilibrium systems:

a. Recipe $(Na_2HPO_4)_T = 10^{-3} M, (HCl)_T = 10^{-3} M$

Species $H_2O, H^+, OH^-, Na^+, Cl^-, H_3PO_4, H_2PO_4^-, HPO_4^{2-}, PO_4^{3-}$

Reactions $H_3PO_4 = H_2PO_4^- + H^+, pK_{a1} = 2.1$
$H_2PO_4^- = HPO_4^{2-} + H^+, pK_{a2} = 7.2$
$HPO_4^{2-} = PO_4^{3-} + H^+, pK_{a3} = 12.3$

b. Recipe $(NH_4Cl)_T = 10^{-3} M, (NaOH)_T = 10^{-3} M$

Species $H_2O, H^+, OH^-, Na^+, Cl^-, NH_4^+, NH_3(aq)$

Reactions $NH_4^+ = NH_3(aq) + H^+; pK_a = 9.2$

c. Recipe Same as system b but add $P_{NH_3} = 10^{-3}$ at

Species Same as system b but add $NH_3(g)$

Reactions Same as system b but add $NH_3(g) = NH_3(aq)$; $pK_H = -1.8$.

3.3 Using a log C-pH diagram, find the pH of vinegar. You may model vinegar as a 5% volume/volume solution of acetic acid.

3.4 Solve the following equilibrium problems:

a. Recipe $(Na_2CO_3)_T = 10^{-3} M, (HCl)_T = 2 \times 10^{-3} M, (NaCl)_T = 10^{-2} M, (NH_4Cl)_T = 10^{-4} M$

Species and reactions are the same as in Problems 3.1 and 3.2, but there is no gas or solid phase.

b. Recipe $P_{CO_2} = 10^{-3.5}$ at, $P_{NH_3} = 10^{-6}$ at, $(NaOH)_T = 10^{-3} M$, $(Na_2HPO_4)_T = 10^{-5} M, (NaCl)_T = 10^{-2} M$

Species and reactions are the same as in Problems 3.1 and 3.2 but without a solid phase.

3.5 Given the following thermodynamic data, calculate the equilibrium composition for the two recipes that follow:

	μ^0 (kcal mol^{-1})
H^+	0
OH^-	-37.59
H_2O	-56.69
Na^+	-62.59
Cl^-	-31.35
$NaCl(aq)$	$> -90.$
$HCl(aq)$	$\cong -27.$
$NaOH(aq)$	$\cong -100.$
CN^-	$+39.6$
$HCN(aq)$	$+26.8$

	Recipe 1	Recipe 2
$(HCN)_T$	$10^{-3}\ M$	$10^{-3}\ M$
$(NaCl)_T$	$10^{-1}\ M$	$10^{-1}\ M$
$(NaOH)_T$	0	$10^{-3}\ M$

a. Calculate equilibrium constants, and decide what species should be considered in the system.
b. Make appropriate ionic strength corrections.
c. Choose appropriate components, write tableaux, and solve.

CHAPTER FOUR

ACIDS AND BASES
ALKALINITY AND pH IN
NATURAL WATERS

The composition of natural waters is controlled by both geochemical and geobiological processes. From a chemical standpoint we can regard the exogenic cycle—the process by which mountains are slowly dissolved and transported to the bottom of the oceans—as consisting of a gigantic acid-base reaction in the flowing water, continuously retransported atop the hills by a Sisyphean sun. The water, made corrosive by its acid content, mostly CO_2, dissolves the basic rock minerals it encounters and in this way acquires most of its solutes which are ultimately precipitated in the ocean sediments. The pH of the water then is primarily determined by a balance between the dissolution of weakly acidic carbon dioxide and basic rocks, alumino-silicates and carbonates. More precisely, the pH is determined by the extent of dissociation of the dissolved carbonic acid, and other weak acids such as water, whose net negative charge (HCO_3^-, CO_3^{2-}, OH^-) has to balance exactly the net positive charge from the strong mineral bases (Na^+, K^+, Ca^{2+}). Each of these two equal quantities provides a definition of alkalinity, one of the most central though perhaps not the best understood concept in aquatic chemistry.

By making a formal distinction between the weak and the strong acids and bases, the concept of alkalinity allows us to study in two steps the mechanisms that control pH in natural waters. In this chapter we examine only the weak acid-base side of the alkalinity equation, considering the other side of the equation as given by some excess of strong base—or excess strong acid in a few cases such as acid mine drainage or acid rain. The dissolution-precipitation mechanisms controlling that concentration of excess strong base—the alkalinity—are examined in the next chapter.

Because in most natural waters carbon dioxide far exceeds the other weak acids, this chapter deals primarily with the carbonate system. A complication that cannot be avoided is that of CO_2 exchange between the aqueous and the gas phases, between water and the atmosphere. This we treat in three stages: first in a closed aqueous systems with no gas phase and thus no CO_2 exchange, second, in a system consisting of an atmosphere of fixed CO_2 partial pressure in equilibrium with the water, and third, as the kinetics of CO_2 exchange at the air-water interface.

Before considering the kinetics at the end of the chapter, we shall take advantage of the simplicity of equilibrium models to obtain some understanding of homeostatic processes in natural waters. Having examined the processes controlling the pH in aquatic systems, we ask the question: How resistant to change is that pH? The answer is merely a generalization of the familiar subject of pH buffers, and we derive a general formula applicable to both simple solutions and complex model systems.

If we consider not only carbonate and water but also the other less important weak acid-base components of natural waters such as phosphate, silicate, ammonia, sulfide, and borate, it is striking that all but borate contain elements essential to the formation of living matter (C, H, O, P, Si, N, S). The cycle of these elements is partially controlled by organisms, and the corresponding weak acids and bases are thus involved in numerous biological processes. As a simple case study, and as a link to the living world, we shall examine the effect of some of these processes on the alkalinity and pH of natural waters.

1. NATURAL WEAK ACIDS AND BASES

Natural waters contain a number of weak acids and bases from a variety of sources. As seen in Table 4.1, the most abundant by far is carbonate, which

TABLE 4.1

Weak Acids and Bases in Natural Waters[1-5]

	Freshwater[1]	Seawater[2]		
	Mean	Warm Surface	Deep Atlantic	Deep Pacific
Carbonate	0.97 mM	2.1 mM	2.3 mM	2.5 mM
Silicate	220 μM	<3 μM	30 μM	150 μM
Ammonia	0–10 μM	<0.5 μM	<0.5 μM	<0.5 μM
Phosphate	0.7 μM	<0.2 μM	1.7 μM	2.5 μM
Borate	1 μM	0.4 mM	0.4 mM	0.4 mM
	Typical Anoxic Hypolimnion[3]	Black Sea (Deep Water)[4]		Cariaco Trench[5]
Sulfide	50–150 μM	330 μM		20 μM
Ammonia	10–40 μM	53 μM		10 μM

originates from the dissolution of carbonate rock, atmospheric CO_2 exchange and respiration of aquatic organisms. By comparison to the total carbonate concentration which averages about one millimolar in freshwater and more than twice that in seawater, all other natural weak acids and bases are relatively unimportant except in particular aquatic systems. Dissolution of silicate minerals contributes a total silicate concentration to freshwaters that is five to ten times lower than the total carbonate. Both concentrations tend to be relatively higher in arid regions. Like other essential plant nutrients (phosphorus and nitrogen), silicate is very depleted in warm surface oceanic waters and increases in concentration from the deep Atlantic to the deep Pacific. This is also true, to a less dramatic degree, for carbon. In temperate freshwaters, uptake by plants typically keeps both ammonia and phosphate surface concentrations below detectable limits in the summer. Apart from their regeneration by the decomposition of organic matter, these two algal nutrients originate

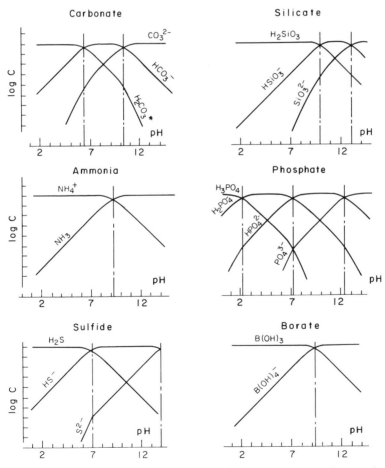

Figure 4.1   Log C-pH diagrams for weak acids-bases that are most common in natural waters.

from fixation of atmospheric nitrogen and dissolution of phosphate rocks, respectively.

Hydrogen sulfide, which is nonexistent in oxygenated systems, can be found at concentrations up to one millimolar in anoxic basins such as the hypolimnions of stratified lakes due to the activity of sulfate-reducing bacteria. In groundwaters contacting sulfide minerals (e.g., iron mines), the concentration of sulfide typically does not exceed 10 μM.

Borate, which is present in trace quantities in continental rock and as a result in freshwaters as well, is the second most abundant weak base in the oceans. This great concentration effect from rivers to seawater ($\times 400$) is apparently due to inefficient and poorly understood removal processes in the oceans.

The log concentration versus pH diagrams for these various weak acid-weak base systems are shown in Figure 4.1 Bicarbonate is the dominant carbonate species near neutral pH, while silicate and borate are essentially undissociated and the ammonium ion is in great excess of ammonia. Both phosphate and

TABLE 4.2

Acid-Base Reactions
(Mixed Acidity Constants)[6]

	$-\log K$	
Reactions	$I = 0$	$I = 0.5\ M$
$H_2O = H^+ + OH^-$	14.00	13.89
$CO_2(g) + H_2O = H_2CO_3^*$	1.46	1.51
$H_2CO_3^* = HCO_3^- + H^+$	6.35	6.30
$HCO_3^- = CO_3^{2-} + H^+$	10.33	10.15
$H_2SiO_3 = HSiO_3^- + H^+$	9.86	9.61
$HSiO_3^- = SiO_3^{2-} + H^+$	13.1	12.71
$H_3PO_4 = H_2PO_4^- + H^+$	2.15	1.87
$H_2PO_4^- = HPO_4^{2-} + H^+$	7.20	6.72
$HPO_4^{2-} = PO_4^{3-} + H^+$	12.35	11.89
$NH_3(g) = NH_3(aq)$	-1.87	-1.64
$NH_4^+ = NH_3(aq) + H^+$	9.24	9.47
$H_2S(g) = H_2S(aq)$	0.99	0.99
$H_2S(aq) = HS^- + H^+$	7.02	6.98
$HS^- = S^{2-} + H^+$	13.9	13.45
$B(OH)_3 + H_2O = B(OH)_4^- + H^+$	9.24	8.97

sulfide have acidity constants near 7. The graphs are constructed using mixed acidity constants applicable at an ionic strength of $10^{-3} M$ ($P = 1$ at, $T = 25°C$). Along with rounded-off pK_a's of 6.3 and 10.3 for carbonate, these are the constants that we shall use arbitrarily throughout this chapter unless otherwise indicated. Other constants are given in Table 4.2 which lists the various acid-base reactions for each system. Note that for convenience the aqueous concentration of carbon dioxide and that of carbonic acid are always added up:

$$(H_2CO_3^*) = (CO_2 \cdot aq) + (H_2CO_3) \tag{1}$$

and that equilibrium constants are defined for the inclusive species $H_2CO_3^*$.

Note. The soluble hydrolysis species of many metals $[Fe(OH)_2^+, Fe(OH)_4^-,$ $PbOH^+$, etc.] also behave as weak acids and bases, and they may be important for the pH control of some aquatic systems. However, because of the dominant importance of the insoluble hydroxide and oxide forms $[Fe(OH)_3(s), Fe_2O_3(s),$ $Pb(OH)_2(s)$, etc.] these particular weak acids and bases are studied in Chapter 5 which deals with solids.

2. ALKALINITY AND RELATED CONCEPTS

2.1 Pure Solutions of CO_2: The Equivalence Point

Before considering natural waters in all their complexity, let us examine the simplest possible case of a carbonate-bearing water, that of a pure CO_2 solution which includes a background electrolyte.

Example 1

Consider a $10^{-3} M$ NaCl solution into which we introduce $10^{-4} M$ CO_2 by bubbling. The system is considered closed to the atmosphere (no gas exchange), and we are interested in its equilibrium composition.

Recipe $(NaCl)_T = 10^{-3} M$
              $(CO_2)_T = 10^{-4} M$

Species (H_2O), H^+, OH^-, Na^+, Cl^-, $H_2CO_3^*$, HCO_3^-, CO_3^{2-}

Reactions

$$H_2O = H^+ + OH^-; \quad pK_w = 14.0 \tag{2}$$

$$H_2CO_3^* = H^+ + HCO_3^-; \quad pK_1 = 6.3 \tag{3}$$

$$HCO_3^- = H^+ + CO_3^{2-}; \quad pK_2 = 10.3 \tag{4}$$

The corresponding log C-pH diagram is given in Figure 4.2. Anticipating that $H_2CO_3^*$ is the dominant carbonate species, we choose it as one of the components. The rest of the principal components are obvious: H^+, Na^+, and

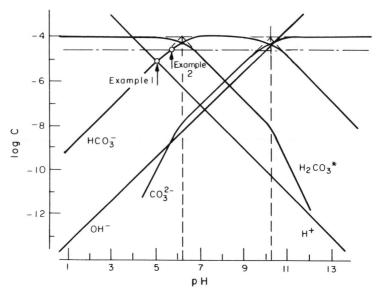

Figure 4.2 Log C-pH diagram for carbonate; graphical solutions of Example 1 (equivalence point) and Example 2.

Cl$^-$. The pH of the system is obtained by solving the mole balance equation for H$^+$ (see Tableau 4.1):

$$TOTH = (H^+) - (OH^-) - (HCO_3^-) - 2(CO_3^{2-}) = 0 \qquad (5)$$

The solution to this equation is given by

$$(H^+) = (HCO_3^-) \gg (OH^-), (CO_3^{2-}) \qquad (6)$$

The pH can be read from the graph (Figure 4.2); it can also be obtained algebraically. In this pH region the mole balance for carbonate simplifies to

$$TOTH_2CO_3 = (H_2CO_3^*) + (HCO_3^-) = (CO_2)_T = 10^{-4} \qquad (7)$$

Introduction of Equation 6 and the mass law for (H$_2$CO$_3$*) into Equation 7 yields

$$10^{6.3}(H^+)^2 + (H^+) = (CO_2)_T = 10^{-4}$$

therefore

$$(H^+) = 10^{-5.17} \ M$$

$$(HCO_3^-) = 10^{-5.17} \ M$$

$$(CO_3^{2-}) = 10^{-10.3} \ M$$

$$(OH^-) = 10^{-8.83} \ M$$

$$(H_2CO_3^*) = 10^{-4.04} \ M$$

Also

$$(Na^+) = (Cl^-) = 10^{-3} \ M$$

TABLEAU 4.1

	H^+	Na^+	Cl^-	$H_2CO_3^*$	pK
H^+	1				
OH^-	-1				14.0
Na^+		1			
Cl^-			1		
$H_2CO_3^*$				1	
HCO_3^-	-1			1	6.3
CO_3^{2-}	-2			1	16.6
Example 1					
NaCl		1	1		$10^{-3} M$
CO_2				1	$10^{-4} M$
Example 2					
NaOH	-1	1			$10^{-4.5} M$

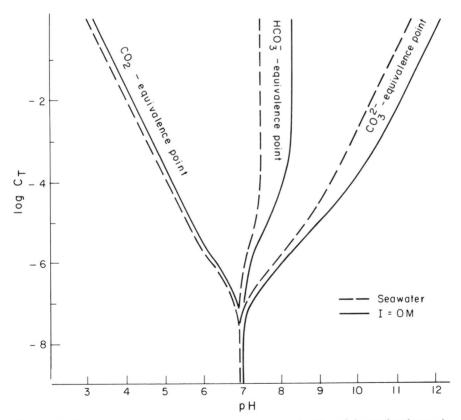

Figure 4.3 Equivalence points of the carbonate system as a function of the total carbonate in solution. The seawater lines are calculated using apparent acidity constants that include the effects of ion pairing; see Chapter 6.

The pH of such a solution containing no strong acid or strong base is called the CO_2 *equivalence point. It is defined as the pH of a pure solution of CO_2 in water* and is given mathematically by the solution of Equation 5, which usually simplifies to $(H^+) = (HCO_3^-)$. The equivalence point depends on the total carbonate concentration and, to a lesser degree, on the ionic strength of the solution. Figure 4.3 gives the CO_2 equivalence point for two different ionic strengths, as a function of the total concentration of carbonate in solution. [The symbol C_T will be used to designate this concentration. It differs from $TOTCO_3$ by including neither the concentration of the gas phase ($CO_2 \cdot g$) nor the concentration of the various carbonate solids when those are considered.]

2.2 Alkalinity: Preliminary Notion

As mentioned earlier, natural waters acquire strong base from the dissolution of rocks. *The resulting net concentration of strong base in excess of strong acid is, by definition, the alkalinity of the water.* For didactic purposes let us simply add NaOH to our closed carbonate system of Example 1.

Example 2

Recipe
$$(NaCl)_T = 10^{-3} \, M$$
$$(CO_2)_T = 10^{-4} \, M$$
$$(NaOH)_T = 10^{-4.5} \, M$$

The reactions and species are identical to those in the previous system. The addition of NaOH at the bottom of Tableau 4.1 yields only two changes in the mole balances:

$$TOTH = (H^+) - (OH^-) - (HCO_3^-) - 2(CO_3^{2-})$$
$$= -(NaOH)_T = -10^{-4.5} \tag{8}$$

$$TOTNa = (Na^+) = (NaCl)_T + (NaOH)_T = 10^{-2.99} \tag{9}$$

According to the definition, the alkalinity of this system, $(NaOH)_T$, is then equal to the negative value of $TOTH$ as expressed in Equation 8:

$$Alk = -(H^+) + (OH^-) + (HCO_3^-) + 2(CO_3^{2-}) = (NaOH)_T \tag{10}$$

It is also equal to the difference between the sodium and the chloride mole balances:

$$TOTNa - TOTCl = Alk = (Na^+) - (Cl^-) = (NaOH)_T \tag{11}$$

The presence of the strong base NaOH produces an excess of positive charge from the strong electrolytes $[(Na^+) > (Cl^-)]$, which is balanced by the net negative charge from the dissociation of carbon dioxide and water. Although not strictly correct, this interpretation of alkalinity as a balance of charges provides a convenient preliminary notion:

$$Alk = (\text{excess negative charge from weak acids})$$

$$= (\text{excess positive charge from strong bases}) \qquad (12)$$

Both the left- and the right-hand sides of this formula (top and bottom) correspond to useful conceptual and mathematical definitions of alkalinity. Since in most natural waters carbonate is much more abundant than other weak acids and bases, and Na^+, K^+, Ca^{2+}, Mg^{2+}, Cl^-, and SO_4^{2-} are the other major ions, the alkalinity equation can usually be expressed explicitly as

$$Alk = (OH^-) - (H^+) + (HCO_3^-) + 2(CO_3^{2-})$$

$$= (Na^+) + (K^+) + 2(Ca^{2+}) + 2(Mg^{2+}) - (Cl^-) - 2(SO_4^{2-}) \quad (13)$$

The separation of the weak and strong acids and bases on each side of the equal sign illustrates the pH buffering of the aqueous phase in natural waters: as more and more strong base is added to the system, the weak acids dissociate increasingly ($H_2CO_3^* \rightarrow HCO_3^- \rightarrow CO_3^{2-}$) to balance the excess positive charge.

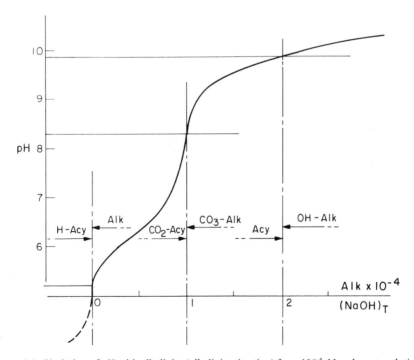

Figure 4.4 Variation of pH with alkalinity (alkalinity titration) for a 10^{-4} M carbonate solution. The arrows indicate the endpoints of the alkalimetric and acidimetric titrations corresponding to the various base and acid neutralizing capacities of the solution: H-Acy = mineral acidity, CO_2-Acy = CO_2 acidity, Acy = acidity, OH-Alk = caustic alkalinity, CO_3-Alk = carbonate alkalinity, Alk = alkalinity.

The degree of dissociation of these weak acids is what determines the pH of the water (as given by the mass laws for HCO_3^- and CO_3^{2-}).

In our example, the addition of NaOH raises the pH, and (HCO_3^-) becomes larger than (H^+). The approximate solution to Equation 10 is given by

$$(HCO_3^-) = (NaOH)_T = 10^{-4.5} \quad \text{(see Figure 4.2)} \tag{14}$$

Introducing Equation 14 and the mass law for HCO_3^- into the carbonate mole balance yields

$$TOTH_2CO_3 = 10^{6.3}(H^+)10^{-4.5} + 10^{-4.5} + \frac{10^{-10.3} \times 10^{-4.5}}{(H^+)} = 10^{-4.0}$$

therefore

$$pH = 5.97$$

As a generalization of this problem, Figure 4.4 shows how the pH of our model system varies when the alkalinity, $(NaOH)_T$, is increased from zero (Alk = 0 is the condition of the CO_2 equivalence point) to $3 \times 10^{-4}\ M$. Note that, to obtain conveniently the pH in the ranges 6.3 to 10.3 and above 10.3, HCO_3^- and CO_3^{2-} should be respectively chosen as the principal components in these pH ranges.

2.3   Alkalinity: Mathematical Definition

Although convenient, the conceptualization of alkalinity as a charge balance is not strictly correct; the proper concept is that of an acid-base balance. The actual definition of alkalinity is thus founded on a conservation equation for hydrogen ions $(TOTH)$, which is not necessarily identical to the electroneutrality equation.

The alkalinity of a solution is the negative of the TOTH expression when the components are the principal components of the solution at the CO_2 equivalence point $(H^+, H_2CO_3^*, Na^+, Cl^-,$ etc., since the equivalence point is typically in the range of pH 4 to 5). Weak acids and bases whose principal components are charged at the CO_2 equivalence point (e.g., NH_4^+ or $H_2PO_4^-$) make the difference between this precise definition of alkalinity and the previous notion of balance of charges. The more exact concept is that of a proton deficiency (with respect to a zero level defined by the principal components at the CO_2 equivalence point) for weak acids and bases which is balanced (or caused) by an excess of strong base over strong acid. In fact, to define alkalinity more precisely, strong acids and strong bases can be characterized as H^+ and OH^- containing compounds that are totally dissociated at the CO_2 equivalence point. For example, as can be seen in Figure 4.1, H_3PO_4 ($pK_{a1} = 2.2$, $pK_{a2} = 7.2$, $pK_{a3} = 12.4$) is a strong monoprotic acid and a weak diprotic acid, and NH_3 ($= NH_4OH$; $pK_a = 9.2$) is a strong base.

The definition of alkalinity is normally restricted to the aqueous phase. However, it can be easily extended to heterogeneous systems by considering the proper principal components in any of the phases of the system, including gases or solids, at the CO_2 equivalence point.

Focusing on the left-hand side of the $TOTH$ equation, we find for a simple carbonate system

$$\text{Alk} = -(H^+) + (OH)^- + (HCO_3^-) + 2(CO_3^{2-}) \tag{15}$$

and for a system that contains other natural weak acids and bases

$$\text{Alk} = -(H^+) + (OH)^- + (HCO_3^-) + 2(CO_3^{2-}) + (NH_3) + (HS^-) + 2(S^{2-})$$
$$+ (HSiO_3^-) + (B(OH)_4^-) - (H_3PO_4) + (HPO_4^{2-}) + 2(PO_4^{3-}) \tag{16}$$

The species, H^+, $H_2CO_3^*$, NH_4^+, H_2S, H_2SiO_3, $B(OH)_3$ and $H_2PO_4^-$ are the principal components at the CO_2 equivalence point since they are the major soluble species for these components in the pH range 4 to 6 (see Figure 4.1 and Tableau 4.2).

The contribution of complexes such as $NaCO_3^-$ or $CaCO_3$ to the alkalinity expression is easily obtained from the knowledge of the principal components at the CO_2 equivalence point (H^+, $H_2CO_3^*$, Na^+, HPO_4^{2-}; see Tableau 4.2):

$$\text{Alk} = -(H^+) + (OH)^- + (HCO_3^-) + 2(CO_3^{2-}) + 2(NaCO_3^-) + 2(CaCO_3)$$
$$- (H_3PO_4) + (HPO_4^{2-}) + 2(PO_4^{3-}) + (CaHPO_4) + (CaOH^+) \tag{17}$$

TABLEAU 4.2

	H^+	CO_2	NH_4^+	H_2S	H_2SiO_3	$B(OH)_3$	$H_2PO_4^-$	Na^+	Ca^{2+}
H^+	1								
OH^-	-1								
$CaOH^+$	-1								1
$H_2CO_3^*$		1							
HCO_3^-	-1	1							
CO_3^{2-}	-2	1							
$NaCO_3^-$	-2	1						1	
$CaCO_3$	-2	1							1
NH_4^+			1						
NH_3	-1		1						
H_2S				1					
HS^-	-1			1					
S^{2-}	-2			1					
H_2SiO_3					1				
$HSiO_3^-$	-1				1				
$B(OH)_3$						1			
$B(OH)_4^-$	-1					1			
H_3PO_4	1						1		
$H_2PO_4^-$							1		
HPO_4^{2-}	-1						1		
PO_4^{3-}	-2						1		
$CaHPO_4$	-1						1		1

A completely systematic method for writing out the alkalinity expression is thus provided by its mathematical definition, even for very complex chemical systems. If the definition of alkalinity were purely mathematical, however, it would be of limited usefulness; an essential property of alkalinity is that it is a measurable quantity.

2.4   Alkalinity: Experimental Definition

The alkalinity of a solution is its acid-neutralizing capacity when the end point of the titration is the CO_2 equivalence point.

This definition clearly focuses on the right-hand side of the $TOTH$ equation and is most easily understood by referring to our previous example: if HCl is added to the system at a concentration equal to that of NaOH, the alkalinity goes to zero which is the condition for the CO_2 equivalence point. Conversely, if HCl is added until the pH reaches the CO_2 equivalence point, the original alkalinity of the system is equal to the concentration of HCl added. This provides an experimental method for determining the alkalinity of a solution by acidimetric titration. The end point is usually recognizable as a sharp inflection in the titration curve (see Figure 4.4) and small errors in the end point affect the alkalinity determination negligibly. The major source of inaccuracy is commonly due to the loss of CO_2 to the atmosphere when the titration is carried to a fixed pH as the end point. (Remember that the CO_2 equivalence point is a function of C_T and that alkalinity is defined for the aqueous phase exclusively!) This can be minimized in various ways, and several techniques have been developed to measure alkalinity with great accuracy. Note that the presence of a weak acid with a pK_a near the CO_2 equivalence point will create discrepancies between the mathematical and experimental definitions of alkalinity (the acid will be incompletely titrated) and complicate its experimental determination by buffering the pH near the end point.

2.5   Other Related Definitions and Quantities

In the same manner that we have defined the CO_2 equivalence point as the pH of a pure CO_2 solution, we can define the bicarbonate and carbonate equivalence points as the pH's of pure bicarbonate (e.g., $NaHCO_3$) and carbonate (e.g., Na_2CO_3) salt solutions, respectively. Mathematically, these equivalence points are most conveniently obtained by equating to zero the proton balance equations corresponding to the respective principal components.

Bicarbonate Equivalence Point (Pure NaHCO₃ Solution)

The components H^+, HCO_3^-, and Na^+ yield Tableau 4.3, from which we obtain the $TOTH$ equation:

$$TOTH = (H^+) - (OH^-) + (H_2CO_3^*) - (CO_3^{2-}) = 0 \qquad (18)$$

TABLEAU 4.3

	H^+	HCO_3^-	Na^+
H^+	1		
OH^-	-1		
$H_2CO_3^*$	1	1	
HCO_3^-		1	
CO_3^{2-}	-1	1	
Na^+			1
$NaHCO_3^-$		1	1

For example, for $(NaHCO_3)_T = 10^{-3}$ M, the $TOTH$ equation simplifies approximately to

$$(H_2CO_3^*) = (CO_3^{2-}) \tag{19}$$

therefore

$$pH = 8.3 \quad \text{(see Figures 4.3 and 4.5)}$$

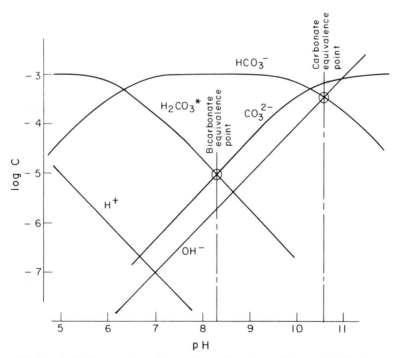

Figure 4.5 Log C-pH diagram for carbonate illustrating the conditions of the bicarbonate and carbonate equivalence points.

Carbonate Equivalence Point (Pure Na_2CO_3 Solution)

The components H^+, CO_3^{2-}, and Na^+ yield Tableau 4.4.

TABLEAU 4.4

	H^+	CO_3^{2-}	Na^+
H^+	1		
OH^-	-1		
$H_2CO_3^*$	2	1	
HCO_3^-	1	1	
CO_3^{2-}		1	
Na^+			1
Na_2CO_3		1	2

$$TOTH = (H^+) - (OH^-) + 2(H_2CO_3^*) + (HCO_3^-) = 0 \qquad (20)$$

For example, for $(Na_2CO_3) = 10^{-3}$ M, the $TOTH$ equation simplifies to

$$(OH^-) = (HCO_3^-) \qquad (21)$$

therefore

$$pH = 10.55 \quad \text{(see Figures 4.3 and 4.5)}$$

Acid and Base Neutralizing Capacity

We have defined the alkalinity of a solution as its acid neutralizing capacity when the end point of the titration is the CO_2 equivalence point. Similar quantities can be defined as the acid or base neutralizing capacity of a solution for HCO_3^- and CO_3^{2-} equivalence points as titration end points. The mathematical expressions of these quantities are the $TOTH$ equations corresponding to the respective principal components.

For CO_2 equivalence point as end point

$$TOTH_1 = (H^+) - (OH^-) - (HCO_3^-) - 2(CO_3^{2-}) \qquad (22)$$

For HCO_3^- equivalence point as end point

$$TOTH_2 = (H^+) - (OH^-) + (H_2CO_3^*) - (CO_3^{2-}) \qquad (23)$$

For CO_3^{2-} equivalence point as end point

$$TOTH_3 = (H^+) - (OH^-) + 2(H_2CO_3^*) + (HCO_3^-) \qquad (24)$$

Each of these quantities is given a different name depending on whether it is positive or negative, that is, whether the corresponding end point is obtained by acid or base addition.

For acid-neutralizing quantities ($TOTH < 0$)

$$\text{Alkalinity} = -TOTH_1$$

$$\text{Carbonate alkalinity} = -TOTH_2$$

$$\text{Caustic alkalinity} = -TOTH_3$$

For base neutralizing quantities ($TOTH > 0$)

$$\text{Mineral acidity} = TOTH_1$$

$$CO_2 \text{ acidity} = TOTH_2$$

$$\text{Acidity} = TOTH_3$$

Each of these quantities is illustrated in the calculated titration curve of Figure 4.4.*

Note 1. In simple carbonate systems the various acid and base neutralizing quantities are related to each other by the equation

$$TOTH_3 = TOTH_2 + C_T = TOTH_1 + 2C_T \tag{25}$$

This equation is used to obtain numerical values of $TOTH$ when switching from one carbonate component to another.

Note 2. The definitions of all the acid and base neutralizing quantities can be generalized to complex systems by considering the $TOTH$ equations corresponding to the principal components of all acids and bases at the various equivalence points. For example, in the presence of small concentrations of ammonia, sulfide, silicate, borate, and phosphate, the carbonate alkalinity is obtained by choosing the principal components H^+, HCO_3^-, NH_4^+, HS^-, H_2SiO_3, $B(OH)_3$, and HPO_4^{2-}:

$$\begin{aligned} \text{Carb-Alk} = -TOTH_2 = {} & -(H^+) + (OH^-) - (H_2CO_3^*) + (CO_3^{2-}) \\ & + (NH_3) - (H_2S) + (S^{2-}) + (HSiO_3^-) + (B(OH)_4^-) \\ & - 2(H_3PO_4) - (H_2PO_4^-) + (PO_4^{3-}) \end{aligned} \tag{26}$$

and the acidity is obtained by choosing the principal components H^+, CO_3^{2-}, NH_3, HS^-, $HSiO_3^-$, $B(OH)_4^-$, and HPO_4^{2-}:

$$\begin{aligned} \text{Acy} = TOTH_3 = {} & (H^+) - (OH^-) + 2(H_2CO_3^*) + (HCO_3^-) \\ & + (NH_4^+) + (H_2S) - (S^{2-}) + (H_2SiO_3) + (B(OH)_3) \\ & + 2(H_3PO_4) + (H_2PO_4^-) - (PO_4^{3-}) \end{aligned} \tag{27}$$

---

* Oceanographers use the term "carbonate alkalinity" to designate the carbonate contribution to the alkalinity: $(HCO_3^-) + 2(CO_3^{2-})$. To avoid confusion, the term p-alkalinity is now used for $-TOTH_2$ in some textbooks.

3. EQUILIBRIUM WITH THE GAS PHASE

3.1 Carbon Dioxide

So far we have considered an aqueous phase that is closed to the atmosphere with no gas exchange taking place at all. Later we shall see that gas exchange kinetics at the air-water interface result in CO_2 equilibration times on the order of days. Equilibrium aqueous systems of fixed total carbonate concentration are thus useful models of natural waters for "fast" processes taking place on a time scale of hours or less. At the other extreme, if we are interested in long time scales (i.e., average annual composition of surface waters), it is then a good approximation to consider CO_2 to be in equilibrium between the water and the atmosphere. In the first case the one-phase model system is defined by two mole balance equations, Alk and C_T, which determine the pH as illustrated in Figure

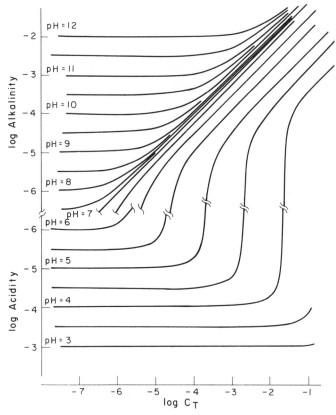

Figure 4.6 General relationship between total carbonate (C_T), alkalinity ($= -$ acidity), and pH in natural waters when the acid-base chemistry is dominated by the carbonate system. (Many graphs similar in concept to this one and others in this chapter and the next were pioneered by various authors including notably Sillén, Butler, Deffeyes, and Stumm and Morgan.)

4.6. Any two of the three parameters Alk, C_T, and pH provide the third one. In the second case the acid-base composition of the aqueous phase (its pH) is determined by the alkalinity and the equilibrium partial pressure of CO_2 in the atmosphere. The three interdependent parameters are then Alk, P_{CO_2}, and pH. The mole balance equation for carbonate is no longer useful and is replaced by the mass law solubility equation (Henry's law) for CO_2 dissolution:

$$CO_2(g) + H_2O = H_2CO_3^*; \quad K_H = 10^{-1.5} \tag{28}$$

$$(H_2CO_3^*) = 10^{-1.5}P_{CO_2} \tag{29}$$

As is apparent from its definition, *the alkalinity of a solution is independent of any gain or loss of CO_2 by the solution* (CO_2 is a component and thus has a coefficient of zero in the H^+ column). Conceptually, this may be seen most easily by noting that at the right-hand side of the *TOTH* expression the excess

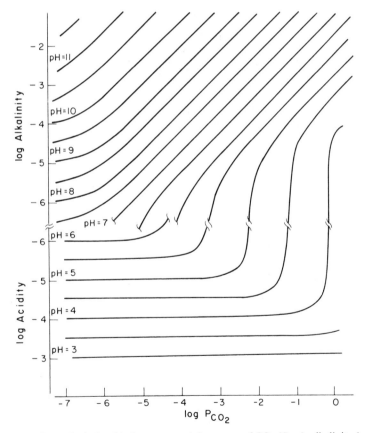

Figure 4.7 General relationship between partial pressure of CO_2 (P_{CO_2}), alkalinity ($= -$ acidity), and pH in natural waters at equilibrium with a gas phase when the acid-base chemistry is dominated by the carbonate system.

of strong bases is unaffected by the CO_2 exchange. The concentrations of each of the species on the left-hand side of the $TOTH$ expression (H^+, OH^-, HCO_3^-, etc.) do vary however; it is only their algebraic combination, the alkalinity, that is invariant upon CO_2 exchange. The total soluble carbonate (C_T) is of course directly dependent on CO_2 gain or loss by the solution.

In Example 4 of Chapter 3 we calculated the pH of a carbonate solution for a fixed partial pressure of CO_2 and—although we did not call it that—the alkalinity of the system, given then by $(NaOH)_T$. The $TOTH$ expression we obtained was precisely the alkalinity equation since CO_2 was chosen automatically as a component.

A solution to the general problem of finding the pH, knowing the P_{CO_2} and the alkalinity of a system where carbonate is the only weak acid, is given directly by the alkalinity equation. The various species concentrations are replaced by their mass law expressions as a function of H^+ and P_{CO_2}:

$$Alk = -(H^+)$$

$$+ \underbrace{K_w(H^+)^{-1}}_{(OH^-)} + \underbrace{K_H K_{a1} P_{CO_2}(H^+)^{-1}}_{(HCO_3^-)} + \underbrace{2K_H K_{a1} K_{a2} P_{CO_2}(H^+)^{-2}}_{(CO_3^{2-})} \quad (30)$$

K_H, K_{a1}, K_{a2} are respectively the Henry's law and the first and second acidity constants of the carbonate system.

This is an explicit third-degree polynomial of (H^+), and it can be solved for any combination of P_{CO_2} and Alk, including negative alkalinities. To obtain a general graph of the relation between the three parameters P_{CO_2}, Alk, and pH, as illustrated in Figure 4.7, it is of course more convenient to calculate the alkalinity, given (H^+) and P_{CO_2}.

3.2   Other Volatile Species

If a body of water is considered totally at equilibrium with the atmosphere, other volatile species with weak acid-base properties should be taken into account to assess their effect on the alkalinity and the pH of the solution. Two such gases that are important are hydrogen sulfide and ammonia:

$$H_2S(g) = H_2S(aq); \quad K_H = 10^{-1.0} \quad (31)$$

$$NH_3(g) = NH_3(aq); \quad K_H = 10^{+1.8} \quad (32)$$

For hydrogen sulfide the situation is much like that for carbon dioxide since the principal component of the sulfide system at the CO_2 equivalence point is H_2S. Loss of H_2S to the atmosphere (which has a very low partial pressure of the gas; approximately 10^{-10} at) will not affect the alkalinity of the water. The total soluble concentration of sulfide $(H_2S)_T$ will decrease, and the pH will increase accordingly.

Example 3

Consider a water whose carbonate system is initially at equilibrium with the atmosphere ($P_{CO_2} = 10^{-3.5}$ at). This water contains sulfide: $(H_2S)_T = 2 \times 10^{-4}$ M, and its initial pH is 8.0. What will be the final pH of this water when hydrogen sulfide is lost to the atmosphere, if $CO_2(g)$ remains at equilibrium and $P_{H_2S} = 0$?

1. Initially

$$P_{CO_2} = 10^{-3.5} \text{ at}$$

$$S_T = 10^{-3.7} M$$

$$pH = 8.0$$

The concentrations of the various species in the system are easily computed:

$$(H_2CO_3^*) = 10^{-1.5} \times 10^{-3.5} = 10^{-5} M$$

$$(HCO_3^-) = \frac{10^{-6.3} \times 10^{-5}}{10^{-8}} = 10^{-3.3} M$$

$$(CO_3^{2-}) = \frac{10^{-10.3} \times 10^{-3.3}}{10^{-8}} = 10^{-5.6} M$$

Since the pK_a for the H_2S–HS^- reaction is 7.0, at pH 8 most of the sulfide will be HS^- and about one-tenth will be H_2S (see Figure 4.1):

$$S_T = (H_2S) + (HS^-) + (S^{2-}) = 10^{-3.7}$$

$$= 10^7 \times 10^{-8}(HS^-) + (HS^-) + 10^{-13.9}\left(\frac{(HS^-)}{10^{-8}}\right) = 10^{-3.7} \quad (33)$$

therefore

$$(HS^-) = 10^{-3.74} M$$

$$(H_2S) = 10^{+7} \times 10^{-8} \times 10^{-3.74} = 10^{-4.74} M$$

$$(S^{2-}) = \frac{10^{-13.9} \times 10^{-3.74}}{10^{-8}} = 10^{-9.64} M$$

We can obtain the value of the alkalinity for this system:

$$\text{Alk} = -(H^+) + (OH^-) + (HCO_3^-) + 2(CO_3^{2-}) + (HS^-) + 2(S^{2-}) \quad (34)$$

$$= (HCO_3^-) + (HS^-) = 10^{-3.17} M$$

2. Eventually

$P_{CO_2} = 10^{-3.5}$ at   (unchanged)

$S_T = 0$ (all sulfide must escape from the system since we assume $P_{H_2S} = 0$)

$\text{Alk} = 10^{-3.17}$ (remains unchanged as CO_2 and H_2S exchange with the atmosphere)

Developing the alkalinity expression in the absence of sulfide

$$Alk = -(H^+) + (OH^-) + (HCO_3^-) + 2(CO_3^{2-}) = 10^{-3.17} \qquad (35)$$

Assuming all terms to be negligible except (HCO_3^-):

$$(HCO_3^-) = 10^{-3.17} = \frac{10^{-6.3} \times 10^{-1.5} \times 10^{-3.5}}{(H^+)}$$

$$(H^+) = 10^{-8.13} \, M$$

After verifying that other terms in the alkalinity expression are indeed negligible, we have a final pH = 8.13. The volatilization of H_2S has resulted in only a small pH increase as CO_2 has replaced the lost H_2S.

The situation for ammonia is different since NH_4^+, not NH_3, is the principal component at the CO_2 equivalence point. Loss of ammonia to the atmosphere ($P_{NH_3} \cong 10^{-8}$ at) thus results in a decrease of the alkalinity of the system. Concomitantly, the pH of the solution decreases. Note, however, that ammonia is orders of magnitude more soluble than carbon dioxide and hydrogen sulfide and that ammonia losses are accordingly much slower.

4. CASE STUDY: MIXING OF TWO WATERS

In estuaries, river confluences, wastewater disposal sites, and so on, waters of different chemical characteristics (e.g., pH and Alk) are mixed. If provided with sufficient information on the chemical characteristics of the original waters and a description of the mixing process, one should be able to predict the chemical characteristics of the mixed water. The basic approach to solving such a problem is to consider the mixing of "conservative" quantities, namely, the mole balances. If two solutions of 1 liter each are mixed, the recipe for the mixture is obviously the sum of the two original recipes divided by two; the pH of the mixture is *not* the average of the two original pH's. To perform such a "mixing" operation, the numerical values (RHS) of the mole balances describing the original waters have to be obtained based on the same set of components. Depending on the original and final constraints (e.g., P_{CO_2} vs. C_T) that are considered for the definition of the system, it might be necessary to consider, serially, different sets of components.

Figure 4.8 illustrates such a solution scheme as applied to the pH of mixtures and to the carbonate system that usually controls it. The conservative quantities, Alk and C_T, are obtained from the given chemical definition of the original waters. As there is no immediate CO_2 exchange with the atmosphere, these quantities are "mixed" and yield the initial composition of the mixture. Over long times of equilibration with the atmosphere, CO_2 equilibrium can be assumed, and the mixture is then defined by a fixed partial pressure of CO_2 and the "mixed" alkalinity. These two equilibrium models idealize gas exchange processes as complete or infinitely slow and mixing processes as instantaneous in given proportions. A more detailed description of the hydrodynamics and

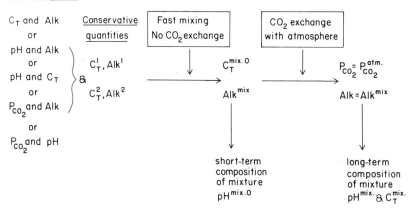

Figure 4.8 General factors controlling the pH in mixtures of carbonate-bearing waters, and schematic of the general method of calculation.

of the gas exchange would permit a chemical definition of the system that varies in time and space:

$$\text{Alk}\,(x, y, z, t); \quad C_T(x, y, z, t) \rightarrow \text{pH}\,(x, y, z, t)$$

Example 4

River A pH = 8.2   Alk = 10^{-3} eq liter^{-1}
River B pH = 5.7   Alk = 10^{-5} eq liter^{-1}
mixture   2 volumes of A for 3 volumes of B.

In order to obtain the total carbonate content of the water after mixing, we need to calculate C_T for each of the rivers.

River A. At a pH of 8.2, HCO_3^- is roughly 100 times larger than the other carbonate species:

$$\text{Alk}^A = -(H^+) + (OH^-) + (HCO_3^-) + 2(CO_3^{2-}) = 10^{-3} \qquad (36)$$

therefore

$$(HCO_3^-) + 2 \times 10^{-10.3} \times 10^{+8.2}(HCO_3^-) = 10^{-3}$$

From the mass laws we then obtain

$$(HCO_3^-) = 10^{-3.01}$$

$$(H_2CO_3^*) = 10^{+6.3} \times 10^{-8.2} \times 10^{-3.01} = 10^{-4.91}$$

$$(CO_3^{2-}) = 10^{-10.3} \times 10^{+8.2} \times 10^{-3.01} = 10^{-5.11}$$

$$C_T^A = (HCO_3^-) + (H_2CO_3^*) + (CO_3^{2-}) = 10^{-3.00} \; M$$

$$(37)$$

River B.   At a pH of 5.7, $H_2CO_3^*$ is the principal carbonate component:

$$Alk^B = -(H^+) + (OH^-) + (HCO_3^-) + 2(CO_3^{2-}) = 10^{-5} \qquad (38)$$

From this we obtain

$$(HCO_3^-) = 10^{-5} + 10^{-5.7} = 10^{-4.92}$$

$$(H_2CO_3^*) = 10^{+6.3} \times 10^{-5.7} \times 10^{-4.92} = 10^{-4.32} \qquad (39)$$

$$C_T^B = (H_2CO_3^*) + (HCO_3^-) = 10^{-4.22}$$

Initial mix.   Before CO_2 exchange

$$Alk^{mix} = \tfrac{2}{5} Alk^A + \tfrac{3}{5} Alk^B = 10^{-3.39} \text{ eq liter}^{-1} \qquad (40)$$

$$C_T^{mix} = \tfrac{2}{5} C_T^A + \tfrac{3}{5} C_T^B = 10^{-3.36} \; M \qquad (41)$$

The pH of the mixture is expected to be approximately neutral, and HCO_3^- is thus expected to be the principal component for the carbonate system. Choosing HCO_3^- as a component,

$$TOTH = (H^+) - (OH^-) + (H_2CO_3^*) - (CO_3^{2-})$$

$$= C_T - Alk = 10^{-4.54} \qquad (42)$$

$$TOTHCO_3 = (H_2CO_3^*) + (HCO_3^-) + (CO_3^{2-}) = C_T = 10^{-3.36} \qquad (43)$$

In the neutral pH range, $(H_2CO_3^*)$ is the major positive term in the $TOTH$ equation. Neglecting the other terms as a first approximation,

$$(H_2CO_3^*) = 10^{-4.54}$$

and

$$(HCO_3^-) = C_T - (H_2CO_3^*) = 10^{-3.39}$$

therefore

$$(H^+) = \frac{(H_2CO_3^*)}{(HCO_3^-)} 10^{-6.3} = 10^{-7.45}$$

$$pH_{mix} = 7.45, \quad \text{initially.}$$

The approximations can be verified to be valid at this point.

Mix Equilibrated with Atmosphere.   After CO_2 equilibration

$$Alk^{mix} = 10^{-3.39} \text{ eq liter}^{-1}$$

$$P_{CO_2} = 10^{-3.5} \text{ at}$$

With the components H^+ and CO_2, the $TOTH$ equation is

$$TOTH = (H^+) - (OH^-) - (HCO_3^-) - 2(CO_3^{2-}) = -Alk = -10^{-3.39}$$

Trial and error, or graphical examination, shows that (HCO_3^-) is the major term of the $TOTH$ equation:

$$(HCO_3^-) = 10^{-3.39} = \frac{10^{-1.5} \times 10^{-6.3} P_{CO_2}}{(H^+)}$$

therefore

$$(H^+) = \frac{10^{-1.5} \times 10^{-6.3} \times 10^{-3.5}}{10^{-3.39}} = 10^{-7.91}$$

$pH_{mix} = 7.91$, eventually.

The solution of a mixing problem such as this one is made particularly convenient by repeated use of generalized graphs relating C_T, Alk, and pH (Figure 4.6) and P_{CO_2}, Alk, and pH (Figure 4.7).

Note. The main point to be remembered from this example is that only conservative quantities can be added together. To find the composition of a mixture, concentrations of particular species cannot be averaged out, much less their logarithms. (For example, one has to be careful in defining an average pH for rain water.) However, any set of components can be chosen to perform the "mixing operation" as long as these are the same for each part of the system. In the preceding example we could have calculated $TOTH$ for each of the rivers by using HCO_3^- as a component and thus obtained directly a convenient $TOTH$ equation for the mix without ever explicitly calculating its alkalinity. It is mandatory, however, to use $H_2CO_3^*$ as a component and alkalinity as the proton balance equation whenever the system is controlled by a fixed partial pressure of CO_2 rather than a total carbonate concentration.

5.   EFFECTS OF BIOLOGICAL PROCESSES ON pH AND ALKALINITY

5.1   Photosynthesis and Respiration

So far we have considered only the atmosphere to be a source or a sink for CO_2 in aquatic systems. In fact photosynthesis and respiration taking place within the water column and in the benthos often dominate the utilization and production of carbon dioxide in natural waters. A very simplified chemical representation of these processes is given by the reaction

$$CO_2 + H_2O \xrightleftharpoons[\text{respiration}]{\text{photosynthesis}} \text{``CH}_2O\text{''} + O_2 \tag{44}$$

where "CH_2O" is taken as a symbol for organic matter. Despite its crudity this chemical representation gives a reasonable approximation of the stoichiometry of total photosynthetic and respiratory activities as they affect water chemistry.

According to Reaction 44 photosynthesis and respiration do not, in a first approximation, affect the alkalinity of the water. Whether the CO_2 is gained or lost in autochthonous processes or by exchange with the atmosphere the alkalinity of the system remains unchanged. (Consider the RHS of the $TOTH$ equation, with $H_2CO_3^*$ as the carbonate component.) Also in either case, the pH decreases with increasing total carbonate. To obtain quantitatively the effect of photosynthesis and respiration on pH, the net loss or gain of CO_2 has to be measured, and the new pH is then calculated, given a constant alkalinity and the new total carbonate concentration

$$C_T^{new} = C_T^{old} + \text{gain of } CO_2 \tag{45}$$

A semiquantitative estimate of the effect of CO_2 gain or loss on pH is obtained by considering the reactions written with the principal components of the system:

$$\text{pH} < 6.3 \qquad\qquad CO_2 + H_2O = \text{``}CH_2O\text{''} + O_2 \tag{46}$$

$$6.3 < \text{pH} < 10.3 \qquad HCO_3^- + H^+ = \text{``}CH_2O\text{''} + O_2 \tag{47}$$

$$\text{pH} > 10.3 \qquad\qquad CO_3^{2-} + 2H^+ = \text{``}CH_2O\text{''} + O_2 \tag{48}$$

At low pH the effect of photosynthesis and respiration on pH is minimal (below pH = 6.3 the reaction shows no proton to be consumed), while at high pH the effect is large (above pH = 10.3 two protons are consumed per fixed carbon). In the carbonate and the bicarbonate regions, CO_2 acts effectively as a strong diprotic and monoprotic acid, respectively. This intuitive observation can be rationalized mathematically by considering the $TOTH$ equation corresponding to the principal components of the system: what is basic or acidic at any pH can be defined as a negative or positive contribution to the $TOTH$ equation. In a first approximation the effect of a change in composition on pH is given by the sign and the magnitude of the change in the numerical value of the $TOTH$ equation (RHS) corresponding to the principal components of the system (see Section 6).

Plant growth requires more than just carbon, water, and light. Other nutrients that are necessary in relatively large quantities are phosphorus and nitrogen, chiefly provided in aquatic systems by phosphate and nitrate or ammonium. The average proportions of these major elements in algal biomass is described by the Redfield formula:[7]

$$\text{Protoplasm} = C_{106}H_{263}O_{110}N_{16}P_1 \tag{49}$$

A more complete stoichiometric description of photosynthetic and respiratory processes in natural waters is thus provided by the following reactions:

$$106CO_2 + 16NO_3^- + H_2PO_4^- + 122H_2O + 17H^+$$

$$\overset{P}{\underset{R}{\rightleftharpoons}} \text{Protoplasm} + 138O_2 \tag{50}$$

$$106CO_2 + 16NH_4^+ + H_2PO_4^- + 106H_2O$$

$$\overset{P}{\underset{R}{\rightleftharpoons}} \text{Protoplasm} + 106O_2 + 15H^+ \tag{51}$$

These overall reactions permit an accurate study of the role of photosynthesis and respiration in modifying the pH and the alkalinity of natural waters despite their inapplicability to particular organisms under particular conditions. As written, they provide a direct measure of alkalinity changes since the reactants are the principal components at the CO_2 equivalence points, H^+, $H_2CO_3^*$, NO_3^-, $H_2PO_4^-$, and NH_4^+:

1 When nitrate is the nitrogen source, the alkalinity increases by 0.16 ($=17/106$) equivalent per mole of carbon fixed.
2 When ammonium is the nitrogen source, the alkalinity decreases by 0.14 ($=15/106$) equivalent per mole of carbon fixed.

The inclusion of nitrogen and phosphorus in the reactions for photosynthesis and respiration does not modify markedly our previous results with respect to the net pH effects of these processes. Consider, for example, the reactions written with the principal components in the pH range 7 to 9:

$$106HCO_3^- + 16NO_3^- + HPO_4^{2-} + 16H_2O + 124H^+$$

$$\underset{R}{\overset{P}{\rightleftarrows}} \text{Protoplasm} + 138O_2 \quad (52)$$

$$106HCO_3^- + 16NH_4^+ + HPO_4^{2-} + 92H^+ \underset{R}{\overset{P}{\rightleftarrows}} \text{Protoplasm} + 106O_2 \quad (53)$$

Clearly, in all situations the uptake of CO_2 dominates the acid-base effect of photosynthesis. For example, if bicarbonate is the major inorganic carbon species, uptake of CO_2 results in an increase in pH. This result is independent of whether $CO_2(aq)$, $H_2CO_3^*$, or HCO_3^- is the actual species taken up by the algae as long as the stoichiometry of the reactions is correct (i.e., H^+ is the counter ion for HCO_3^-). The magnitude of the pH effect is influenced by the nature of the nitrogen source, however, photosynthesis drives the pH higher when nitrate rather than ammonium is the nitrogen source for aquatic plants. At low pH where $H_2CO_3^*$ is the major carbonate species, the nitrogen uptake can dominate the acid-base chemistry of photosynthesis. For example, uptake of ammonia in bog water (pH $\cong$ 4) can result in a net pH decrease, as seen in Reaction 51.

5.2 Other Microbial Processes

Microbial processes other than photosynthesis and respiration are also involved in the cycle of elements. Sulfate reduction, nitrogen fixation, nitrification, denitrification, methane production, and others, are all locally and globally important in governing the cycles of sulfur, nitrogen, and carbon.

Although the energetics of these oxidation-reduction processes are examined in Chapter 7, the stoichiometric approach we have taken for photosynthesis and respiration permits us to examine here their effect on the pH and alkalinity of aquatic systems. As before, the methodology consists of writing overall reactions describing the stoichiometry of the various processes as a function of the principal components for the alkalinity expression (i.e., at the CO_2 equivalence point; pH $\cong$ 5). The production or consumption of protons in such reactions then yields the alkalinity changes. The effects on pH are obtained by considering the gains or losses of protons when reactants and products are written as the principal components under ambient conditions.

Table 4.3 lists such reactions and the corresponding qualitative effects on alkalinity and pH. When chemical conditions are specified, the pH change can be calculated by considering the effects of the various reactions on the mole balances. Note that in each reaction of Table 4.3 the change in alkalinity is precisely the change in the electrical charge of the species of specific interest (S, N, C). For example, from an acid-base perspective denitrification and sulfate reduction are strictly equivalent to removal of nitric and sulfuric acid, respectively.

TABLE 4.3

Effects of Aquatic Microbial Processes on Alkalinity and pH in Closed Systems

Sulfate reduction
$$SO_4^{2-} + 2``CH_2O" + 2H^+ \rightarrow$$
$$H_2S + 2H_2O + 2CO_2$$

Alkalinity increases by 2 eq per (1) mole of sulfate reduced; pH < 6.3 → pH increases; 7.0 > pH > 6.3 → pH ≅ constant; pH > 7.0 → pH decreases.

Nitrogen fixation (with concomitant photosynthesis)
$$106CO_2 + 8N_2 + H_2PO_4^- + 130H_2O + H^+ \rightarrow$$
$$\text{Protoplasm} + 118O_2$$

Alkalinity increases by 0.13 eq per (2) mole of nitrogen fixed but this effect is due to P uptake and is in fact negligible; pH increases throughout pH range.

Nitrification
$$NH_4^+ + \tfrac{3}{2}O_2 \rightarrow NO_2^- + 2H^+ + H_2O$$

Alkalinity decreases by 2 eq per (3) mole of ammonium oxidized; pH decreases throughout pH range.

$$NO_2^- + \tfrac{1}{2}O_2 \rightarrow NO_3^-$$

No effect on alkalinity or pH. (4)

Denitrification
$$4NO_3^- + 5``CH_2O" + 4H^+ \rightarrow$$
$$2N_2 + 5CO_2 + 7H_2O$$

Alkalinity increases by 1 eq per (5) mole of nitrate reduced; pH < 6.3 → pH increases; pH > 6.3 → pH decreases slightly.

Methane fermentation
$$``CH_2O" + ``CH_2O" \rightarrow CH_4(g) + CO_2$$

Alkalinity remains constant; (6) pH < 6.3 → pH ≅ constant; pH > 6.3 → pH decreases.

Note: "CH$_2$O" represents organic matter.

Example 5

Consider the effect on pH and alkalinity of reducing all the sulfate to sulfide (without any loss of gas) in a lake characterized initially by

$$Alk^0 = 10^{-3} \, M$$

$$(SO_4^{2-})_T^0 = 10^{-4} \, M$$

$$pH^0 = 7.5$$

Choice of Components
The principal components at a pH of 7.5 are

$$H^+, HCO_3^-, SO_4^{2-}, HS^- \quad \text{(see Figure 4.1)}$$

Initial Composition
We need to obtain the right-hand side of the mole balance equations corresponding to our principal components. The last two columns of Tableau 4.5 give

$$TOTSO_4 = 10^{-4}$$

$$TOTHS = 0$$

To obtain the values of $TOTH$ and $TOTHCO_3$, we have to calculate the concentrations of all the carbonate species, given the alkalinity (10^{-3}) and the pH (7.5):

$$Alk = -(H^+) + (OH^-) + (HCO_3^-) + 2(CO_3^{2-}) = 10^{-3} \quad (54)$$

$$= -10^{-7.5} + 10^{-6.5} + (HCO_3^-) + 2\left(\frac{10^{-10.3}}{10^{-7.5}}\right)(HCO_3^-) = 10^{-3} \quad (55)$$

TABLEAU 4.5

	H^+	HCO_3^-	SO_4^{2-}	HS^-
H^+	1			
OH^-	-1			
$H_2CO_3^*$	1	1		
HCO_3^-		1		
CO_3^{2-}	-1	1		
SO_4^{2-}			1	
H_2S	1			1
HS^-				1
S^{2-}	-1			1
Initial Composition	$10^{-4.21}$	$10^{-2.97}$	$10^{-4.0}$	0
Change in Composition	$+10^{-4.0}$	$+10^{-3.7}$	$-10^{-4.0}$	$+10^{-4.0}$
Final Composition	$10^{-3.79}$	$10^{-2.90}$	0	$10^{-4.0}$

therefore

$$(HCO_3^-) = 10^{-3} \, M$$

$$(CO_3^{2-}) = \frac{10^{-10.3}}{10^{-7.5}} \, 10^{-3} = 10^{-5.8}$$

$$(H_2CO_3^*) = 10^{6.3} \times 10^{-7.5} \times 10^{-3} = 10^{-4.2}$$

From this calculation we obtain (see Tableau 4.5)

$$TOTH = 10^{-7.5} - 10^{-6.5} + 10^{-4.2} - 10^{-5.8} = 10^{-4.21}$$

$$TOTHCO_3 = 10^{-4.2} + 10^{-3.0} + 10^{-5.8} = 10^{-2.97}$$

Change in Composition
Let us write the reaction for sulfate reduction with the principal components of the system (see Table 4.3):

$$SO_4^{2-} + 2\text{``CH}_2O\text{''} \rightarrow HS^- + 2HCO_3^- + H^+ \tag{56}$$

Given $\Delta TOTSO_4^{2-} = -10^{-4}$ (sulfate entirely reduced), we obtain according to the stoichiometry of Reaction 56

$$\Delta TOTH = +10^{-4}$$

$$\Delta TOTHCO_3 = +2 \times 10^{-4} \, M$$

$$\Delta TOTHS = +10^{-4} \, M$$

therefore

$$\text{new } TOTH = 10^{-3.79}$$

$$\text{new } TOTHCO_3 = 10^{-2.90}$$

$$\text{new } TOTHS = 10^{-4}$$

Final Composition
Following the columns of Tableau 4.5, we can write the mole balance equations:

$$TOTH = (H^+) - (OH^-) + (H_2CO_3^*) - (CO_3^{2-}) + (H_2S)$$

$$- (S^{2-}) = 10^{-3.79} \tag{57}$$

$$TOTHCO_3 = (H_2CO_3^*) + (HCO_3^-) + (CO_3^{2-}) = 10^{-2.90} \tag{58}$$

$$TOTHS = (H_2S) + (HS^-) + (S^{2-}) = 10^{-4.0} \tag{59}$$

Since the right-hand side of the $TOTH$ equation has increased compared to its original value, the pH is expected to be lower than before, and (CO_3^{2-}) and (S^{2-}) can be neglected in all the mole balances, as can (H^+) and (OH^-) in Equation 57:

$$TOTH = (H_2CO_3^*) + (H_2S) = 10^{6.3}(H^+)(HCO_3^-)$$

$$+ 10^{7.0}(H^+)(HS^-) = 10^{-3.79}$$

$$TOTHCO_3 = (H_2CO_3^*) + (HCO_3^-) = [10^{6.3}(H^+) + 1](HCO_3^-) = 10^{-2.90}$$
$$TOTHS = (H_2S) + (HS^-) = [10^{7.0}(H^+) + 1](HS^-) = 10^{-4.0}$$

Substituting (HCO_3^-) and (HS^-) into the first equation yields an explicit equation for (H^+):

$$\frac{10^{3.4}(H^+)}{1 + 10^{6.3}(H^+)} + \frac{10^{3.0}(H^+)}{1 + 10^{7.0}(H^+)} = 10^{-3.79}$$

The resulting quadratic can be solved:

$$(H^+) = 10^{-7.25} \ M$$

Substituting back into the other mole balances

$$(HCO_3^-) = 10^{-2.95} \ M$$
$$(HS^-) = 10^{-4.19} \ M$$

The other species concentrations are obtained from the mass laws:

$$(H_2CO_3^*) = 10^{-3.90} \ M; \quad (CO_3^{2-}) = 10^{-6.00} \ M$$
$$(H_2S) = 10^{-4.44} \ M; \quad (S^{2-}) = 10^{-10.84} \ M$$

Note that the reduction of the sulfate has resulted in a slight increase in the alkalinity of the system:

$$Alk' = -TOTH + TOTHCO_3 + TOTHS = 10^{-2.92}$$

Despite this consumption of sulfuric acid, the pH actually decreases from 7.5 to 7.25. This is due to the accompanying production of CO_2 in the sulfate reduction process.

6. BUFFERING CAPACITY

From the point of view of *homo polluens* what matters most among the characteristics of the environment is its resistance to change, its homeostasis, its buffering processes. Staying in the chemical realm, the problem is to understand how and by how much the composition of the environment will change for a given input. Attacking such a problem is quite a leap from the traditional study of buffers that are used to stabilize pH in reaction vessels. Yet, if we narrow our focus to the question of pH in natural waters—a particularly topical question in view of a widespread acid rain problem in the northern hemisphere—the same principles that apply in the laboratory can be used to ascertain the behavior of lakes, rivers, groundwaters, and oceans.

Before we develop a general theorem for the "buffer capacity" of complex chemical system, it is worth remembering the limitations of our approach. Although the systems that we consider may be quite complex indeed, they are but very simplified idealizations of real systems; the most tenuous simplification is assuredly that of chemical equilibrium. We have to recall here our earlier arguments of time scales and *partial* equilibrium. To the traditional buffering questions about how much the pH will change for a given addition of acid or base and what species or reactions control the pH, we have to add: Over what time scale? The homeostatic behavior of the real system is not the same over hours as it is over geological times; it is dependent on a number of relatively slow reactions such as equilibration with gas or solid phases. We shall see that our corresponding partial equilibrium models have different buffering capacities. These buffering capacities pertain to reality only to the extent that the equilibrium models do.

6.1 Definition

The buffer capacity of a chemical system is typically defined as the ratio of an infinitesimal base addition to the resulting infinitesimal increase in pH. It is given by the slope of the titration curve—base concentration minus acid concentration versus pH—which may be obtained theoretically or experimentally:

$$\beta_H = -\frac{\Delta TOTH}{\Delta pH} \tag{60}$$

The buffer capacity so defined has dimensions of moles per liter and provides a measure of the ability of the system to resist pH changes against strong acid or base additions. For example, if we consider the titration curve of Figure 4.4, we can see that the buffer capacity of the carbonate system is largest when the pH is close to the pK_a's (6.3 and 10.3).

To proceed with calculations of buffer capacities in complex systems, we need a more precise mathematical definition. Let us define an "ideal" strong acid or base as one that affects a chemical system exclusively through hydrogen or hydroxyl ion additions (for example, an acid HA that dissociates completely into H^+ and A^-, the latter being strictly unreactive). In practice, most strong acids and bases are good approximations of ideal ones. Consider now the addition of an infinitesimal quantity $(\partial TOTH)$ of an ideal strong acid or base, and the resulting change (∂pH) in pH. A strict mathematical definition of the buffer capacity β_H of the system is given by

$$\beta_H = -\left(\frac{\partial pH}{\partial TOTH}\right)^{-1} \tag{61}$$

The partial derivative indicates that all total component concentrations, except H^+, are kept constant. Note that the order of the differentiation is consistent with the causality of the real or imaginary titration process: the pH change results from an acid or base addition, and not vice versa as implied in the preceding formula.

6.2 Minor Species Theorem[8]

An approximate value of the pH buffering capacity of any chemical system is given by

$$\beta_H = 2.3 \sum_i \lambda_i^2(S_i) \tag{62}$$

where λ_i is the coefficient of the species S_i in the $TOTH$ equation when the components are the principal components of the system. This can be rephrased in a more compact, but perhaps more esoteric, form: where λ_i is the proton level of i compared to the principal components of the system.

A simplified demonstration of this theorem can be given as a direct differentiation of the $TOTH$ equation. Consider the generalized tableau involving the principal components of the system (Tableau 4.6). From the coefficients in the H^+ column, the proton conservation equation can be written:

$$TOTH = \sum_i \lambda_i(S_i) \tag{63}$$

All the components (here the principal components) but H^+ have identically null λ_i. Consider the mass laws of the other species (the "minor species") as a function of the components:

$$(S_i) = \frac{K_i}{\gamma_i} \times \{H^+\}^{\lambda_i} \times (a_1)^{v_{i1}} \times (a_2)^{v_{i2}} \ldots \tag{64}$$

K_i are the equilibrium constants, and γ_i the activity coefficients. If we suppose the ionic strength to be given, the activity coefficients are fixed. The activities (a_k) of the principal components, except that of H^+, are also approximately independent of $TOTH$:

1 For solids $a_k = 1$.
2 For gases $a_k = P_k$ always taken as constant.
3 For solutes $a_k = \gamma_k(C_k)$, where the activity coefficient γ_k is constant for a given ionic strength.

Consider the mole balance equation corresponding to C_k (see Tableau 4.6):

$$TOTC_k = (C_k) + v_{3k}(S_3) + v_{4k}(S_4) + \ldots \tag{65}$$

By definition of the principal components, (C_k) is the largest concentration in this equation. If all the other concentrations can be neglected

$$(C_k) \cong TOTC_k = \text{constant}$$

From our own experience (Chapter 3) we know that this approximation holds very well under most circumstances. There are of course some cases where two concentrations have approximately the same numerical value, for example, near

TABLEAU 4.6

Principal Components

	Hydrogen Ion H$^+$ {H$^+$}	Solids Z$_1$ $a_1 = 1$	Z$_2$ $a_2 = 1$ $\ldots$	Gases G$_1$ P$_1$	G$_2$ $\ldots$ P$_2$ $\ldots$	C$_1$	Soluble (Major) Species C$_2$ $\ldots$	C$_k$ $a_k = \gamma_k(C_k)\ldots$
Component Activities								
Major Species								
S_1	0	0$\ldots$		0$\ldots$		1	0$\ldots$	0
S_2	0	0$\ldots$		0$\ldots$		0	1 0$\ldots$	
$\vdots$	$\vdots$	$\vdots$		$\vdots$		$\vdots$	0	
	1							1
Minor Species								
$S_1 = \mathrm{H}^+$	$\lambda_1 = 1$	$v_{11} = 0$	$v_{12} = 0\ldots$					$v_{1k} = 0$
$S_2 = \mathrm{OH}^-$	$\lambda_2 = -1$	$v_{21} = 0$	$v_{22} = 0$					$v_{2k} = 0$
S_3	λ_3	$v_{31} = 0$	$v_{32} = 0$					v_{3k}
$\vdots$	$\vdots$	$\vdots$	$\vdots$					
S_i	λ_i	v_{i1}	v_{i2} $\ldots$					$v_{ik}\ldots$
$\vdots$	$\vdots$	$\vdots$	$\vdots$					

the pK_a of a weak acid. This is why it is necessary to use the adjective "approximate" in stating the theorem. We shall examine later the magnitude of the errors that result from it.

Grouping all constant terms into one, X_i, the mass law expression for the minor species S_i becomes a function of $\{H^+\}$ only:

$$(S_i) = X_i\{H^+\}^{\lambda_i} \tag{66}$$

where X_i is a constant independent of $TOTH$ (and pH). The $TOTH$ equation is then given as an explicit function of $\{H^+\}$ only, and the partial derivatives can be replaced by total derivatives:

$$TOTH = \sum_i X_i\lambda_i\{H^+\}^{\lambda_i} \tag{67}$$

therefore

$$\beta_H = -\left(\frac{\partial pH}{\partial TOTH}\right)^{-1} = -\left(\frac{dpH}{dTOTH}\right)^{-1}$$

$$= -\frac{dTOTH}{dpH} = +2.3\{H^+\}\frac{dTOTH}{d\{H^+\}} \tag{68}$$

The factor of 2.3 is introduced by the differentiation of a base 10 logarithm.

$$\beta_H = 2.3\{H^+\}\sum_i X_i\lambda_i^2\{H^+\}^{\lambda_i-1} = 2.3\sum_i \lambda_i^2 X_i\{H^+\}^{\lambda_i} \tag{69}$$

$$= 2.3\sum_i \lambda_i^2(S_i) \qquad \text{Q.E.D.} \tag{70}$$

This intuitive proof of the minor species theorem provides also a definition of what is acidic (proton donor: $\Delta TOTH > 0$) or basic (proton acceptor: $\Delta TOTH < 0$) in a given system. An added chemical constituent behaves as an acid (or a base) in a chemical system if it has a positive (or negative) proton level compared to the principal components of the system. Thus the application of the minor species theorem need not be restricted to strong acids and bases; it can be used to calculate small pH variations brought about by any change in the composition of a system:

$$\Delta pH = -\frac{\Delta TOTH}{\beta_H}$$

In this formula $\Delta TOTH$ is calculated as the change in the right-hand side of the $TOTH$ equation corresponding to the principal components of the system. In effect the minor species theorem provides a justification, a generalization, and a quantification of the qualitative reasoning that we carried out in our discussion of the effects of biological processes on pH (Section 5). This is elaborated further in Section 6.5.

6.3 Buffer Capacity of Weak Acid-Base Solutions

Consider a solution of a weak acid:

$$HA = H^+ + A^-; \quad K_a \tag{71}$$

For pH's below the pK_a the major species, and hence the principal component, is HA. The buffer capacity is then given according to the minor species theorem:

$$pH < pK_a : \beta_H = 2.3\{(A^-) + (H^+)[+ (OH^-); \text{ small at low pH}]\} \tag{72}$$

Conversely, for pH's above the pK_a the buffer capacity is given by

$$pH > pK_a : \beta_H = 2.3\{(HA) + (OH^-)[+ (H^+); \text{ small at high pH}]\} \tag{73}$$

For pH equal to the pK_a either HA or A^- can be taken as the major or minor species:

$$pH = pK_a : \beta_H = 2.3[(HA) + (H^+) + (OH^-)] = 2.3[(A^-) + (H^+) + (OH^-)] \tag{74}$$

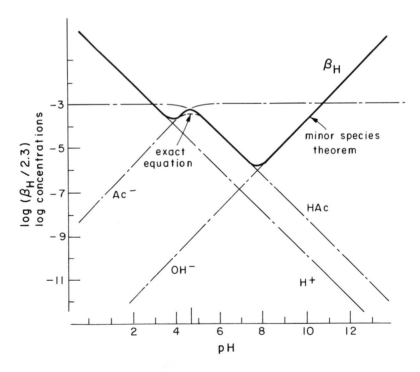

Figure 4.9 Buffer capacity of the acetate system. Except at the pK_a, where the buffer capacity is maximum, the minor species theorem gives values of β_H that are indistinguishable from those of the exact equation.

For such a simple system an exact expression of β_H can actually be derived, valid at any pH:

$$\beta_H = 2.3\left[(H^+) + (OH^-) + \frac{(HA)(A^-)}{(HA) + (A^-)}\right] \qquad (75)$$

As can be seen in Figure 4.9, where both this equation and the approximation given by the minor species theorem are plotted on a log C-pH graph, the minor species theorem provides an excellent approximation of the buffer capacity over the whole pH range. The fit of the two formulae is worst when there is no dominant species, when the pH is equal to the pK_a. In this case the minor species theorem overestimates the buffer capacity by a factor of two. This is true even in very complex cases. For applications where more precision is needed, a correction factor can be applied to the value of β_H obtained from the theorem. Note that the buffer capacity of a weak acid-base solution is maximum at its pK_a where it is equal to one quarter the total acid-base concentration ($\times 2.3$). One pH unit away from the pK_a the buffer capacity is reduced to one-tenth of the total acid-base concentration ($\times 2.3$). This provides a convenient rule of thumb for choosing appropriate buffers in the laboratory.

6.4 Buffer Capacity of Heterogeneous Systems

The buffer capacity of heterogeneous systems has been studied relatively little, partly because slow reactions make experimental studies difficult, partly because the mathematical complexity of the theoretical calculations appears formidable. Paradoxically, the minor species theorem is especially easy to apply to such systems as the choice of principal components is made more obvious by the presence of solids and gases.

Consider, for example, a carbonate-bearing water in equilibrium with a fixed partial pressure of CO_2. Taking CO_2 as principal component, the buffering capacity of such a system is given directly by

$$\beta_H = 2.3[(HCO_3^-) + 4(CO_3^{2-}) + (H^+) + (OH^-)] \qquad (76)$$

For many natural waters this quantity is practically equal to the alkalinity ($\times 2.3$). It is much larger than the buffering capacity of the aqueous phase alone which, in the neutral pH range, is obtained by applying Equation 62 to the carbonate system with HCO_3^- as the principal component:

$$\beta_H = 2.3[(H_2CO_3^*) + (CO_3^{2-}) + (H^+) + (OH^-)] \qquad (77)$$

At pH $= 8$ (approximate seawater pH) the buffer capacity of the carbonate system in equilibrium with the gas phase is roughly 100 times greater than that of the isolated aqueous phase. Note that the buffer capacity of the system is also increased by the presence of carbonate complexes, since the concentration of such species are also included in the expression of β_H.

Consider now a carbonate system in equilibrium not only with a fixed partial pressure of CO_2 but also with a calcium carbonate solid. Taking both CO_2 and $CaCO_3$ as principal components, we have

$$\beta_H = 2.3[4(Ca^{2+}) + (HCO_3^-) + 4(CO_3^{2-}) + (H^+) + (OH^-)] \tag{78}$$

If more heterogeneous reactions are considered, the buffering capacity of the system increases as more soluble species are included in the expression of β_H, in this case (Ca^{2+}).

At the limit, if *all* soluble species in an aquatic system are controlled by the solubility of solid and gas phases, an expression of the maximum possible buffering capacity is obtained. Consider a complete set of principal components including only H^+, gases, and solids. (It is shown in Chapter 5 that this set contains the maximum number of solid and gas phases that can be in equilibrium with the solution). Then the proton level λ_i of each species is exactly equal to its electrical charge, and β_H is given by

$$\beta_H^{max} = 4.6 \times I \tag{79}$$

where $I = \frac{1}{2} \sum \lambda_i^2(S_i)$ is the ionic strength of the system. Of course this theoretical maximum value of the buffering capacity of an aquatic system may not be attained in fact, even if we consider the geochemical processes that control the composition of the water over geological time. For example, in most waters the concentration of the chloride ion is probably not controlled by the solubility of any mineral phase. Nonetheless, the expression of β_H^{max} shows that the buffer capacity of a system cannot be larger than its ionic content ($\times 4.6$), and this value is a rough estimate of the buffer capacity of the system in which all heterogeneous geochemical processes are considered to be operative.

6.5   Applications and Limitations of the Minor Species Theorem

Recall Examples 3, 4, and 5 where we calculated changes in pH due to changes in the composition of a system. We shall apply the minor species theorem to these examples to illustrate the methodology and gain some insight into the applicability of this differential approach to calculating pH changes.

In **Example 3** the system is at equilibrium with the $CO_2(g)$ in the atmosphere but not the $H_2S(g)$. The principal components at pH $= 8$, H^+, CO_2, HS^-, and the initial composition, $(HCO_3^-) = 10^{-3.3}$ M, $(CO_3^{2-}) = 10^{-5.6}$ M, $(H_2S) = 10^{-4.74}$ M, $(S^{2-}) = 10^{-9.44}$ M, yield the buffer capacity:

$$\beta_H = 2.3[(HCO_3^-) + 4(CO_3^{2-}) + (H_2S)] = 1.22 \times 10^{-3} M$$

The reaction of escape of hydrogen sulfide gas written with the principal components

$$HS^- + H^+ \rightarrow H_2S(g)$$

yields

$$\Delta TOTH = -S_T = -2 \times 10^{-4} \, M$$

$$\Delta pH \cong +\frac{2 \times 10^{-4}}{1.22 \times 10^{-3}} \cong +0.16$$

By comparing this result with that of Example 3 ($\Delta pH = +0.13$), we can see that the application of the minor species theorem gives a reasonably accurate answer in this case.

To calculate the changes in pH from the buffer capacity of the system in **Example 4**, it is convenient to consider that the initial system consists of River A and that both the dilution of A and the addition of B correspond to changes in the composition of A.

System Isolated from the Atmosphere

With the principal components H^+ and HCO_3^-, the minor species theorem yields

$$\beta_H = 2.3[(H_2CO_3^*)_A + (CO_3^{2-})_A] = 4.6 \times 10^{-5} \, M$$

$$\Delta TOTH = -\tfrac{3}{5} TOTH_A + \tfrac{3}{5} TOTH_B$$

$$= -\tfrac{3}{5}[(H_2CO_3^*)_A - (CO_3^{2-})_A] + \tfrac{3}{5}[(H_2CO_3^*)_B - (CO_3^{2-})_B]$$

$$= +2.6 \times 10^{-5} \, M$$

$$\Delta pH = -\frac{2.6 \times 10^{-5}}{4.6 \times 10^{-5}} = -0.57$$

This change is large enough not to be very accurately predicted by this differential calculation: the actual pH change is $\Delta pH = -0.75$.

System at Equilibrium with the Atmosphere

The initial composition of A at equilibrium with the atmosphere would be

$$pH = 8.3$$

$$(HCO_3^-)_A = 10^{-3} \, M$$

$$(H_2CO_3^*)_A = 10^{-5} \, M$$

$$(CO_3^{2-})_A = 10^{-5} \, M$$

With the principal components H^+ and CO_2, the buffer capacity is given by

$$\beta_H = 2.3[(HCO_3^-)_A + 4(CO_3^{2-})_A] = 2.31 \times 10^{-3} \, M$$

$$\Delta TOTH = +\tfrac{3}{5} \, Alk_A - \tfrac{3}{5} \, Alk_B = 5.94 \times 10^{-4} \, M$$

$$\Delta pH = -\frac{5.94 \times 10^{-4}}{2.31 \times 10^{-3}} = -0.26$$

The calculated final pH is thus $8.3 - 0.26 = 8.04$, which is a bit higher than that obtained in the detailed calculation of Example 4 (7.91).

In **Example 5** where the pH change is brought about by the reduction of sulfate to sulfide, at the initial pH of 7.5, the principal components of the system are H^+, HCO_3^-, HS^-, SO_4^{2-} (see Tableau 4.5). Thus

$$\beta_H = 2.3[(H_2CO_3^*) + (CO_3^{2-}) + \text{no sulfide initially}] = 1.49 \times 10^{-4} M$$

The sulfate reduction reaction with the principal components

$$SO_4^{2-} + 2\text{"CH}_2O\text{"} \rightarrow HS^- + 2HCO_3^- + H^+$$

yields

$$\Delta TOTH = -\Delta(SO_4^{2-})_T = +10^{-4} M$$

and

$$\Delta pH = -\frac{10^{-4}}{1.49 \times 10^{-4}} = -0.67$$

This calculated change in pH is markedly larger than the actual one (-0.25), a reflection of the fact that a sizable fraction of the sulfide formed remains undissociated as H_2S (hence $\Delta TOTH$ is effectively smaller than $10^{-4} M$) and that the buffering capacity that is provided by the sulfide itself as it forms has not been accounted for (i.e., β_H is effectively larger than $1.49 \times 10^{-4} M$).

Overall it is apparent that the calculation of pH changes on the basis of the buffer capacity of a system provides accurate answers only when these pH changes are relatively small, say, $|\Delta pH| < 0.2$. One should also be wary of changes in compositions when weak acids and bases (e.g., H_2S) are introduced into a system because of their effect on the system's buffer capacity. Nonetheless, the direction of pH change and its approximate magnitude can be readily calculated by applying the methodology outlined here. Of course buffer capacity calculations especially provide a quick method for estimating whether or not the pH will change appreciably.

7.   EXCHANGE OF GASES AT THE AIR-WATER INTERFACE

At any given time most natural waters are not in equilibrium with the atmosphere because the aquatic processes that consume or produce volatile compounds are often faster than are gas exchange processes. For example, on a summer day, the rate of photosynthesis by phytoplankton is usually fast enough to result in supersaturation of oxygen and undersaturation of carbon dioxide in the euphotic zones of lakes and oceans. Conversely, the rate of respiration during the night can cause O_2 depletion and CO_2 supersaturation. As a result of these and other processes (microbial activity, temperature changes, etc.) there is normally a diurnal and seasonal cycle of gas exchange in natural water bodies.

The general expression for the exchange of a nonreactive volatile compound between a gaseous and liquid phase is of the form

$$J_g = k_g(C_g^s - C_g)$$

where J_g is the rate of transfer of the compound per unit area (mol cm^{-2} s^{-1}), k_g is the rate transfer coefficient (cm s^{-1}), C_g^s is the soluble concentration of the compound (mol liter^{-1}) in equilibrium with the partial pressure P_g of the compound in the gas phase, and C_g is the concentration of the compound in the liquid phase (mol liter^{-1}).

This empirical rate law can be rationalized by physical models of the exchange process. Various models provide different interpretations for the transfer coefficient k_g as a function of the properties of the volatile compound (primarily its molecular diffusion coefficient D_g) and the characteristics of the fluid motion, usually restricted to the liquid phase unless the compound is extremely volatile. These models are principally of two general types:

1   The surface film model where the transfer coefficient is interpreted as the ratio of the molecular diffusion coefficient and the thickness Z of a diffusion boundary layer:

$$k_g = \frac{D_g}{Z}$$

2   The surface renewal model(s) where the transfer coefficient is interpreted as the square root of the ratio of the molecular diffusion coefficient D_g and a characteristic time θ, representing the frequency of renewal of the surface film by bulk solution, due to turbulence:

$$k_g = \left(\frac{D_g}{\theta}\right)^{1/2}$$

Various surface renewal models differ primarily in the underlying assumptions for the distribution of surface renewal times, and consequently they have different formulae for the effective parameter θ.

Conclusive experimental evidence proving the correctness or incorrectness of either model is still lacking. Although available data may be construed to favor slightly the surface renewal model, the great conceptual simplicity of the surface film model makes it a better heuristic tool, and it is the one that we are presenting succinctly in the following sections.

The exchange rate of a reactive volatile species can be affected by reactions in solution, and in addition to the physical transfer coefficient k_g, the rate law may include a chemical enhancement factor E_g:

$$J_g = E_g k_g(C_g^s - C_s)$$

As discussed next, such a chemical enhancement factor is seldom important in natural waters where the degree of turbulence is usually high and physical transfer processes are efficient. Because the molecular diffusion coefficients of typical aquatic solutes span a relatively narrow range of values (2 to 5×10^{-5} cm^2 s^{-1}), the transfer of gases between the atmosphere and surface waters is dominated by the hydrodynamic characteristics of the water and is almost independent of the nature of the volatile compounds. Chemical oceanographers who have a propensity for simple physical images like to speak of the transfer coefficient k_g as a "piston velocity," with values in the range of 10^{-4} to 10^{-2} cm s^{-1}. For surface water depths of the order of 1 to 10 m, this represent a characteristic exchange time of the order of 10^4 to 10^7 seconds, or 2 hr to 100 d.

□ □ □

7.1   Governing Equation

By assuming that the transfer of a gas between water and atmosphere is limited by molecular diffusion through a water interfacial laminar layer, a mathematical description of gas transfer kinetics can be obtained. For steady-state conditions under a number of simplifying assumptions, the rate of gas transfer may be described by an equation of the form

$$J_g = 10^3 \frac{D}{Z}(C_g^s - C_g) \tag{80}$$

where J_g is the rate of transfer (flux) of the gas per unit area (mol cm^{-2} s^{-1}), D is the molecular diffusion coefficient of the gas (cm^2 s^{-1}), C_g^s is the soluble concentration (mol liter^{-1}) of the gas in equilibrium with the given partial pressure P_g, C_g is the concentration of the gas in the bulk aqueous phase (mol liter^{-1}), and Z is the average depth of the water laminar boundary layer through which the diffusion is taking place (cm). The factor of 10^3 is included for consistency among the various units.

With the exception of NH_3 and SO_2 which are extremely soluble, the gases of interest in aquatic systems (O_2, N_2, CO_2, H_2S, CH_4, NO_x) are sufficiently volatile so that no boundary layer in the gas phase need be considered. For example, in the case of transfer from air to water, there is no microzone of gas depletion on the air side of the interface, as the rate of mixing in the air is much faster than the rate of dissolution into the water.

Many of the assumptions leading to Equation 80 are rarely, if ever, satisfied in natural waters. Still, the form of the equation is worth keeping as it conveniently separates parameters describing the hydrodynamic regime of the system (Z) from those pertaining to the particular gas of interest. Although the parameter Z may not strictly correspond to the thickness of an actual laminar boundary layer, it describes the influence of the mixing regime—the turbulence of the water—on the gas exchange kinetics as a single parameter: the thickness of an *equivalent* boundary layer. In principle, Z should be the same for all gases in a given body of water at a particular point in time. For a given gas only D/Z is measurable directly as the ratio of the gas flux to the concentration gradient, as described by Equation 80.

7.2 Chemical Enhancement

Equation 80 provides a convenient method of describing gas exchange kinetics for unreactive gases such as O_2, N_2, or CH_4. However, for gases that dissociate into weak acids and bases (CO_2, H_2S, NH_3), one sometimes observes experimentally that the gas transfer is more rapid than expected on the basis of that equation, given an independent estimate of the boundary layer thickness. This is interpreted as the effect of chemical reactions within the boundary layer. To take such an effect into account, one can define a chemical enhancement factor (E_g)[9-11] so that the actual gas flux is given by

$$J_g = 10^3 E_g \frac{D}{Z}(C_g^s - C_g) \tag{81}$$

Theoretical values of E_g can be obtained by considering the kinetics of the various chemical reactions and the diffusion of all chemical species in infinitesimal sublayers within the laminar boundary layer. The problem is particularly complex as several competing reactions have to be considered simultaneously, and such effects as the electrical interactions among ions of different mobilities have to be taken into account. Such sophistication does not seem warranted in face of the intrinsic crudeness of the laminar boundary layer model itself and the imprecision of the experimental measurements. In addition the parameter E_g is dependent upon the particular chemistry of the system; it varies as the water equilibrates with the atmosphere.

We shall examine here limiting cases to this problem, focusing on the question of CO_2 transfer across the air-water interface. Consider the usual boundary layer model (see Figure 4.10) consisting of an unmixed water layer of thickness Z between an atmosphere of a given fixed P_{CO_2} and a well-mixed bulk solution of known chemistry. As a molecule of CO_2 is dissolved into the upper boundary of the layer, it can either diffuse downward or react with water according to

$$CO_2 + H_2O \underset{}{\overset{k_1}{\rightleftharpoons}} HCO_3^- + H^+ \tag{82}$$

or

$$CO_2 + OH^- \underset{}{\overset{k_2}{\rightleftharpoons}} HCO_3^- \tag{83}$$

The forward kinetics of these two reactions, which are known to be the slowest of all those involved in the carbonate system, are described by the equation[12]

$$\frac{d(CO_2)}{dt} = -k_1(CO_2) - k_2(CO_2)(OH^-) + \dots \tag{84}$$

where $k_1 = 3 \times 10^{-2}\ s^{-1}$ and $k_2 = 8.5 \times 10^3\ M^{-1}\ s^{-1}$.

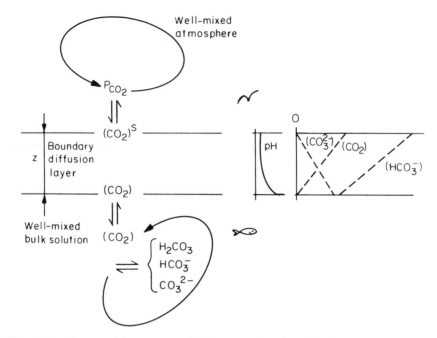

Figure 4.10 Diagram of the transport of CO_2 across a boundary diffusion layer at the air-water interface. The carbonate species concentration profiles are drawn assuming constant alkalinity throughout the diffusion layer. In this illustration the water is undersaturated with respect to the atmosphere, and the net carbon flux is toward the water. Note that the (CO_3^{2-}) gradient is opposite to those of (CO_2) and (HCO_3^-) corresponding to the increase in pH with depth.

If the resulting reaction rate is much slower than the diffusion process across the layer, we can effectively ignore the reactivity of CO_2 and treat the gas exchange problem according to Equation 80. If, at the other extreme, the chemical reactions are much faster than diffusion ($D_{CO_2} = 2 \times 10^{-5}$ cm^2 s^{-1}), then we can treat the problem by considering equilibrium at every point in the layer (equilibrium enhancement). If the chemical and transport rates are comparable, we then have no choice but to consider both simultaneously in a rather complex model. The critical parameter determining which of these cases is applicable is the depth Z of the boundary layer:

1. The characteristic time for diffusion through the boundary layer is given by the ratio of the total number of moles of CO_2 in a given volume of the boundary layer to the flux through that volume:

$$t_{\text{diffusion}} \cong \frac{(CO_2)Z \text{ (unit area)}}{J_{CO_2} \text{ (unit area)}} \cong \frac{Z^2}{D} \tag{85}$$

2. The characteristic time for chemical reaction of CO_2 is given directly by the forward kinetic coefficients:

$$t_{\text{reaction}} \cong \left[k_1 + k_2(OH^-) \right]^{-1} \tag{86}$$

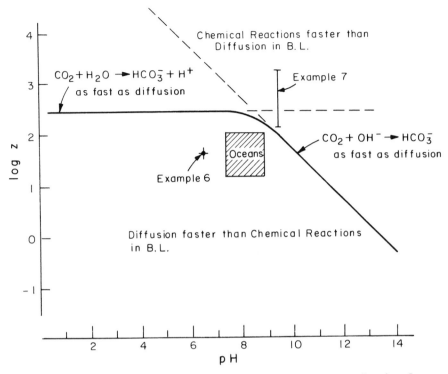

Figure 4.11 Chemical enhancement of CO_2 transport across the air-water interface. In most bodies of water the effective thickness of the diffusion boundary layer (given here as log Z, Z in μm) is sufficiently small that CO_2 diffusion is rapid compared to the kinetics of HCO_3^- formation and chemical enhancement is negligible.

3. Equating the two time scales ($t_{\text{diffusion}} = t_{\text{reaction}}$) yields the critical depth of the diffusion boundary layer:

$$Z \cong \left[\frac{D}{k_1 + k_2(OH^-)} \right]^{1/2} = \left[\frac{1}{1 + 10^{-8.55}/(H^+)} \right]^{1/2} \times 250 \ \mu m \qquad (87)$$

A logarithmic plot of this equation appears in Figure 4.11. As shown in the figure only for very quiescent lakes with deep surface boundary layers or for alkaline waters are the chemical reactions as fast as, or faster than, the diffusion through the surface layer. In most natural water the enhancement of the exchange kinetics of CO_2 due to chemical reactions can be neglected.

7.3 Kinetics of CO_2 Equilibrium in a Water Column

Let us consider the time course of CO_2 equilibration in a well-mixed water column with the typical conditions of negligible chemical enhancement. Carbon dioxide is then the only species transported across the boundary layer; chemical reactions in the bulk solution are much faster than the CO_2 transport, and equilibrium can be assumed among the carbonate species.

Example 6

Given a water body of average depth, $h = 10$ m, with sufficient mixing to insure uniform concentrations throughout, and the following initial conditions how will the carbonate concentration in the water change as a function of time?

$$P_{CO_2} = 10^{-3.5} \text{ at}$$

$$\text{Alk} = 10^{-3.5} \; M$$

$$pH^0 = 6.5$$

Boundary layer thickness: $Z = 40 \; \mu m$ (typical range being 20–1000 μm):

$$D_{CO_2} = 2 \times 10^{-5} \text{ cm}^2 \text{ s}^{-1}$$

Using our previous notation to study the chemistry, let us make the approximation

$$(CO_2)_{aq} \cong (H_2CO_3^*)$$

In this system for pH < 9, the following approximations are easily justified:

$$\text{Alk} = (HCO_3^-)$$

$$C_T = (H_2CO_3^*) + (HCO_3^-)$$

$$C_T^0 = \text{Alk}\,(10^{6.3} \times 10^{-6.5} + 1) = 10^{-3.3}$$

$$(CO_2) = (H_2CO_3^*) = C_T - \text{Alk}$$

Using dm ($= 10^{-1}$ m) throughout as unit of length since 1 dm^3 = 1 liter,

$$\frac{dC_T}{dt} = J_{CO_2} \frac{\text{Area}}{\text{Volume}} = \frac{J_{CO_2}}{h} = \frac{J_{CO_2}}{100}$$

therefore

$$\frac{dC_T}{dt} = 10^{-2} \frac{D_{CO_2}}{Z} [(CO_2)^s - (CO_2)]$$

where

$$10^{-2} \frac{D_{CO_2}}{Z} = 10^{-5.3} \text{ s}^{-1} = 10^{-1.75} \text{ h}^{-1}$$

$$(CO_2)^s = 10^{-1.5} P_{CO_2} = 10^{-5} \; M$$

$$(CO_2) = C_T - \text{Alk} = C_T - 10^{-3.5}$$

$$\frac{dC_T}{dt} = 10^{-1.75}[10^{-3.5} - C_T]$$

The solution of this differential equation for the initial condition $C_T^0 = 10^{-3.3} \; M$ is given by

$$C_T = 10^{-3.5} + 10^{-3.7} e^{-0.018t} \quad (t \text{ in hours})$$

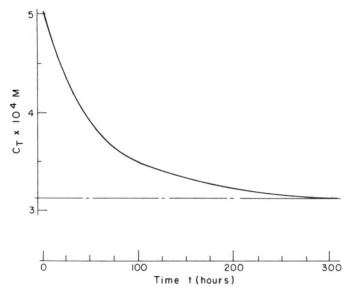

Figure 4.12 Kinetics of CO_2 equilibration (with no chemical enhancement) in a well-mixed 10-m water column with an effective diffusion boundary layer of 40 μm (Example 6).

This solution is plotted in Figure 4.12, where it can be seen that about five days are necessary to approach equilibrium with the atmosphere. At each point in time the concentration of all solution species (including H^+) can be computed from C_T and Alk by assuming chemical equilibrium in the bulk solution phase.

7.4 Equilibrium Enhancement of CO_2 Exchange

As seen in Figure 4.11, it is possible to find conditions where the chemical reactions of the carbonate system in the diffusion layer occur faster than the diffusion itself. A very simplified treatment is to assume chemical equilibrium at each boundary of the layer and add together the diffusion fluxes of each of the carbonate species:

$$J_i = \frac{D_i}{Z}\left[(S_i)^s - (S_i)\right] \tag{88}$$

where i refers to the various carbonate species.

If we consider the constraint of electroneutrality and neglect the problem of electrostatic interactions among diffusing ions, we can reasonably assume that the alkalinity is constant throughout the diffusion layer and equal to the alkalinity of the bulk solution. This assumption is particularly good if the ionic strength is markedly higher than the carbonate species concentrations.

By taking all the diffusion coefficients to be approximately equal ($\cong 2 \times 10^{-5}$ cm^2 s^{-1}), the enhancement factor is simply the ratio of the C_T gradient to the CO_2 gradient:

$$J_T = \sum_i J_i = \frac{D}{Z}\left[C_T^s - C_T\right] \tag{89}$$

$$J_{CO_2} = \frac{D}{Z}\left[(CO_2)^s - (CO_2)\right] \tag{90}$$

$$E_g = \frac{J_T}{J_{CO_2}} = \frac{C_T^s - C_T}{(CO_2)^s - (CO_2)} \tag{91}$$

Example 7

Consider the conditions in a quiescent lake:

$$P_{CO_2} = 10^{-3.5} \text{ at}$$
$$\text{Alk} = 10^{-4} \ M$$
$$\text{pH} = 9.5$$

The boundary conditions at the air interface are defined by

$$P_{CO_2} = 10^{-3.5} \text{ at}$$
$$\text{Alk} = 10^{-4} \ M$$

therefore

$$\text{pH}^s = 7.3$$
$$(H_2CO_3^*)^s = 10^{-5} \ M$$
$$(HCO_3^-)^s = 10^{-4} \ M$$
$$(CO_3^{2-})^s = 10^{-7} \ M$$

On the bulk water side of the interface, the boundary conditions are given by

$$(H^+) = 10^{-9.5} \ M$$
$$(\text{Alk}) = 10^{-4} \ M$$

therefore

$$(H_2CO_3^*) = 10^{-7.49} \ M$$
$$(HCO_3^-) = 10^{-4.29} \ M$$
$$(CO_3^{2-}) = 10^{-5.09} \ M$$

Note that the CO_3^{2-} gradient is opposite to the $H_2CO_3^*$ and HCO_3^- gradients. The total carbonate and carbon dioxide gradients are obtained from the differences

$$C_T^s - C_T = 5 \times 10^{-5}$$
$$(CO_2)^s - (CO_2) = (H_2CO_3^*) - (H_2CO_3^*) = 10^{-5}$$

yielding a maximum enhancement factor $E_g = 5$. Actual experimental laboratory measurements of such a system[10] have yielded $E_g \cong 3.7 - 6.5$.

Although this discussion is limited to the coupling of chemical reactions with the surface film model, it has been demonstrated that the influence of chemical reactions on the rate of gas transfer are practically the same in all physical models of gas transfer.[13] In the equilibrium enhancement case where the chemical reactions are considered infinitely fast, the very same result is applicable regardless of the model: the volatile species gradient can simply be replaced by the total concentration gradient in the transfer rate equation. This provides a convenient way to evaluate an upper limit on the possible role of chemical reactions in the transport of reactive solutes.

□ □ □

REFERENCES

1 H. D. Holland, *The Chemistry of the Atmosphere and Oceans*, Wiley, New York, 1978.
2 J. P. Riley and G. Skirrow, Eds., *Chemical Oceanography*, Academic, New York, 1965.
3 G. E. Hutchinson, *A Treatise on Limnology*, Wiley, New York, 1974.
4 J. P. Riley and C. Skirrow, *Chemical Oceanography*, 2nd ed., Academic, London, 1975.
5 F. A. Richards and B. B. Benson, *Deep Sea Res.*, **7**, 254–264 (1961).
6 R. M. Smith and A. E. Martell, *Critical Stability Constants*, Vol. 4, *Inorganic Ligands*, Plenum, New York, 1976.
7 A. C. Redfield, *James Johnston Memorial Volume*, Liverpool Univ. Press, 1934.
8 F. M. M. Morel, R. E. McDuff, and J. J. Morgan, *J. Mar. Chem.*, **4**, 1 (1976).
9 J. A. Quinn and N. C. Otto, *J. Geophys. Res.*, **76**, 1539–1549, 1971.
10 S. Emerson, *Limnol. Oceanogr.*, **20**, 743–753, 1975.
11 S. Emerson, *Limnol. Oceanogr.*, **20**, 754–761, 1975.
12 W. Stumm and J. J. Morgan, *Aquatic Chemistry*, 2nd ed., Wiley, New York, 1981, p. 211.
13 P. V. Danckwerts, *Gas-Liquid Reactions*, McGraw-Hill, New York, 1970.

PROBLEMS

4.1 What is the alkalinity in each of the following systems?

   a. $(NaOH)_T = 10^{-3}\ M, (NaHCO_3)_T = 10^{-2}\ M, P_{CO_2} = 10^{-2.5}$ at

   b. $(Na_2CO_3)_T = 10^{-4}\ M, (NH_3)_T = 10^{-2}\ M, P_{CO_2} = 10^{-2.5}$ at

   c. Carbonate system: $P_{CO_2} = 10^{-3.5}$ at, pH = 7.3

   d. $(H_2SO_4)_T = (NaHCO_3)_T = 10^{-3}\ M$

4.2 Calculate the pH given:

   a. Alk $= C_T = 10^{-3}\ M$

   b. $(NaOH)_T = (CO_2)_T = 10^{-3}\ M$

   c. $P_{CO_2} = 10^{-3.5}$ at, Alk $= 10^{-3}\ M$

   d. Alk $= -10^{-2}\ M$ (mineral acidity)

4.3 Consider the approximate photosynthetic reaction: $CO_2 + H_2O \rightarrow$ "CH_2O" + O_2. What are the qualitative changes in pH in the following systems?

a. $C_T = 10^{-3}$ M, pH = 7.3
b. $P_{CO_2} = 10^{-3.5}$ at, pH = 7.3
c. $CaCO_3(s)$ at equilibrium, $P_{CO_2} = 10^{-3.5}$ at
d. $CaCO_3(s)$ at equilibrium, pH = 11.0, $(Ca^{2+}) = 10^{-3}$ M

4.4 What are the alkalinity changes introduced by each of the following reactions, each proceeding to the right by 10^{-3} M of the first reactant (solids precipitate and gases escape the system)?

a. $HS^- + H^+ \rightarrow H_2S(g)$
b. $SO_4^{2-} + 2"CH_2O" \rightarrow HS^- + 2HCO_3^- + H^+$
c. $NH_4^+ \rightarrow NH_3(g) + H^+$
d. $Ca^{2+} + 2HCO_3^- \rightarrow CaCO_3(s) + CO_2(g) + H_2O$

4.5 a. Using the appropriate data from Table 5.4, provide the alkalinity expression (LHS) for a carbonate system containing soluble aluminum and ferric hydroxide species.
   b. What are the effects of $Al(OH)_3(s)$ and $Fe(OH)_3(s)$ precipitation on alkalinity? What are the qualitative effects on pH, for pH $\cong$ 7.0? For pH $\cong$ 9.0?

4.6 Planet Thethys has a nitrogen and hydrogen sulfide atmosphere ($P \cong 1$ at), liquid water, and photosynthetically supported life. Thethysian aquatic chemists have thus defined the alkalinity expression with respect to the H_2S equivalence point.

   a. Write this alkalinity expression (LHS) including necessary carbonate and phosphate species.
   b. What is the alkalinity of the system:

$$(FeS)_T = 10^{-3} \ M, \ (CaCO_3)_T = 10^{-4} \ M$$

   c. What is the effect on alkalinity of $CO_2(g)$ exchange with the atmosphere?
   d. What is the effect on alkalinity of the photosynthetic reaction

$$CO_2(g) + \tfrac{2}{3}H_2S(g) + \tfrac{1}{3}H_2O \rightarrow \text{``}CH_2O\text{''} + \tfrac{2}{3}SO_2(g)$$

   e. For alkalinity $\cong 10^{-3}$ M and $P_{H_2S} = 10^{-3}$ at. What is the average pH in Thethysian lakes? What are the qualitative effects of $CO_2(g)$ exchange and photosynthesis on this pH?

4.7 An acid mine drainage $\left[(H_2SO_4)_T = 10^{-4} \ M, \ P_{CO_2} = 10^{-1.5} \ \text{at}\right]$ mixes 1:2 with a stream characterized by $P_{CO_2} = 10^{-1.5}$ at, pH = 6.0. What are the alkalinity and the pH of (a) the acid mine drainage, (b) the stream, (c) the mix at the point of confluence, (d) the mix after cascading down the hill and equilibrating with the atmosphere, and (e) the mix after standing in a small lake and supporting the growth of 3 mg liter^{-1} of algal biomass (NO_3^- as N source)?

4.8 Yesterday you prepared a 10^{-2} M NaOH solution in distilled water.

a. What pH did you measure right away?

b. The pH measured today is only 10; why?

c. What will be the ultimate pH?

4.9 Using apparent constants appropriate for seawater (see Chapter 6), derive a plot of pH versus C_T, given Alk $= 2.5 \times 10^{-3}$ M (C_T from 0 to 3×10^{-3} M). How good a measure of photosynthesis and respiration would a surface seawater pH be? Discuss.

4.10 Calculate the small change in pH due to the addition of 10^{-5} M FeCl$_3$ in a system at equilibrium with Fe(OH)$_3$(s) at pH $= 7.3$ and $C_T = 10^{-3}$ M.

4.11 Your backyard pond (a nice little water body with a stable ground water source) had a pH of 8.2 in April. The pH measured now (November) is 7.5, and when you bubble a sample with air, the pH goes to 7.7.

a. Why these pH differences?

b. What would have been the measured pH if you had bubbled a sample in April?

c. Calculate the alkalinity and the total carbonate content of the pond in April and November.

4.12 An industry discharges 700 kg of ammonium chloride (NH$_4$Cl) per hour into a stream. For the purpose of this problem, the stream is divided into three regions (flow rate $= 2$ m^3 s^{-1}):

1 Upstream from the discharge where the water is in equilibrium with the atmosphere (0.032% CO$_2$; 0.0000% NH$_3$).

2 Some distance downstream from the discharge where the waste has been mixed and equilibrated with the water, but air-water equilibrium has not had time to be reestablished.

3 Farther from the discharge where air-water equilibrium has been reestablished.

Questions

a. The pH measured in region 1 is 9.4. Supposing that sodium is the main cation in the stream, what is the concentration of sodium?

b. Which way will the pH vary as we measure farther downstream: region 1? 2? 3?

c. List species, write appropriate mole balance equations, and compute pH in region 2.

d. Compute pH in region 3.

4.13 The epilimnion of your backyard pond has a pH $\cong 7.0$ and alkalinity $\cong 10^{-3}$ M.

a. Assuming the carbonate system to dominate the chemistry of your pond, calculate the concentrations of all species. Is your pond in equilibrium with the atmosphere? Explain.

b. You just washed the bricks of your house walls with HCl. How much acid can you add to your pond before changing the pH by more than 0.2 units (get an approximate answer by considering the buffering capacity).

c. The hypolimnion of your pond is anoxic. Its composition is derived from that of the epilimnion by reducing $3 \times 10^{-4} M$ of SO_4^{2-} to H_2S. What are the alkalinity, the pH, and the concentrations of all species in the hypolimnion?

d. In the fall the epilimnion (1-m depth) and the hypolimnion (2-m depth) mix with minimal gas exchange. What is the composition (Alk, pH, concentrations) of the resulting water?

e. What would be the composition of the mixed water if you forced mixing by bubbling air at the bottom of the pond (assume gas exchange, not oxidation).

4.14 Consider a carbonate solution, pH $= 8.0$; $C_T = 5.0 \times 10^{-4} M$.

a. Calculate the alkalinity and the detailed composition of the system.

b. Calculate and draw a precise acidimetric titration curve (pH versus HCl_T), considering $C_T =$ constant.

c. Calculate and draw a precise acidimetric titration curve for $P_{CO_2} = 10^{-3.51}$ at $=$ constant.

d. Suppose that the solution is contained in a 1-liter stoppered bottle and that the pressure of the enclosed gas (approximately 10 mliter) is monitored. Draw an acidimetric titration curve (P_{CO_2} vs. HCl_T).

e. Discuss the relative merits of the three corresponding methods for alkalinity determination.

4.15 The following data were obtained from analysis of a rainwater sample collected in Pittsburgh, Pennsylvania, in September 1979:

		Concentration
Ion	mg liter^{-1}	mol liter^{-1} ($\times 10^5$)
H^+	(pH $= 4.19$)	6.46
NH_4^+	1.75	12.5
Na^+	0.337	1.46
K^+	0.741	1.90
Ca^{2+}	3.00	7.50
Mg^{2+}	0.855	3.56
SO_4^{2-}	13.0	13.5
NO_3^-	2.15	15.4
Cl^-	0.44	1.23

    a. Compute the Σ cation equivalent/Σ anion equivalent; what does this reveal?

    b. Compute the mineral acidity of this sample.

    c. Sketch an *approximate* titration curve assuming:

       That the titration is carried out with 0.01 N NaOH on a 100 milliliter sample in a closed jacketed beaker.

       That high-purity nitrogen gas is bubbled through the sample so that the titration is performed under a nitrogen atmosphere.

       That the given pH remains essentially unchanged after the onset of $N_2(g)$ bubbling and prior to the addition of NaOH.

       Indicate the points of inflection and explain their significance.

    d. Explain the purpose of performing the titration in a closed jacketed beaker under a nitrogen atmosphere.

4.16 For three diffusion boundary layer thicknesses (30–60–120 μm) calculate the rate of CO_2 exchange across the surface of the ocean ($T = 20°C$) as a function of pH in the range of 7.5 to 9 (use "apparent" carbonate acidity constants: $K_{a1} = 10^{-5.9}$; $K_{a2} = 10^{-9.1}$).

CHAPTER FIVE

SOLID DISSOLUTION AND PRECIPITATION
ACQUISITION AND CONTROL OF ALKALINITY

With an average runoff of 30 cm y^{-1}, the hydrologic cycle erodes the continents at an overall rate of about 60 μm y^{-1}. Approximately 0.5 g liter^{-1} of weathered continental rock is thus transported by rivers to the oceans, roughly 80% as suspended solids (physical erosion) and only about 20% as dissolved species (chemical erosion).[1] Although the suspended material is not completely inert chemically (for example, ion exchange in detrital clays may be an important sink for Na^+ in seawater), we largely ignore its existence in this chapter and focus on the smaller part of the exogenic material flux from the continents to the ocean, that which undergoes phase changes from the continental rock to the aqueous solution and finally to the sediments. Dissolution and precipitation reactions control these phase changes; they are the chemical processes that we wish to study here as an idealization of weathering and sedimentation phenomena.

There are wide differences in the conditions under which dissolution of continental rocks and precipitation of oceanic sediments take place. At one extreme very low ionic strength rainwater gets into the soil where the effective partial pressure of CO_2 is high (10^{-3} to 10^{-1} at; typically $10^{-2.5}$ at) due to intense microbial activity. Some of the solids are highly unstable in this corrosive, rapidly renewed groundwater environment; the chemical driving forces are rather large and reasonably well understood. At the other extreme there is seawater which contains about half the chloride of the earth's surface and is a medium of high ionic strength (about 0.7 M); the residence time of seawater is long (about 30,000 y), and the oceans are practically in equilibrium with the atmosphere ($P_{CO_2} = 10^{-3.5}$ at). The solids that control the solubility of the seawater constituents are for the most part barely, and perhaps only locally, oversaturated. In many instances we are just beginning to discover the nature of the controlling phases that serve as sinks for the minor and major soluble constituents of the oceans.

In Chapter 4 we examined the concept of alkalinity in part as a convenient way to study the chemistry of dissolved weak acids—particularly CO_2—without worrying about the source of the balancing strong bases. The question of how natural waters acquire their alkalinity by dissolution of continental rock is one that we wish to address here. Owing to the complexity and multiplicity of rock and soil types and to the sometimes sluggish and poorly characterized kinetics, we may not be able to develop precise quantitative models of alkalinity in freshwaters. Yet we can use limiting cases to understand chemically how the nature of the rocks controls the alkalinity of the water.

Our understanding of alkalinity control in the oceans is both stronger and more tenuous. On relatively short time scales the alkalinity of the oceans is rather constant in time and space, and the small variations that exist are well studied and easily accounted for. Over geological times the difficult question is paradoxically not why seawater alkalinity is so high (about 2.3 meq liter^{-1} compared to 0.1 meq liter^{-1} for the average river) but why it is so low. If seawater were simply river water concentrated by evaporation, its alkalinity would be almost a thousand times what it is (by normalization to chloride concentrations). In order to understand the control of alkalinity in seawater, we thus need to understand quantitatively the removal processes for the major cations Ca^{2+}, Na^+, K^+, and Mg^{2+}. These do not involve simply equilibration with solid phases in the sediments; ion exchange in detrital clays, high and low temperature reactions with volcanic material at oceanic ridges, and biological activity all seem to play an important role as well.

Due to limitations in space and knowledge, the actual questions addressed in this chapter are somewhat narrower and simpler than those posed by the study of the exogenic cycle. After a brief examination of the chemical nature of rocks, we develop some simple methodologies—diagrams and calculations—to study equilibrium between solids and the aqueous phase. This is useful for understanding if a particular constituent is saturated with respect to a particular phase and what concentrations of dissolved constituents a certain solid may contribute to a contacting water. Generalizing our approach to include several chemical compounds, we then examine the question of the coexistence of solid phases and determine the conditions under which solids with common constituents can coexist at equilibrium and which solid is thermodynamically most stable when they cannot coexist. These questions naturally bring us to discuss Gibbs' phase rule in view of the mathematical and conceptual framework we have developed and its significance in natural waters. Going back to a more direct study of the exogenic cycle, we then examine a few simple examples of weathering reactions to show how freshwaters acquire their alkalinity, how the composition of the water is controlled by the types of rocks it encounters. We also look at the mechanisms that control the alkalinity of seawater. Since equilibrium concepts are clearly insufficient in these studies of the acquisition and control of alkalinity, we close the chapter with a brief discussion of precipitation-dissolution kinetics in aquatic systems.

A number of caveats are in order before embarking on the study of solubility relationships in aquatic systems:

1. Many precipitation-dissolution reactions are sluggish. For example, a large supersaturation of $CaCO_3$ is often observed in natural waters before precipitation of the solid actually occurs. High temperature, the presence of nucleating surfaces, and biological activity can all dramatically enhance precipitation-dissolution kinetics. The proper application of equilibrium relationships depends on one's knowledge of such processes.

2. Even when the kinetics of precipitation are fast, the solid formed is often not the most stable solid thermodynamically. For example, opal—a cryptocrystalline form of silica—is typically precipitated by organisms where quartz (crystalline silica) is thermodynamically the stable form of SiO_2. The evolution of the solid to its more stable form (through dehydration, recrystallization, etc.) is usually very slow and often requires high temperatures.

3. The metastable solids that are initially precipitated are often "nonstoichiometric." For example, a common form of manganese oxide, "γ-MnO_2," has the approximate stoichiometry of $MnO_{1.3}$. Many natural solids also typically contain impurities; foreign ions are incorporated in the matrix. The equilibrium constants to be used for such solids have operational value but probably little true thermodynamic significance.

4. The formation of pure solid phases is not the only, or perhaps even the dominant, process by which many solutes are removed from solution. We shall see in Chapter 8 how surface adsorption and solid solution formation can remove a solute from solution much below saturation conditions for pure solids. Such processes can be viewed as a way to describe thermodynamically the formation of solid species with activities different from unity.

1. THE CHEMICAL NATURE OF ROCKS

In order to understand the composition of natural waters, it is indispensable to understand something of the composition of the rocks that they contact, transform, and partially dissolve. The necessary excursions into the field of geology are often particularly frustrating to the aquatic chemist. It sometimes seems as if xenophobic mineralogists had arranged their multidimensional taxonomy with the intention of forming a maze impenetrable to outsiders.

At the risk of tediousness and oversimplification let us recall here a few fundamentals of geological terminology. Rocks (which are assemblages of minerals) are broadly classified into igneous, sedimentary, and metamorphic. Igneous rocks are solidified magma, the molten material from below the earth's crust. Their weathering produces sedimentary rocks which cover some 80% of the surface of the continents. Metamorphic rocks result from pressure and heat transformations of the other rocks and are often classified, as they are here, with their parent material by geochemists who are more interested in composition than appearances. Sketchily, igneous rocks come in two basic shades (though many colors), black and gray. The darker shade is characteristic of lavas and described by an assortment of adjectives: dark, basaltic, extrusive, volcanic, and mafic (rich in magnesium and iron). The light colored material is typically formed by intrusions of cooled magma into the crust and variously described as light, granitic, intrusive, plutonic, or sialic (rich in silica and aluminum). Sedimentary rocks comprise (1) sandstones, the familiar stuff on beaches; (2) shales, which represent more than 50% of all exposed rocks and include the clays; (3) evaporites, which are salt deposits formed by evaporation of seawater; and (4) limestones such as chalk and marbles.

Only four elements account for about 89% of the igneous rock mass: oxygen, silicon, aluminum, and iron (Table 5.1). The four elements that provide the major cation content of natural waters, calcium, sodium, potassium, and magnesium, roughly account for the remaining 11%. The elemental composition of average sedimentary rock is of course very similar to that of the parent igneous rock, the major difference being the addition of some CO_2 (0.05 g/g), H_2O (0.04 g/g), and HCl (0.03 g/g). Although their elemental composition is similar, the mineralology of igneous and sedimentary rocks is quite different. Tables 5.2 and 5.3 list minerals—taken here as defined chemical solid phases—

TABLE 5.1

Major Elements in Continental Rock[a,b]

	Igneous Rock	Sedimentary Rock
O	46.8	49.0
Si	29.7	27.4
Al	8.4	7.6
Fe	4.6	4.4
Ca	3.5	3.3
Mg	1.8	1.6
Na	2.5	0.7
K	2.7	2.6
C	—	1.3
H	—	0.4
Cl	—	1.7

[a] Source: Adapted from Garrels and MacKenzie (1971).[2]
[b] Note: The figures given are in weight percent; rocks dried to 110°C.

TABLE 5.2
Some Important Minerals in Continental Igneous Rocks

	Silicates	Alumino-Silicates	
		Micas	Feldspars
Olivines[a]			**Plagioclases**[c]
Fosterite	$2MgO \cdot SiO_2$		Anorthite $\quad CaO \cdot Al_2O_3 \cdot 2SiO_2$
Fayalite	$2FeO \cdot SiO_2$		
Pyroxenes			
Enstatite	$MgO \cdot SiO_2$		
Wollastonite	$CaO \cdot SiO_2$		
Hedenbergite	$CaO \cdot FeO \cdot 2SiO_2$		
Diopside	$CaO \cdot MgO \cdot 2SiO_2$	Biotite or Phlogopite	Albite $\quad Na_2O \cdot Al_2O_3 \cdot 6SiO_2$
Amphiboles[b]		$K_2O \cdot 6MgO \cdot Al_2O_3 \cdot 6SiO_2 \cdot 2H_2O$	
Tremolite	$2CaO \cdot 5MgO \cdot 8SiO_2 \cdot H_2O$	Muscovite	**Orthoclases**
Iron tremolite	$2CaO \cdot 5FeO \cdot 8SiO_2 \cdot H_2O$	$K_2O \cdot 3Al_2O_3 \cdot 6SiO_2 \cdot 2H_2O$	K. feldspar $\quad K_2O \cdot Al_2O_3 \cdot 6SiO_2$
Quartz			
α-quartz	SiO_2		
Cristobalite	SiO_2		

Basaltic ←——————————————————————→ Granitic

[a] Continuously variable composition from Mg to Fe.
[b] Variable composition; Na often part of composition; in the hornblende variety, Al substitutes for part of the Si.
[c] Continuously variable composition from Ca to Na.

TABLE 5.3

Some Important Minerals in Continental Sedimentary Rocks

Silicates (Predominant in Sandstones)		Alumino-Silicates (Predominant in Shales: Clays)	Carbonates (Predominant in Limestone)		Chlorides (Predominant in Evaporites)		Sulfates (Predominant in Evaporites)	
Quartz		**Montmorillonites**[a]	Calcite	$CaCO_3$	Halite	NaCl	Gypsum	$CaSO_4 \cdot 2H_2O$
α-quartz	SiO_2	$Na_2O \cdot 7Al_2O_3 \cdot 22SiO_2 \cdot nH_2O$	Aragonite	$CaCO_3$			Anhydrite	$CaSO_4$
Cristobalite	SiO_2	$CaO \cdot 7Al_2O_3 \cdot 22SiO_2 \cdot nH_2O$	Dolomite	$CaMg(CO_3)_2$				
Amorphous Silica	$SiO_2\text{-}nH_2O$	**Illite** (Hydrated Mica)	Magnesite	$MgCO_3$				
		$3K_2O \cdot 2MgO \cdot 9Al_2O_3 \cdot 28SiO_2 \cdot 8H_2O$						
		Chlorites						
		$5MgO \cdot Al_2O_3 \cdot 3SiO_2 \cdot 4H_2O$						
		Kaolinite						
		$Al_2O_3 \cdot 2SiO_2 \cdot 2H_2O$						

[a] Extremely variable composition; often contains K and Mg.

183

that predominate in the various types of rocks. Silicates and silicate-rich alumino-silicates (chiefly feldspars) are characteristic of igneous rocks, while carbonates and silicate-poor alumino-silicates (mostly clays) are abundant in sedimentary material. Note that abraded but unweathered igneous rocks also constitute a variable fraction of the sedimentary rock mass and that quartz is rather ubiquitous. In these tables the formulae of complex alumino-silicates are given as the "oxide formulae" to emphasize the basic nature of the minerals and to simplify their stoichiometric decomposition into the components SiO_2 and Al_2O_3 [or any combination thereof such as kaolinite: $Al_2O_3 \cdot 2SiO_2 \cdot 2H_2O = Al_2Si_2O_5(OH)_4$].

By itself the stoichiometric information provides a rough understanding of the chemistry of weathering. The major constituents of rocks are rather insoluble: aluminum and iron oxides contribute no significant solute concentrations to natural waters, while dissolved silica concentrations average only 0.15 mM. In addition to the relative enrichment of aluminum over silicon in individual rocks (not in the average rock composition), weathering is thus largely the process of progressively stripping the four major cations, Ca^{2+}, Na^+, K^+, and Mg^{2+}, from the alumino-silicates. Note the relative sodium impoverishment in sedimentary rock. In the process the water acquires part of its dissolved load, most of which actually originates from the dissolution of carbonates, sulfates, sulfides, and chlorides in sedimentary material (plus gases and aerosols).

To go further in our understanding of weathering processes and to make it more quantitative, thermodynamic and/or kinetic information is necessary. Table 5.4 provides solubility constants for most of the major minerals listed in Tables 5.2 and 5.3 and for some other minerals such as sulfides and hydroxides that are not particularly abundant in continental rocks but play an important

TABLE 5.4

Solubility Products of Various Minerals

		Log K^a	Reference
Chlorides			
Halite	$NaCl(s) = Na^+ + Cl^-$	1.54	3
Sylvite	$KCl(s) = K^+ + Cl^-$	0.98	3
Chlorargyrite	$AgCl(s) = Ag^+ + Cl^-$	-9.74	4
Sulfates			
Gypsum	$CaSO_4 \cdot 2H_2O(s) = Ca^{2+} + SO_4^{2-} + 2H_2O$	-4.62	4
Celestite	$SrSO_4(s) = Sr^{2+} + SO_4^{2-}$	-6.50	4
Barite	$BaSO_4(s) = Ba^{2+} + SO_4^{2-}$	-9.96	4
Oxides and Hydroxides[b]			
Calcium Hydroxide	$Ca(OH)_2(s) = Ca^{2+} + 2OH^-$	-5.19	4, 5
	$CaOH^+ = Ca^{2+} + OH^-$	-1.15	5

TABLE 5.4 (Continued)

		Log K^a	Reference
Brucite	$Mg(OH)_2(s) = Mg^{2+} + 2OH^-$	-11.1	4, 5
	$MgOH^+ = Mg^{2+} + OH^-$	-2.6	5
	$Mg_4(OH)_4^{4+} = 4Mg^{2+} + 4OH^-$	-16.3	4, 5
Gibbsite	$Al(OH)_3(s) = Al^{3+} + 3OH^-$	-33.5	4, 5
	$AlOH^{2+} = Al^{3+} + OH^-$	-9.0	5
	$Al(OH)_2^+ = Al^{3+} + 2OH^-$	-18.7	5
	$Al(OH)_3 = Al^{3+} + 3OH^-$	-27.0	5
	$Al(OH)_4^- = Al^{3+} + 4OH^-$	-33.0	5
	$Al_2(OH)_2^{4+} = 2Al^{3+} + 2OH^-$	-20.3	5
	$Al_3(OH)_4^{5+} = 3Al^{3+} + 4OH^-$	-42.1	5
Manganous	$Mn(OH)_2(s) = Mn^{2+} + 2OH^-$	-12.8	5
Hydroxide	$MnOH^+ = Mn^{2+} + OH^-$	-3.4	5
Ferrous Hydroxide	$Fe(OH)_2(s) = Fe^{2+} + 2OH^-$	-15.1	5
	$FeOH^+ = Fe^{2+} + OH^-$	-4.5	5
	$Fe(OH)_2 = Fe^{2+} + 2OH^-$	-7.4	5
	$Fe(OH)_3^- = Fe^{2+} + 3OH^-$	-11.0	5
	$Fe(OH)_4^{2-} = Fe^{2+} + 4OH^-$	-10.0	5
Goethite	$\alpha \cdot FeOOH(s) + H_2O = Fe^{3+} + 3OH^-$	-41.5	4, 5
Ferric Hydroxide	$am \cdot Fe(OH)_3(s) = Fe^{3+} + 3OH^-$	-38.8	4
Hematite	$\frac{1}{2}\alpha \cdot Fe_2O_3(s) + \frac{3}{2}H_2O = Fe^{3+} + 3OH^-$	-42.7	4
	$FeOH^{2+} = Fe^{3+} + OH^-$	-11.8	5
	$Fe(OH)_2^+ = Fe^{3+} + 2OH^-$	-22.3	5
	$Fe(OH)_4^- = Fe^{3+} + 4OH^-$	-34.4	5
	$Fe_2(OH)_2^{4+} = 2Fe^{3+} + 2OH^-$	-25.1	5
	$Fe_3(OH)_4^{5+} = 3Fe^{3+} + 4OH^-$	-49.7	5
Tenorite	$CuO(s) + H_2O = Cu^{2+} + 2OH^-$	-20.4	5
Cupric Hydroxide	$Cu(OH)_2(s) = Cu^{2+} + 2OH^-$	-19.4	5
	$Cu(OH)_4^{2-} = Cu^{2+} + 4OH^-$	-16.4	5
	$Cu_2(OH)_2^{2+} = 2Cu^{2+} + 2OH^-$	-17.6	5
Litharge (Red)	$PbO(s) + H_2O = Pb^{2+} + 2OH^-$	-15.3	5
Massicot (Yellow)	$PbO(s) + H_2O = Pb^{2+} + 2OH^-$	-15.1	5
	$PbOH^+ = Pb^{2+} + OH^-$	-6.3	5
	$Pb(OH)_2 = Pb^{2+} + 2OH^-$	-10.9	5
	$Pb(OH)_3^- = Pb^{2+} + 3OH^-$	-13.9	5
	$Pb_2(OH)^{3+} = 2Pb^{2+} + OH^-$	-7.6	5
	$Pb_3(OH)_4^{2+} = 3Pb^{2+} + 4OH^-$	-32.1	5
	$Pb_4(OH)_4^{4+} = 4Pb^{2+} + 4OH^-$	-35.1	5
	$Pb_6(OH)_8^{4+} = 6Pb^{2+} + 8OH-$	-68.4	5
Carbonates			
Aragonite	$CaCO_3(s) = Ca^{2+} + CO_3^{2-}$	-8.22	4
Calcite	$CaCO_3(s) = Ca^{2+} + CO_3^{2-}$	-8.35	4
Magnesite	$MgCO_3(s) = Mg^{2+} + CO_3^{2-}$	-7.46	4

(continued next page)

TABLE 5.4 (Continued)

		Log K^a	Reference
Nesquehonite	$MgCO_3 \cdot 3H_2O(s) = Mg^{2+} + CO_3^{2-} + 3H_2O$	-4.67	4
Dolomite	$CaMg(CO_3)_2(s) = $ Calcite $+$ Magnesite	-1.70	3
Disordered Dolomite	$CaMg(CO_3)_2(s) = $ Calcite $+$ Magnesite	-0.08	3
Strontianite	$SrCO_3(s) = Sr^{2+} + CO_3^{2-}$	-9.0	4
Rhodochrosite	$MnCO_3(s) = Mn^{2+} + CO_3^{2-}$	-10.4	4
Siderite	$FeCO_3(s) = Fe^{2+} + CO_3^{2-}$	-10.7	4
Malachite	$Cu_2CO_3(OH)_2(s) =$ $2Cu^{2+} + CO_3^{2-} + 2OH^-$	-33.8	4
Azurite	$Cu_3(CO_3)_2(OH)_2(s) =$ $3Cu^{2+} + 2CO_3^{2-} + 2OH^-$	-46.0	4
Cerussite	$PbCO_3(s) = Pb^{2+} + CO_3^{2-}$	-13.1	4
Phosphates			
Brushite	$CaHPO_4 \cdot 2H_2O(s) =$ $Ca^{2+} + HPO_4^{2-} + 2H_2O$	-6.6	4
Hydroxylapatite	$Ca_5(PO_4)_3OH(s) = 5Ca^{2+} + 3PO_4^{3-} + OH^-$	-55.6	8
Newberyite	$MgHPO_4 \cdot 3H_2O(s) =$ $Mg^{2+} + HPO_4^{2-} + 3H_2O$	-5.8	4
Bobierrite	$Mg_3(PO_4)_2 \cdot 8H_2O(s) =$ $3Mg^{2+} + 2PO_4^{3-} + 8H_2O$	-25.2	4
Vivianite	$Fe_3(PO_4)_2 \cdot 8H_2O(s) =$ $3Fe^{2+} + 2PO_4^{3-} + 8H_2O$	-36.0	4
Strengite	$FePO_4 \cdot 2H_2O(s) = Fe^{3+} + PO_4^{3-} + 2H_2O$	-26.4	4
Berlinite	$AlPO_4(s) = Al^{3+} + PO_4^{3-}$	-20.6	6
Sulfides			
Pyrrhotite	$FeS(s) = Fe^{2+} + S^{2-}$	-18.1	4
Pyrite	$FeS_2(s) = $ Pyrrhotite $+ S^0$	-10.4	3
Alabandite	$MnS(s) = Mn^{2+} + S^{2-}$	-13.5	4
Covellite	$CuS(s) = Cu^{2+} + S^{2-}$	-36.1	4
Galena	$PbS(s) = Pb^{2+} + S^{2-}$	-27.5	4
Chalcopyrite	$CuFeS_2(s) = $ Covellite $+$ Pyrrhotite	-6.0	3
Silicates			
Quartz $(\alpha + \beta)$	$SiO_2(s) + H_2O = H_2SiO_3$	-4.00	3
Cristobalite $(\alpha + \beta)$	$SiO_2(s) + H_2O = H_2SiO_3$	-3.45	3
Amorphous silica	$SiO_2(s) + H_2O = H_2SiO_3$	-2.71	3
Fosterite	$\frac{1}{4}[2MgO \cdot SiO_2](s) + H^+ =$ $\frac{1}{2}Mg^{2+} + \frac{1}{4}H_2SiO_3 + \frac{1}{4}H_2O$	7.11	3
Fayalite	$\frac{1}{4}[2FeO \cdot SiO_2](s) + H^+ =$ $\frac{1}{2}Fe^{2+} + \frac{1}{4}H_2SiO_3 + \frac{1}{4}H_2O$	4.21	3
Enstatite	$\frac{1}{2}[MgO \cdot SiO_2](s) + H^+ = \frac{1}{2}Mg^{2+} + \frac{1}{2}H_2SiO_3$	5.82	3
Wollastonite	$\frac{1}{2}[CaO \cdot SiO_2](s) + H^+ = \frac{1}{2}Ca^{2+} + \frac{1}{2}H_2SiO_3$	6.82	3

TABLE 5.4 (Continued)

		Log K^a	Reference
Hedenbergite	$\frac{1}{4}[CaO \cdot FeO \cdot 2SiO_2](s) + H^+ =$ $\frac{1}{4}Ca^{2+} + \frac{1}{4}Fe^{2+} + \frac{1}{2}H_2SiO_3$	4.60	3
Diopside	$\frac{1}{4}[CaO \cdot MgO \cdot 2SiO_2](s) + H^+ =$ $\frac{1}{4}Ca^{2+} + \frac{1}{4}Mg^{2+} + \frac{1}{2}H_2SiO_3$	5.30	3
Tremolite	$\frac{1}{14}[2CaO \cdot 5MgO \cdot 8SiO_2 \cdot H_2O](s) + H^+ =$ $\frac{1}{7}Ca^{2+} + \frac{5}{14}Mg^{2+} + \frac{4}{7}H_2SiO_3$	4.46	3
Alumino-Silicates			
Gibbsite	$Al(OH)_3(s) + H_2SiO_3 = \frac{1}{2} \text{ Kaolinite} + \frac{3}{2}H_2O$	4.25	3
Phlogopite	$\frac{1}{14}[K_2O \cdot 6MgO \cdot Al_2O_3 \cdot 6SiO_2 \cdot 2H_2O](s)$ $+ H^+ = \frac{1}{14} \text{ Kaolinite} + \frac{1}{7}K^+$ $+ \frac{3}{7}Mg^{2+} + \frac{2}{7}H_2SiO_3 + \frac{3}{14}H_2O$	5.01	3
Muscovite	$\frac{1}{2}[K_2O \cdot 3Al_2O_3 \cdot 6SiO_2 \cdot 2H_2O](s) + H^+$ $+ \frac{3}{2}H_2O = \frac{3}{2} \text{ Kaolinite} + K^+$	3.51	3
Anorthite	$\frac{1}{2}[CaO \cdot Al_2O_3 \cdot 2SiO_2](s) + H^+ + \frac{1}{2}H_2O =$ $\frac{1}{2} \text{ Kaolinite} + \frac{1}{2}Ca^{2+}$	9.83	3
Albite	$\frac{1}{2}[Na_2O \cdot Al_2O_3 \cdot 6SiO_2](s) + H^+ + \frac{5}{2}H_2O =$ $\frac{1}{2} \text{ Kaolinite} + 2H_2SiO_3 + Na^+$	−0.68	3
K-feldspar	$\frac{1}{2}[K_2O \cdot Al_2O_3 \cdot 6SiO_2](s) + H^+ + \frac{5}{2}H_2O =$ $\frac{1}{2} \text{ Kaolinite} + 2H_2SiO_3 + K^+$	−3.54	3
Na- Montmorillonite[c]	$\frac{1}{2}[Na_2O \cdot 7Al_2O_3 \cdot 22SiO_2 \cdot 6H_2O](s) + H^+$ $+ \frac{15}{2}H_2O = \frac{7}{2} \text{ Kaolinite} + 4H_2SiO_3 + Na^+$	−9.1	8
Ca- Montmorillonite[c]	$\frac{1}{2}[CaO \cdot 7Al_2O_3 \cdot 22SiO_2 \cdot 6H_2O](s) + H^+$ $+ \frac{15}{2}H_2O = \frac{7}{2} \text{ Kaolinite} + 4H_2SiO_3$ $+ \frac{1}{2}Ca^{2+}$	−7.7	8

[a] Except where noted all constants are valid for 25°C and ionic strength $I = 0$ M.

[b] See also oxides formed in redox reactions, Table 7.1.

[c] As noted by Helgeson et al.,[3] thermodynamic properties of minerals with continuously variable stoichiometric composition (e.g., montmorillonites and illites) are not simply defined. The values reported here allow for representative calculations of water composition not rigorous thermodynamic geochemical analysis. Note also that the solubilities of anorthite and albite reported by Helgeson et al.[3] are markedly higher than those reported by Stumm and Morgan[8] and that they may thus be inconsistent with the tabulated values for Na- and Ca-montmorillonites.

role in aquatic chemistry. Also listed in Table 5.4 are some copper and lead solid phases to exemplify the chemical control of trace elements in natural waters. It should be noted that thermodynamic data for solids that react very slowly at ordinary pressures and temperatures are difficult to obtain (they are often extrapolated from high P and T conditions) and may thus be unreliable. Also minerals that exist in continuously variable compositions (e.g., montmorillonites) are in principle difficult to define thermodynamically; the corresponding constants listed in Table 5.4 are provided for illustrative purposes, not for exact thermodynamic calculations. Finally, the presence of foreign ions may affect markedly the stability of natural minerals compared to their idealized pure chemical counterparts.

2. SOLUBILITY OF SOLIDS: EFFECTS OF COMPOSITION

In Chapter 3 we briefly discussed the problem of determining whether a particular solid is or is not present at equilibrium, given a particular recipe for the system. This problem can be solved by originally assuming the solid to be present, carrying out the calculation of equilibrium composition, and finally checking that all components are present in sufficient concentrations to ensure the presence of the solid; that is, the *solid* is calculated to have a positive *concentration* in the system. Alternatively, the equilibrium composition can be calculated by assuming that the solid does not form in the system and then verifying that the solubility product is not exceeded.

In this section we want to generalize the problem and determine under what conditions (e.g., concentration and pH) a given solid may form or what is the composition of a solution in equilibrium with a given solid phase.

2.1 Solubility of Chlorides and Sulfates

Precipitation and dissolution of chlorides and sulfates are seemingly simple to study since Cl^- and SO_4^{2-} have no interesting acid-base chemistry over the pH range of natural waters. However, in nature the precipitation or dissolution of such solids occurs only in the processes of formation or weathering of evaporites. The multiplicity of solutes and solid phases that can form (e.g., complex chlorides and sulfates) and the complexity of the ionic interactions in concentrated brines (which are highly nonideal solutions) in fact make the problem of predicting the precipitation of chlorides and sulfates one of the most difficult in aquatic chemistry. Nonetheless, halite (NaCl) is the most abundant solid in evaporites and is known to form in the laboratory when seawater is evaporated to about one-tenth its original volume. As a simple example let us calculate the evaporative concentration of seawater theoretically necessary for precipitation of halite.

Example 1. Precipitation of Halite from Seawater

Seawater Concentrations

$$(Na^+)^0 = 0.47 \ M$$

$$(Cl^-)^0 = 0.55 \ M$$

$$I^0 = 0.65 \ M$$

Halite Precipitation

$$NaCl(s) = Na^+ + Cl^-; \quad pK_s = -1.54 \tag{1}$$

Considering here that no reaction other than the precipitation of NaCl is taking place, let us define f as the relative reduction in total volume as the water evaporates. The concentration of each constituent is then increased by f:

$$(Na^+) = f(Na^+)^0$$

$$(Cl^-) = f(Cl^-)^0$$

$$I = fI^0$$

The brink of precipitation for halite is achieved when the activity product equals the solubility product:

$$\gamma_{Na}f(Na^+)^0\gamma_{Cl}f(Cl^-)^0 = K_s$$

therefore

$$2 \log f + \log \gamma_{Na} + \log \gamma_{Cl} = \log K_s - \log(Na^+)^0 - \log(Cl^-)^0 = 2.13 \tag{2}$$

Making the rather gross assumption that the activity coefficients can be described by the Davies equation (see Chapter 2), we obtain an explicit equation in f:

$$2 \log f - \frac{(0.65f)^{1/2}}{1 + (0.65f)^{1/2}} + 0.13f = 2.13 \tag{3}$$

Solving Equation 3 by trial and error yields $f = 7.9$. This concentration factor is somewhat lower than is observed for halite precipitation in seawater (about 10.5), a reflection of the fact that the Davies equation provides a poor representation of ionic interactions in such a complex system.

Similar calculations for gypsum precipitation would be even more off the mark as the activities of calcium and sulfate are even a lesser fraction (and a more complicated function) of their total concentration in evaporated seawater than are Na^+ and Cl^- (see Chapter 6). The problem of predicting the correct sequence of solids in evaporite formation can be solved by iterative computer techniques using thermodynamic formulae that account for first- and second-order interactions among ions (see Chapter 2). On the basis of the activity coefficient formulae of Pitzer,[9,10] Harvie, Weare, and Engster[11,12,13] have developed a chemical model for calculating mineral solubilities in water that varies from zero to high ionic strength in a batch evaporation process. Mineral sequences in evaporite formation can then be predicted by assuming either that chemical equilibrium is achieved at all times (equilibrium path) or that precipitated phases are no longer interacting with the solution (fractionation path). Comparisons of the model predictions with mineral sequences in well-studied evaporite deposits have demonstrated a close correspondence of the data with the equilibrium path calculations. Some existing discrepancies can be rationalized by considering that fractionation occurs between successive batches of equilibrium path evaporation.

2.2 Solubility of Hydroxides (and Oxides)

The formation and dissolution of hydroxide and oxide solids (oxides can be considered simply as dehydrated hydroxides) are important in the aquatic chemistry of many metal ions, particularly iron and manganese. Since the hydroxide ion concentration is given directly as a function of pH (the precipitation reactions can in fact be written with H^+ production rather than OH^- consumption), the mathematical problem is straightforward, the only complication being the existence of hydrolysis species such as $Fe(OH)_2^+$ and $MnOH^+$. These hydrolysis species often play a dominant role in controlling the solubility of hydroxide forming metals, and their equilibrium constants are thus listed along with those of the solids in Table 5.4.

Example 2. Solubility of Fe(OH)$_3$(s)

For a hydroxide solid such as $Fe(OH)_3$ the most convenient way to display the solubility as a function of concentration and pH is in the form of a log C-pH diagram as shown on Figure 5.1. This figure (which is very similar to Figure 3.4 of Chapter 3 where we ignored the polymeric species), is obtained by plotting the logarithmic form of the various mass laws for the hydrolysis of iron III. Choosing as the solubility product that of amorphous $Fe(OH)_3$(s) which is most applicable to a system containing a fresh iron precipitate, we obtain by rearranging the reactions of Table 5.4:

$$Fe^{3+} + 3H_2O = Fe(OH)_3(s) + 3H^+; \quad pK = 3.2 \tag{4}$$

$$FeOH^{2+} + 2H_2O = Fe(OH)_3(s) + 2H^+; \quad pK = 1.0 \tag{5}$$

$$Fe(OH)_2^+ + H_2O = Fe(OH)_3(s) + H^+; \quad pK = -2.5 \tag{6}$$

$$Fe(OH)_4^- + H^+ = Fe(OH)_3(s) + H_2O; \quad pK = -18.4 \tag{7}$$

$$Fe_2(OH)_2^{4+} + 4H_2O = 2Fe(OH)_3(s) + 4H^+; \quad pK = 3.5 \tag{8}$$

$$Fe_3(OH)_4^{5+} + 5H_2O = 3Fe(OH)_3(s) + 5H^+; \quad pK = 3.3 \tag{9}$$

The total iron III in solution at equilibrium with the solid $Fe(OH)_3$(s) is also the maximum concentration of soluble iron at any pH:

$$(Fe\cdot III)_{soluble} = (Fe^{3+}) + (Fe(OH)^{2+}) + (Fe(OH)_2^+) + (Fe(OH)_4^-)$$
$$+ 2(Fe_2(OH)_2^{4+}) + 3(Fe_3(OH)_4^{5+}) \tag{10}$$

For any given pH and analytical concentration of Fe(III), the graph of Figure 5.1 shows whether the solid $Fe(OH)_3$ is saturated or not and gives the concentrations of all Fe(III) species in solution. As can be seen in the figure the hydroxide complexes of Fe(III), not the free ion Fe^{3+}, account for most of the soluble iron concentration throughout the pH range of natural waters. As we shall see in the next chapter, however, other soluble complexes may dominate the hydroxide species. A graph such as Figure 5.1 is thus valid only when there is no important soluble species of Fe(III) other than the hydroxide complexes; in particular, there must be no strong organic Fe(III) complexing agent in the system.

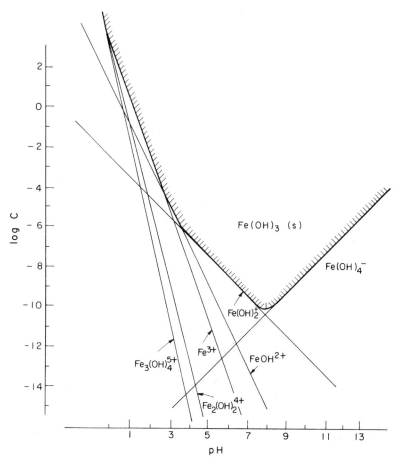

Figure 5.1 Log C-pH diagram for Fe(III) at equilibrium with amorphous $Fe(OH)_3(s)$, obtained by plotting the logarithm of Equations 4 through 9 (Example 2). Any system whose total Fe(III) concentration (log C) and pH yield a point within the hatched area of the graph is supersaturated with respect to am.$Fe(OH)_3$; at equilibrium the soluble Fe(III) and the pH of such a system must yield a point on the saturation line of the diagram. Note the importance of the polymeric hydroxo species at low pH and the increase in solubility at high pH due to $Fe(OH)_4^-$ formation. If a more stable form of ferric hydroxide were considered (see Table 5.3), the whole diagram would be translated to lower log C values.

In utilizing Figure 5.1, one should bear in mind that ferric salts themselves are strong acids. While pH can be varied independently of $(Fe \cdot III)_T$ by addition of strong base or strong acid, an experimental change in $(Fe \cdot III)_T$ typically leads to a pH variation. Consider, for example, the recipe

$$(FeCl_3)_T = 2 \times 10^{-3} \, M$$

For simplicity let us ignore ionic strength effects and possible iron chloride complexes (see Chapter 6). Guessing that ferric hydroxide precipitates, we have a straightforward choice of principal components: H^+, Cl^-, $Fe(OH)_3$. Thus

$$TOTH = (H^+) - (OH^-) + 3(Fe^{3+}) + 2(FeOH^{2+}) + (Fe(OH)_2^+)$$
$$- (Fe(OH)_4^-) + 4(Fe_2(OH)_2^{4+}) + 5(Fe_3(OH)_4^{5+}) = 3(FeCl_3)_T$$
$$= 6 \times 10^{-3} \, M \tag{11}$$

All the terms in this equation have an explicit expression in (H^+):

$$(H^+) - 10^{-14}(H^+)^{-1} + 3 \times 10^{3.2}(H^+)^3 + 2 \times 10^{1.0}(H^+)^2 + 10^{-2.5}(H^+)$$
$$- 10^{-18.4}(H^+)^{-1} + 4 \times 10^{3.5}(H^+)^4 + 5 \times 10^{3.3}(H^+)^5 = 6 \times 10^{-3} \tag{12}$$

Solving this polynomial by trial and error leads to

$$pH = 2.31$$

A pure ferric chloride solution is thus strongly acidic, and at such a low pH the iron is quite soluble (see Figure 5.1):

$$(Fe^{3+}) = 1.86 \times 10^{-4} \, M$$
$$(FeOH^{2+}) = 2.40 \times 10^{-4} \, M$$
$$(Fe(OH)_2^+) = 1.55 \times 10^{-5} \, M$$

therefore

$$(Fe \cdot III)_{soluble} = 4.42 \times 10^{-4} \, M$$

At higher concentrations of ferric salts the pH is in fact so low that the solid actually—and paradoxically—dissolves. The pH below which the solid dissolves can be calculated by taking the difference between the $TOTFe(OH)_3$ and $TOTH$ equations [neglecting (OH^-), $Fe(OH)_4^-$, and $Fe_3(OH)_4^{5+}$ which are small at the pH of interest]:

$$TOTH = (H^+) + 3(Fe^{3+}) + 2(FeOH^+) + (Fe(OH_2)^+) + 4(Fe_2(OH)_2^{4+})$$
$$= 3(FeCl_3)_T$$
$$TOTFe = (Fe(OH)_3 \cdot s) + (Fe^{3+}) + (FeOH^{2+}) + (Fe(OH)_2^+) + 2(Fe_2(OH)_2^{4+})$$
$$= (FeCl_3)_T$$

therefore

$$3TOTFe - TOTH = 3(Fe(OH)_3 \cdot s) + (FeOH^{2+}) + 2(Fe(OH)_2^+)$$
$$+ 2(Fe_2(OH)_2^{4+}) - (H^+) = 0$$

and
$$3(Fe(OH)_3 \cdot s) = (H^+) - (FeOH^{2+}) - 2(Fe(OH)_2^+) - (2Fe_2(OH)_2^{4+})$$

The solid dissolves when the solid concentration becomes negative:

$$(Fe(OH)^{2+}) + 2(Fe(OH)_2^+) + 2(Fe_2(OH)_2^{4+}) \geq (H^+)$$

therefore

$$10^{1.0}(H^+)^2 + 2 \times 10^{-2.5}(H^+) + 2 \times 10^{3.5}(H^+)^4 \geq (H^+)$$

and

$$(H^+) > 10^{-1.35} \, M$$

Introducing the value of (H^+) into the $TOTFe$ equation with substituted mass law expressions gives the maximum $FeCl_3$ concentration above which the $Fe(OH)_3$ dissolves:

$$(FeCl_3)_T \geq 10^{3.2} \times 10^{-4.05} + 10^{1.0} \times 10^{-2.7} + 10^{-2.5} \times 10^{-1.35}$$
$$+ 2 \times 10^{3.5} \times 10^{-5.4}$$

therefore

$$(FeCl_3)_T \geq 0.18 \, M$$

2.3 Solubility of Carbonates, Sulfides, and Phosphates

Studying the precipitation of hydroxide solids as a function of pH is a relatively straightforward exercise since the hydroxide concentration is known immediately at any given pH. Such is not the case with other ligands such as carbonate, sulfide, or phosphate, whose free concentrations depend not only on pH but also on the total ligand concentration and perhaps on the concentration of the precipitating metal (if it is present in excess of the ligand). In the next few examples we examine the different possible situations for precipitation of a metal with a weak acid ligand. For consistency among the examples, calcite $(CaCO_3)$ is used throughout as the precipitating solid, but the concepts and the methodology are applicable to all other carbonates, sulfides, and phosphates. (FeS, a dominant solid in anoxic—i.e., oxygen free—systems is used extensively as an example in the next section.)

   The easiest case of carbonate precipitation to study is that where the system is at equilibrium with a fixed partial pressure of CO_2. In such a case the free ligand concentration is given explicitly as a function of pH, and simple numerical or graphical solutions are obtained.

Example 3.   Solubility of $CaCO_3$ at Fixed P_{CO_2}

Consider the typical partial pressure of carbon dioxide in the atmosphere: $P_{CO_2} = 10^{-3.5}$ at. From the mass laws of the carbonate system, the free carbonate concentration can be expressed as a function of pH:

$$(CO_3^{2-}) = \frac{10^{-6.3}}{(H^+)} \times \frac{10^{-10.3}}{(H^+)} \times 10^{-1.5}P_{CO_2} = 10^{-21.6}(H^+)^{-2} \qquad (13)$$

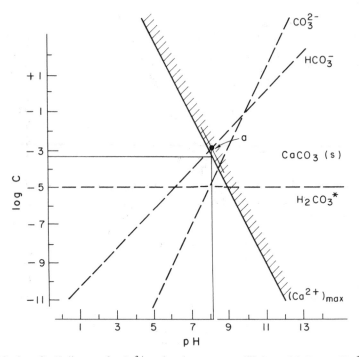

Figure 5.2   Log C-pH diagram for Ca^{2+} and carbonate at equilibrium with $P_{CO_2} = 10^{-3.5}$ at and $CaCO_3(s)$ (Example 3). A system whose calcium concentration and pH yield a point in the hatched area of the graph is supersaturated with respect to $CaCO_3$ (calcite); at equilibrium the free calcium concentration and the pH must yield a point on the saturation line of the diagram. [The solution of Example 3 is given by point a which satisfies the condition $(HCO_3^-) = 2(Ca^{2+})$. The composition of the system is then obtained from the intersection of the corresponding vertical line with the graphs of the various species.]

When calcite precipitates, the free calcium ion also becomes an explicit function of pH:

$$(Ca^{2+})(CO_3^{2-}) = K_s = 10^{-8.3} \tag{14}$$

therefore

$$(Ca^{2+})_{max} = 10^{+13.3}(H^+)^2 \tag{15}$$

For any pH this result, illustrated in Figure 5.2 as a log C-pH diagram, gives the maximum concentration of calcium that can be added to a solution before saturation of calcite under the given partial pressure of CO_2. Note that no soluble calcium species other than Ca^{2+} is assumed to be present here.

For example, consider the recipe

$$P_{CO_2} = 10^{-3.5} \text{ at}$$

$$(CaCO_3)_T = 10^{-2.0} \ M$$

With the assumption that $CaCO_3$ precipitates, the principal components H^+, CO_2, and $CaCO_3$ lead to the equation:

$$TOTH = (H^+) - (OH^-) - (HCO_3^-) - 2(CO_3^{2-}) + 2(Ca^{2+}) = 0 \tag{16}$$

Graphically (see point **a** on Figure 5.2) the solution of this equation is given by

$$2(Ca^{2+}) = (HCO_3^-)$$

so that

$$2 \times 10^{+13.3}(H^+)^2 = \frac{10^{-6.3}}{(H^+)} 10^{-1.5} \times 10^{-3.5}$$

$$pH = 8.3$$

$$(Ca^{2+}) = 10^{-3.3} \ M$$

$$TOTCa = (Ca^{2+}) + (CaCO_3 \cdot s) = 10^{-2.0}$$

and thus

$$(CaCO_3 \cdot s) = 10^{-2.0} - 10^{-3.3} = 9.5 \times 10^{-3} \ M$$

For concentrations of $(CaCO_3)_T$ less than $10^{-3.3} \ M$, the solid would dissolve.

Calcium carbonate systems not in equilibrium with a fixed partial pressure of CO_2 are slightly more difficult to study since the free carbonate concentration is then not a function of pH only; (CO_3^{2-}) depends also on the total soluble carbonate which itself depends on the degree of precipitation. In the next two examples we consider systems with given total calcium and carbonate concentrations and with a fixed pH that may be varied arbitrarily by adding a strong acid or a strong base. The problem is to find out whether there is enough calcium and carbonate for calcite precipitation at any pH and what the resulting composition of the system will be.

Example 4. Solubility of $CaCO_3(s)$ Given $(CO_3)_T > (Ca)_T$

$$(Ca)_T = 10^{-4} \ M$$

$$(CO_3)_T = 10^{-2} \ M$$

Since the total carbonate in this system is 100 times larger than the total calcium, the speciation of carbonate is minimally affected by the presence of calcium, whether calcite precipitates or not. The various carbonate species are thus plotted in the usual manner in Figure 5.3.

For calcium, on the other hand, if the solid does not precipitate,

$$(Ca^{2+}) = (Ca)_T = 10^{-4} \ M \tag{17}$$

If the solid precipitates,

$$(Ca^{2+}) = \frac{K_s}{(CO_3^{2-})} = \frac{10^{-8.3}}{(CO_3^{2-})} \tag{18}$$

Both of the (Ca^{2+}) versus pH graphs corresponding to Equations 17 and 18 are plotted on Figure 5.3. At any pH, the lower of these two graphs [smaller

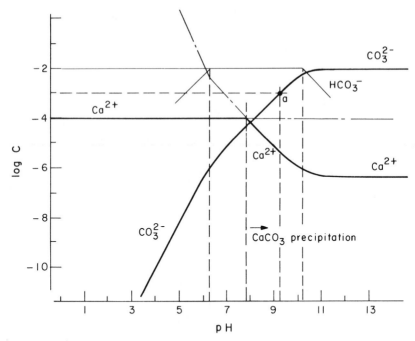

Figure 5.3 Log C-pH diagram for Ca^{2+} and carbonate with $(CO_3)_T = 10^{-2}$ M and $Ca_T = 10^{-4}$ M (Example 4). Two graphs for calcium are obtained, one by considering no solid species and the other by assuming precipitation of $CaCO_3(s)$. The critical pH for $CaCO_3(s)$ precipitation is obtained by the intersection of these two calcium graphs, and the equilibrium free calcium ion concentration as a function of pH is given by the lower of the two graphs (heavy line).

$(Ca^{2+})]$ gives the correct calcium concentration at equilibrium. This is because the solubility product expression,

$$(Ca^{2+}) = \frac{K_s}{(CO_3^{2-})} = \frac{10^{-8.3}}{(CO_3^{2-})}$$

gives both the maximum concentration of calcium in solution and the minimum concentration of calcium necessary to obtain a precipitate at a given free carbonate concentration. Thus at low (CO_3^{2-}) there is insufficient calcium to meet the solubility product, precipitation does not occur, and (Ca^{2+}) is controlled by the total calcium concentration. At higher (CO_3^{2-}), calcium is abundant enough to allow precipitation, and (Ca^{2+}) is controlled by the solubility product expression.

The critical pH above which the solid precipitates is obtained by equating the two expressions for (Ca^{2+}), Equations 17 and 18:

$$\frac{10^{-8.3}}{(CO_3^{2-})} = (Ca^{2+}) = (Ca)_T = 10^{-4} \tag{19}$$

According to Figure 5.3, the two graphs intersect in the region where HCO_3^-

is the major carbonate species, thus providing an expression for (CO_3^{2-}):

$$(CO_3^{2-}) = 10^{-10.3} \frac{(HCO_3^-)}{(H^+)} = 10^{-10.3} \frac{(CO_3)_T}{(H^+)} = \frac{10^{-12.3}}{(H^+)}$$

and

$$\frac{10^{-8.3}}{10^{-12.3}} (H^+) = 10^{-4}$$

$$pH_{crit} = 8.0$$

Given a complete recipe, we can obtain the pH of the solution on the graph by examining the appropriate $TOTH$ equation, that corresponding to the principal components of the system. Consider, for instance, the following recipe for Example 4:

$$(CaCO_3)_T = 10^{-4} \ M$$

$$(Na_2CO_3)_T = 10^{-3} \ M$$

$$(NaHCO_3)_T = 8.9 \times 10^{-3} \ M \quad \left[\text{to give } (CO_3)_T = 10^{-2} \ M \right]$$

Guessing that the solid does precipitate and thus choosing the principal components, H^+, HCO_3^-, $CaCO_3$, and Na^+, we obtain the appropriate $TOTH$ equation:

$$TOTH = (H^+) - (OH^-) + (H_2CO_3^*) - (CO_3^{2-}) + (Ca^{2+})$$

$$= -(Na_2CO_3)_T = -10^{-3} \ M \tag{20}$$

This equation is readily simplified to $(CO_3^{2-}) = 10^{-3} \ M$, which leads to pH $= 9.3$ graphically; see point **a** in Figure 5.3.

The methodology used in this example is typical of that used for all such problems:

1 Graph species for the component in excess (metal or ligand).
2 Graph species for the less abundant components under both hypotheses of the presence and absence of solid.
3 Take the "lowest lines" of the less abundant component species as the correct species concentration graph.
4 Obtain the critical pH (the threshold of precipitation) by equating the expressions for the less abundant component under conditions of precipitation and no precipitation (in the diagram the pH is given by the intersection of the two corresponding graphs).

Note that this methodology breaks down when both components involved in the precipitation of a solid are present in approximately the same concentration. In such a case one may use the methodology to obtain the critical precipitation pH. Below that pH, in the absence of precipitation, the mole balance equations ($TOTCa$ and $TOTCO_3$ in our example) are independent of each other and provide the correct solution regardless of their relative numerical values. However, to obtain the solution composition in the precipitation region where the solid itself must be chosen as a component, it becomes necessary to consider simultaneously several species and their corresponding mass laws, as illustrated in the next example.

Example 5.   Solubility of $CaCO_3(s)$ Given $(Ca)_T \cong (CO_3)_T$

$$(Ca)_T = 10^{-3.2}\ M$$

$$(CO_3)_T = 10^{-3.3}\ M$$

Following the methodology presented earlier, we first plot the free calcium concentration on the graph as if no solid precipitated (Figure 5.4):

$$TOTCa = (Ca^{2+}) = (Ca)_T = 10^{-3.2}\ M \tag{21}$$

For (CO_3^{2-}), if $CaCO_3(s)$ precipitates,

$$(CO_3^{2-}) = \frac{10^{-8.3}}{(Ca^{2+})} = 10^{-5.1}\ M \tag{22}$$

If $CaCO_3(s)$ does not precipitate, (CO_3^{2-}) is given by the usual log C-pH diagram. In particular, in the region where HCO_3^- is the major carbonate species,

$$(CO_3^{2-}) = 10^{-10.3}\frac{(HCO_3^-)}{(H^+)} = 10^{-10.3}\frac{(CO_3)_T}{(H^+)} = \frac{10^{-13.6}}{(H^+)} \tag{23}$$

The lowest of the two (CO_3^{2-}) graphs in Figure 5.4 is the pertinent one. The two lines intersect in the HCO_3^- region; the critical pH is then obtained by equating Equations 22 and 23:

$$10^{-5.1} = \frac{10^{-13.6}}{(H^+)}$$

therefore

$$pH_{crit} = 8.5$$

Above this pH the precipitation of $CaCO_3$ will markedly decrease the concentrations of all species in solution. Consider the $TOTCa$ equation corresponding to the principal components H^+, Ca^{2+}, and $CaCO_3$:

$$TOTCa = (Ca^{2+}) - (CO_3^{2-}) - (HCO_3^-) - (H_2CO_3^*) = (Ca)_T - (CO_3)_T$$

$$= 10^{-3.89} \tag{24}$$

All concentrations can be expressed as a function of (CO_3^{2-}) and (H^+):

$$\frac{10^{-8.3}}{(CO_3^{2-})} - (CO_3^{2-}) - 10^{10.3}(H^+)(CO_3^{2-}) = 10^{-3.89} \tag{25}$$

The graph of this implicit function of (H^+) is shown on Figure 5.4. The equivalent

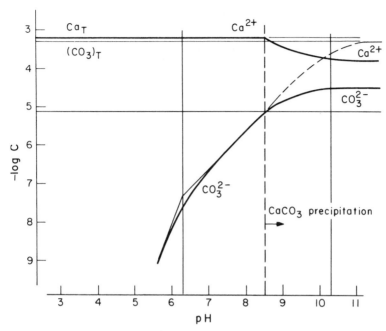

Figure 5.4 Log C-pH diagram for Ca^{2+} and carbonate with $(CO_3)_T = 10^{-3.3}$ M and $Ca_T = 10^{-3.2}$ M (Example 5). The critical pH for $CaCO_3(s)$ precipitation is obtained by the intersection of the two (CO_3^{2-}) graphs according to the general methodology. Above this critical pH a significant fraction of both calcium and carbonate are precipitated as $CaCO_3$; the free calcium and carbonate concentrations are then obtained as the solution of the quadratic equation shown in Example 5.

expression for (Ca^{2+}) is also plotted on the figure:

$$(Ca^{2+}) - \frac{10^{-8.3}}{(Ca^{2+})} - 10^{2.0}\frac{(H^+)}{(Ca^{2+})} = 10^{-3.89} \tag{26}$$

At very high pH, when (HCO_3^-) is negligible, (CO_3^{2-}) and (Ca^{2+}) are obtained by solving a simple quadratic:

$$\frac{10^{-8.3}}{(CO_3^{2-})} - (CO_3^{2-}) = 10^{-3.89}$$

therefore

$$(CO_3^{2-}) = 10^{-4.50}$$

$$(Ca^{2+}) = 10^{-3.80}$$

Although the approximate equality of the total metal and ligand concentrations make this example more complicated than the previous ones, the appropriate choice of principal components does lead to a reasonably simple solution.

Analytical information for natural waters is usually limited to the total concentrations in the aqueous phase, for example, Ca_T, C_T, and Alk. If we want to know whether or not $CaCO_3$ is saturated in such systems, the problem to be solved is slightly different from, and in fact easier than, those we have considered so far. In the following two examples we show how saturating concentrations of calcium can be calculated, given C_T and pH or alkalinity and pH.

Example 6.   Maximum Ca^{2+} Concentration Given C_T

A direct solution to this problem is obtained as a function of pH by using the ionization fraction expression for the free carbonate (see Example 9 of Chapter 3):

$$C_T = (H_2CO_3^*) + (HCO_3^-) + (CO_3^{2-}) \tag{27}$$

$$C_T = [10^{16.6}(H^+)^2 + 10^{10.3}(H^+) + 1](CO_3^{2-}) \tag{28}$$

$$(CO_3^{2-}) = C_T[10^{16.6}(H^+)^2 + 10^{10.3}(H^+) + 1]^{-1} \tag{29}$$

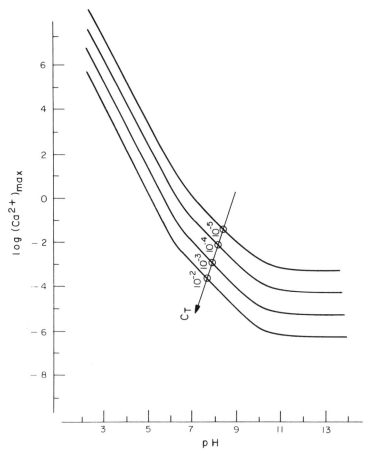

Figure 5.5   Log (Ca^{2+}) versus pH diagram for solutions saturated with calcium carbonate (calcite) at various concentrations of total dissolved carbonate, C_T.

Then using the solubility product of $CaCO_3(s)$, we solve for (Ca^{2+}):

$$(Ca^{2+})_{max} = \frac{10^{-8.3}}{(CO_3^{2-})} \tag{30}$$

$$(Ca^{2+})_{max} = \frac{10^{-8.3}}{C_T} \left[10^{16.6}(H^+)^2 + 10^{10.3}(H^+) + 1 \right] \tag{31}$$

This formula is plotted in Figure 5.5 for various values of C_T.

Example 7.   Maximum Ca^{2+} Concentration Given Alk

This problem also has an explicit solution in (H^+). For carbonate dominated systems

$$Alk = -(H^+) + (OH^-) + (HCO_3^-) + 2(CO_3^{2-}) \tag{32}$$

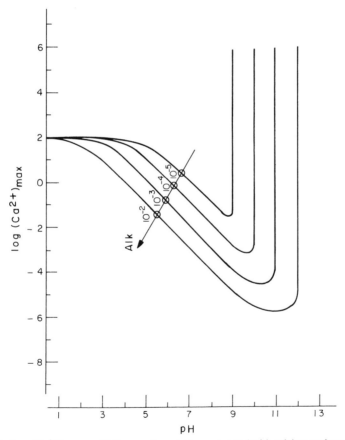

Figure 5.6   Log (Ca^{2+}) versus pH diagram for solutions saturated with calcium carbonate (calcite) at various alkalinities. In this graph the alkalinity is attributed exclusively to the bicarbonate, carbonate, and hydroxide, minus hydrogen, ions. At high pH where the hydroxide accounts entirely for the alkalinity of the system, there is no carbonate in the water, so $CaCO_3(s)$ cannot precipitate.

therefore

$$(CO_3^{2-}) = \frac{Alk + (H^+) - 10^{-14}/(H^+)}{2 + 10^{10.3}(H^+)} \tag{33}$$

Then using the solubility product of $CaCO_3(s)$, we solve for (Ca^{2+}):

$$(Ca^{2+})_{max} = \frac{10^{-8.3}}{(CO_3^{2-})} \tag{34}$$

$$(Ca^{2+})_{max} = \frac{10^{-8} + 10^2(H^+)}{Alk + (H^+) - 10^{-14}/(H^+)} \tag{35}$$

This formula is plotted in Figure 5.6 for various values of Alk. Note that when the alkalinity of the system is entirely accounted for by the hydroxide ion concentration (OH^-), the system contains no carbonate, and the calcium concentration is unbounded.

3. COMPETITION AND COEXISTENCE AMONG SEVERAL SOLID PHASES

At high pH in a system containing calcium and carbonate the precipitation of calcium hydroxide can, in principle, become thermodynamically favorable:

$$Ca^{2+} + 2OH^- = Ca(OH)_2(s); \quad pK_s = 5.2$$

Which of the two possible solid phases considered ($CaCO_3$ and $Ca(OH)_2$) would then control the soluble calcium concentration as pH is increased? Can both solids coexist at equilibrium? Although these are the types of questions that we wish to address here, calcium provides a poor example because the precipitation of $Ca(OH)_2$ takes place at pH values outside of the range of interest for natural waters (pH > 12). As a more realistic simple example, let us consider the iron(II)-sulfide-carbonate system.

Example 8.   Competition between Sulfide and Carbonate Fe(II) Solids

To illustrate how two ligands may compete for precipitation of the same metal, let us consider a situation where both total ligand concentrations are in excess of the metal.

Recipe $(Fe \cdot II)_T = 10^{-6} M$
           $(S \cdot -II)_T = 10^{-5} M$
           $(CO_3)_T = 10^{-3} M$

Reactions
$H_2S = HS^- + H^+$;	$pK = 7.0$	(36)
$HS^- = S^{2-} + H^+$;	$pK = 13.9$	(37)
$H_2CO_3^* = HCO_3^- + H^+$;	$pK = 6.3$	(38)
$HCO_3^- = CO_3^{2-} + H^+$;	$pK = 10.3$	(39)
$FeS(s) = Fe^{2+} + S^{2-}$;	$pK_{s1} = 18.1$	(40)
$FeCO_3(s) = Fe^{2+} + CO_3^{2-}$;	$pK_{s2} = 10.7$	(41)

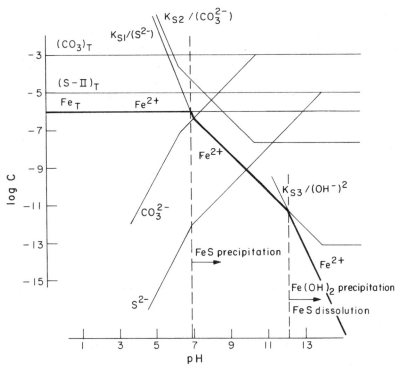

Figure 5.7 Log C-pH diagram for Fe(II) in a carbonate and sulfide bearing water with all ligands in excess: $(CO_3)_T = 10^{-3}$ M, $(S.-II)_T = 10^{-5}$ M, $Fe_T = 10^{-6}$ M (Example 8). The two intersections obtained among the four (Fe^{2+}) graphs [considering no precipitate, $FeCO_3(s)$, $FeS(s)$, and $Fe(OH)_2(s)$] delimit the pH domains of stability of the various solids. In this case $FeCO_3(s)$ is never stable, and the domains of stability of $FeS(s)$ and $Fe(OH)_2(s)$ have no overlap. The equilibrium free ferrous ion concentration is given by the lowest of the four graphs at each pH (heavy line).

Since the total concentrations of sulfide and carbonate are in great excess of the iron, the corresponding free ion concentrations are effectively independent of any precipitation; the graphs for (S^{2-}) and (CO_3^{2-}) can be plotted in the usual manner on a log C-pH diagram (Figure 5.7). Three possible cases then arise for the free ferrous ion concentration:

1 No solid precipitates

$$(Fe^{2+}) = (Fe \cdot II)_T = 10^{-6} \ M \qquad (42)$$

2 FeS(s) precipitates

$$(Fe^{2+}) = \frac{K_{s1}}{(S^{2-})} = \frac{10^{-18.1}}{(S^{2-})} \qquad (43)$$

3 FeCO₃(s) precipitates

$$(Fe^{2+}) = \frac{K_{s2}}{(CO_3^{2-})} = \frac{10^{-10.7}}{(CO_3^{2-})} \qquad (44)$$

Each of the corresponding three graphs is plotted on Figure 5.7. As in the previous examples (4 and 5), the lowest Fe^{2+} concentration graph is the correct one at any pH. As seen on Figure 5.7, a simple case of competition among the two solids arises: the stable solid at a given pH is the one that most "lowers" the ferrous ion concentration. In this particular case the ferrous carbonate solid never forms because the ferrous sulfide is too insoluble and it keeps (Fe^{2+}) below saturation of $FeCO_3$ at all pH's; that is, the graph for Equation 43 is always lower than that for Equation 44.

The critical pH at which FeS precipitates is obtained in the usual way by equating Equations 42 and 43. According to Figure 5.7, this point is in the H_2S region:

$$10^{-6.0} = \frac{10^{-18.1}}{(S^{2-})} = \frac{10^{-18.1} \times (H^+)^2}{10^{-13.9} \times 10^{-7.0} \times (S \cdot -II)_T} \tag{45}$$

therefore

$$pH_{crit} = 6.9$$

This value is a bit low because it is very close to the upper limit of the H_2S region. (If Figure 5.7 were more exact, the graph of Equation 43 would be rounded near the pH = 7 transition.) Expressing the free sulfide by its ionization fraction,

$$(S^{2-}) = (S \cdot -II)_T [10^{20.9}(H^+)^2 + 10^{13.9}(H^+)]^{-1} \tag{46}$$

yields the more exact answer:

$$pH_{crit} = 7.07.$$

For the sake of completeness, let us consider the possible precipitation of the ferrous hydroxide solid that we have neglected so far in this example,

$$Fe(OH)_2(s) = Fe^{2+} + 2OH^-; \quad pK_{s3} = 15.1 \tag{47}$$

When this solid precipitates, the free ferrous ion concentration is given by

$$(Fe^{2+}) = 10^{12.9}(H^+)^2 \tag{48}$$

The corresponding graph shown on Figure 5.7 is lower than any other (Fe^{2+}) graph only at very high pH. The critical pH above which $Fe(OH)_2(s)$ precipitates and FeS(s) dissolves is obtained by equating Equations 43 and 48; in the HS^- region

$$10^{12.9}(H^+)^2 = \frac{10^{-18.1}}{(S^{2-})} = \frac{10^{-18.1}(H^+)}{10^{-13.9}(S \cdot -II)_T} \tag{49}$$

therefore

$$pH'_{crit} = 12.1$$

Example 9.   Coexistence of Sulfide and Carbonate Fe(II) Solids

In Example 8 both solids, FeS(s) and $FeCO_3(s)$, cannot coexist because, from low to high pH, the free ferrous ion (Fe^{2+}) is effectively controlled, first by its total concentration and then by sulfide precipitation below the saturation value for carbonate. In order for the carbonate solid to form, there must be a large (Fe^{2+}) concentration. When FeS precipitates, the carbonate solid will form only if the total iron concentration is in excess of the total sulfide.

Recipe $(Fe \cdot II)_T = 10^{-4} \ M$
                 $(S \cdot -II)_T = 10^{-5} \ M$
                 $(CO_3)_T = 10^{-3} \ M$

Reactions   Same as Example 8.

Following our usual method, we plot the free ion concentration graphs in the order of the total concentrations in the system. First, the free carbonate ion concentration (CO_3^{2-}) will not be affected much by any precipitation since the total carbonate is 10 times in excess of anything else (Figure 5.8). Second, the

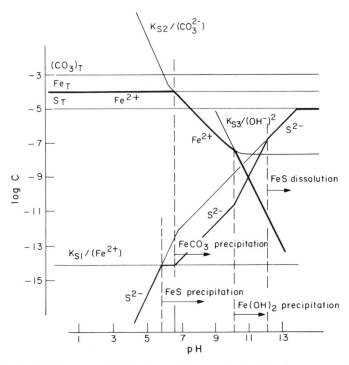

Figure 5.8 Log C-pH diagram for Fe(II) in a carbonate and sulfide bearing water with the metal in excess of the sulfide: $(CO_3)_T = 10^{-3} \ M$, $(S. - II)_T = 10^{-5} \ M$, $Fe_T = 10^{-4} \ M$ (Example 9). The diagram is obtained according to the general methodology by first ignoring the relatively low sulfide concentration. This yields the domains of stability of $FeCO_3(s)$ and $Fe(OH)_2(s)$ (which do not overlap) and the graphs of the free ferrous ion concentration (heavy line). The domain of stability of FeS(s) is then obtained from the intersections of the various (S^{2-}) graphs [considering no solid, FeS(s) and Fe^{2+}, FeS(s) and $FeCO_3(s)$, and FeS(s) and $Fe(OH)_2(s)$]. Because the total iron is in excess of the total sulfide, the domain of stability of FeS(s) can overlap with that of the other two solids.

free ferrous ion can only be slightly affected by sulfide, and the graphs for (Fe^{2+}) are obtained as usual.

1   No precipitate forms

$$(Fe^{2+}) = (Fe \cdot II)_T = 10^{-4} \, M \tag{50}$$

2 $FeCO_3$ precipitates

$$(Fe^{2+}) = \frac{K_{s2}}{(CO_3^{2-})} = \frac{10^{-10.7}}{(CO_3^{2-})} \tag{51}$$

The correct (Fe^{2+}) graph is the lowest one on Figure 5.8, and the critical pH for $FeCO_3$ precipitation is obtained by equating Equations 50 and 51. In the HCO_3^- region

$$10^{-4} = \frac{10^{-10.7}}{(CO_3^{2-})} = \frac{10^{-10.7}(H^+)}{10^{-10.3}(CO_3)_T} \tag{52}$$

therefore

$$pH_{crit} = 6.6$$

We can turn now to the question of the speciation of the trace sulfide that we have heretofore neglected.

If FeS does not precipitate, (S^{2-}) is obtained from the usual sulfide graph (see Figure 5.8). In the H_2S region

$$(S^{2-}) = \frac{10^{-13.9} \times 10^{-7.0}(S \cdot -II)_T}{(H^+)^2} = \frac{10^{-25.9}}{(H^+)^2} \tag{53}$$

If FeS precipitates,

$$(S^{2-}) = \frac{K_{s1}}{(Fe^{2+})} = \frac{10^{-18.1}}{(Fe^{2+})} \tag{54}$$

In Equation 54, (Fe^{2+}) is given by either Equation 50 or 51, whichever yields the lower concentration at any given pH. For pH < 6.6

$$(S^{2-}) = \frac{10^{-18.1}}{(Fe^{2+})} = \frac{10^{-18.1}}{10^{-4.0}} = 10^{-14.1} \tag{55}$$

The correct (S^{2-}) graph is the lowest one in Figure 5.8, and we obtain the critical pH for FeS precipitation by equating Equations 53 and 54 (guessing from the graph that the critical pH is less than 6.6):

$$\frac{10^{-25.9}}{(H^+)^2} = 10^{-14.1} \tag{56}$$

therefore

$$pH'_{crit} = 5.9$$

Above this pH, FeS precipitates and controls the sulfide species concentrations. In the region where the ferrous ion concentration is controlled by precipitation of $FeCO_3$ (pH > 6.6), the free sulfide is then dependent on the free carbonate, in a sort of "second-order interaction":

$$(S^{2-}) = \frac{10^{-18.1}}{(Fe^{2+})} = \frac{10^{-18.1}}{10^{-10.7}}(CO_3^{2-}) = 10^{-7.4}(CO_3^{2-}) \tag{57}$$

Finally, for completeness, let us consider again the precipitation of $Fe(OH)_2(s)$:

$$(Fe^{2+}) = 10^{12.9}(H^+)^2 \tag{58}$$

Equating Equations 51 and 58, we find the critical pH above which $Fe(OH)_2(s)$ precipitates and $FeCO_3(s)$ dissolves; in the HCO_3^- region

$$10^{12.9}(H^+)^2 = \frac{10^{-10.7}}{(CO_3^{2-})} = \frac{10^{-10.7}(H^+)}{10^{-10.3}(CO_3)_T} \tag{59}$$

therefore

$$pH''_{crit} = 10.3$$

Again, this pH is too close to the pK to be accurate; (CO_3^{2-}) cannot be neglected. The more complete formula given by the ionization fraction,

$$(CO_3^{2-}) = (CO_3)_T[1 + 10^{10.3}(H^+)]^{-1} \tag{60}$$

yields

$$pH''_{crit} = 10.1$$

Also, as in Example 8, FeS(s) dissolves for pH above 12.1:

$$pH'''_{crit} = 12.1$$

The graph of Figure 5.8 is thus made up of 5 regions:

pH $<$ 5.9	Fe is totally in solution as Fe^{2+}.
5.9 $<$ pH $<$ 6.6	Fe is mostly present as Fe^{2+}, but FeS(s) precipitates.
6.6 $<$ pH $<$ 10.3	Fe is mostly present as $FeCO_3(s)$ which coexists with FeS(s).
10.3 $<$ pH $<$ 12.1	Fe is mostly present as $Fe(OH)_2(s)$ which coexists with FeS(s).
12.1 $<$ pH	Fe is precipitated as $Fe(OH)_2(s)$.

☐ ☐ ☐

The only difference between the last two examples is the concentration of total Fe(II) in the system. One may obtain a general description of the chemistry of Fe(II) for all combinations of Fe(II) total concentration and pH—all other conditions such as $(CO_3)_T$ and $(S\cdot{-}II)_T$ being fixed and given—in the form of a $\log(Fe\cdot II)_T$ versus pH diagram as shown in Figure 5.9. Such *solubility diagrams* are obtained by a generalization of the methodology described in the preceding examples. The boundaries of the various domains are derived by equating the controlling expressions for the major species in the neighboring areas, a not very difficult but often tedious process. For example, the various lines of Figure 5.9, where $(CO_3)_T = 10^{-3}$ M and $(S\cdot{-}II)_T = 10^{-5}$ M, are obtained in the following eight steps:

1. $Fe^{2+}/FeS(s)$ line (southwestern boundary of FeS stability domain). Equating $(Fe^{2+}) = (Fe\cdot II)_T$ and $(Fe^{2+}) = K_{s1}/(S^{2-})$ and expressing (S^{2-}) as a

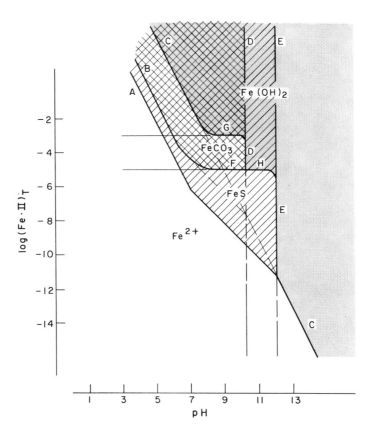

Figure 5.9 Solubility diagram for Fe(II) in the presence of carbonate (10^{-3} M) and sulfide (10^{-5} M). The boundaries of the domains of stability of the various solids are given by FeS(s): A and E; FeCO$_3$(s): B, F, and D; Fe(OH)$_2$(s): C, G, D, H, E, and C. Each solid domain is indicated on the diagram by a particular hatching to illustrate the domains of solid coexistence when the iron is in excess of the precipitating ligands and the strict competition among solids at low iron concentrations.

function of $(S \cdot -II)_T$ and pH yields line A:

$$\text{pH} < 7; \quad (\text{Fe} \cdot \text{II})_T = 10^{7.8}(\text{H}^+)^2 \tag{61}$$

$$\text{pH} > 7; \quad (\text{Fe} \cdot \text{II})_T = 10^{0.8}(\text{H}^+) \tag{62}$$

2. $\text{Fe}^{2+}/\text{FeCO}_3(s)$ line (southwestern boundary of FeCO_3 stability domain). Equating $(\text{Fe}^{2+}) = (\text{Fe} \cdot \text{II})_T$ and $(\text{Fe}^{2+}) = K_{s2}/(\text{CO}_3^{2-})$ and expressing (CO_3^{2-}) as a function of $(\text{CO}_3)_T$ and pH yields line B:

$$\text{pH} < 6.3; \quad (\text{Fe} \cdot \text{II})_T = 10^{8.9}(\text{H}^+)^2 \tag{63}$$

$$\text{pH} > 6.3; \quad (\text{Fe} \cdot \text{II})_T = 10^{2.6}(\text{H}^+) \tag{64}$$

3. $\text{Fe}^{2+}/\text{Fe(OH)}_2(s)$ line (southwestern boundary of Fe(OH)_2 stability domain). Equating $(\text{Fe}^{2+}) = (\text{Fe} \cdot \text{II})_T$ and $(\text{Fe}^{2+}) = K_{s3}/(\text{OH}^-)^2$ and expressing (OH^-) as a function of pH yields line C:

$$(\text{Fe} \cdot \text{II})_T = 10^{12.9}(\text{H}^+)^2 \tag{65}$$

4. $\text{FeCO}_3(s)/\text{Fe(OH)}_2(s)$ line (eastern boundary of FeCO_3 stability domain). Equating $(\text{Fe}^{2+}) = K_{s2}/(\text{CO}_3^{2-})$ and $(\text{Fe}^{2+}) = K_{s3}/(\text{OH}^-)^2$ yields line D:

$$\text{pH} = 10.3 \quad \text{(as in Example 9)} \tag{66}$$

5. $\text{FeS}(s)/\text{Fe(OH)}_2(s)$ line (eastern boundary of FeS stability domain). Equating $(\text{Fe}^{2+}) = K_{s1}/(\text{S}^{2-})$ and $(\text{Fe}^{2+}) = K_{s3}/(\text{OH}^-)^2$ yields line E:

$$\text{pH} = 12.1 \quad \text{(as in Example 8)} \tag{67}$$

The remaining three lines correspond to boundaries where a ligand is exhausted by precipitation and a new iron solid forms. These are more difficult to obtain precisely than the others but easier to determine approximately. The methodology is described in some detail for the $\text{FeCO}_3(s)$ precipitation line corresponding to exhaustion of sulfide.

6. $\text{FeCO}_3(s)/\text{FeS}(s)$ line (southern boundary of FeCO_3 stability domain). In the FeS stability domain with FeS and HS^- as components, the relevant mole balance is written

$$TOTHS = (\text{HS}^-) - (\text{Fe}^{2+}) = (\text{S} \cdot -II)_T - (\text{Fe} \cdot \text{II})_T \tag{68}$$

When FeCO_3 precipitates, (Fe^{2+}) is given by

$$(\text{Fe}^{2+}) = \frac{K_{s2}}{(\text{CO}_3^{2-})} \tag{69}$$

and hence

$$(\text{S}^{2-}) = \frac{K_{s1}}{(\text{Fe}^{2+})} = \frac{(\text{CO}_3^{2-})K_{s1}}{K_{s2}} \tag{70}$$

Expressing (Fe^{2+}) and $(HS^-) = 10^{13.9}(H^+)(S^{2-})$ as a function of $(CO_3)_T$ and (H^+) in the $TOTHS$ equation then yields line F:

$$(Fe \cdot II)_T = 10^{-5.01} + 10^{2.6}(H^+) \tag{71}$$

$(Fe \cdot II)_T = (S \cdot -II)_T = 10^{-5.0}$ is a good intuitive approximation for this boundary since iron becomes available for precipitation with carbonate when it is in excess of the sulfide.

7. $Fe(OH)_2(s)/FeCO_3(s)$ line (southern boundary of $Fe(OH)_2$ stability domain). Using $(Fe^{2+}) = K_{s3}/(OH^-)^2$ and $(CO_3^{2-}) = K_{s2}/(Fe^{2+})$ in the $TOTHCO_3$ equation pertinent in the $FeCO_3$ domain $[(HCO_3^-) - (Fe^{2+}) = (CO_3)_T - (Fe \cdot II)_T]$ yields line G:

$$(Fe \cdot II)_T = 10^{-3.0} - \frac{10^{-13.3}}{(H^+)} + 10^{12.9}(H^+)^2 \tag{72}$$

$(Fe \cdot II)_T = (CO_3)_T = 10^{-3.0}$ is a good intuitive approximation.

8. $Fe(OH)_2(s)/FeS(s)$ line (southern boundary of $Fe(OH)_2$ stability domain). Using $(Fe^{2+}) = K_{s3}/(OH^-)^2$ and $(S^{2-}) = K_{s1}/(Fe^{2+})$ in the $TOTHS$ equation pertinent in the FeS domain $[(HS^-) - (Fe^{2+}) = (S \cdot II)_T - (Fe \cdot II)_T)]$ yields line H:

$$(Fe \cdot II)_T = 10^{-5.0} - \frac{10^{-17.1}}{(H^+)} + 10^{12.9}(H^+)^2 \tag{73}$$

Again $(Fe \cdot II)_T = (S \cdot -II)_T = 10^{-5}$ is a good intuitive approximation.

Note on Figure 5.9 that the stability domains of the various solids do not coincide with the regions of dominance, sections of the log $(Fe \cdot II)_T$ versus pH diagram where a solid accounts for most of the total iron concentration. For example, (Fe^{2+}) is the principal iron species in parts of the domains where $FeCO_3(s)$ and $FeS(s)$ precipitate, and there are several regions where the various solids coexist. The results of Examples 8 and 9 can be checked in Figure 5.9 by following the appropriate horizontal lines on the graph.

In general, if we consider the possible precipitation of one metal with several ligands, it is rather obvious from Example 8 that solids with ligands in excess of the metal cannot coexist at equilibrium. This is true in particular for hydroxides (there is effectively a concentration of 55.4 M of $(OH^-)_T$ in water) and usually for carbonates (with the exception of calcium and magnesium in systems such as seawater). Typically we can expect that either the hydroxide or the carbonate solid of a metal may precipitate but not both.

As illustrated in Examples 8 and 9, in anoxic waters the relative values of pH, $(S \cdot -II)_T$, $(CO_3)_T$, and $(Fe \cdot II)_T$ are of great importance in determining the sulfide chemistry and hence the chemistry of many trace metals that form insoluble sulfides. The free sulfide ion activity may be controlled by (1) $(S \cdot -II)_T$ and pH when there is no precipitate, (2) $(Fe \cdot II)_T$ when FeS precipitates but not $FeCO_3$, and (3) $(CO_3)_T$ and pH when both FeS and $FeCO_3$ precipitate [and assuming $(CO_3)_T > (S \cdot -II)_T$].

4. THE PHASE RULE

Determining the possible coexistence among solid phases in aquatic systems containing a large number of components can become quite complicated. An upper limit on the number of phases that can coexist at equilibrium is given by the "Gibbs phase rule," perhaps the best known but not the best understood basic theorem of chemical thermodynamics. First formulated by J. W. Gibbs in 1875,[14] the phase rule is a mere mathematical consequence of the conservation and energetic constraints in chemical systems at equilibrium. Here we shall state and illustrate the phase rule for the restricted conditions that we have considered so far—dilute aqueous solutions at a given pressure and temperature. In so doing, we hope to enhance our understanding of this old theorem of Gibbs and to widen the field of its applicability in aquatic systems.

Consider the first three rules we have set for choosing principal components: (1) H_2O; (2) H^+; (3) species of fixed activities such as gases at given partial pressures or solid phases. A question that readily arises is: Could it happen that there would be more species with fixed activities than there are independent components in the system? The answer to this question is our first statement of the phase rule: *The number of species with fixed activities* (*solids and gases at fixed partial pressure*) *must be smaller than the number of components in a chemical system at fixed pressure and temperature.* (Note that this statement of the rule is unaffected by whether water is or is not considered as a component; if we choose to consider water as an explicit component, we also have to consider its activity fixed at unity.)

As given here, the phase rule is reasonably intuitive. Since all species are either components or expressed as a function of components by mass laws, a system where all components have fixed activities is strictly invariant. The number of species with fixed activities then certainly cannot exceed the number of components; we can show that it has to be in fact smaller using a simple example to avoid the unnecessary complications and awkward notation of the general case. The demonstration consists of a reasoning *ab absurdo* in which we consider a system with as many solids and gases at fixed partial pressure as components and show that equilibrium cannot be achieved.

Example 10.   Maximum Number of Phases in the Ca-CO₃ System

Using our familiar calcium and carbonate system, let us ask the question: Can the solids calcium carbonate and calcium hydroxide coexist in water under a given partial pressure of carbon dioxide? According to our phase rule, this should be impossible since there would be three fixed activities ($\{CO_2\} = P_{CO_2}$, $\{CaCO_3\} = 1$, and $\{Ca(OH)_2\} = 1$) in a three component system (e.g., H^+, Ca^{2+}, CO_3^{2-}).

Consider, for example, that the system is defined by $P_{CO_2} = 10^{-3.5}$ at and the usual reactions:

$$CO_2(g) + H_2O = H_2CO_3^*; \qquad pK = 1.5 \qquad (74)$$

$$H_2CO_3^* = 2H^+ + CO_3^{2-}; \quad pK = 16.6 \qquad (75)$$

$$CaCO_3(s) = Ca^{2+} + CO_3^{2-}; \quad pK = 8.3 \qquad (76)$$

$$Ca(OH)_2(s) + 2H^+ = Ca^{2+} + 2H_2O; \quad pK = -22.8 \qquad (77)$$

By combination of these four reactions, we can write another reaction involving exclusively solids, gases, and water:

$$CO_2(g) + Ca(OH)_2(s) = CaCO_3(s) + H_2O; \quad pK = -13.0 \qquad (78)$$

It is clear that at the given partial pressure of CO_2 this reaction cannot be at equilibrium:

$$\frac{1}{P_{CO_2}} = 10^{+3.5} \neq 10^{+13.0}$$

The free energy of this reaction, $\Delta G = -RT \ln K - RT \ln P_{CO_2} = -12.9$ kcal mol^{-1}, is always negative, and the reaction must proceed to the right until either $CO_2(g)$ or $Ca(OH)_2(s)$ is exhausted:

$$CO_2(g) + Ca(OH)_2(s) \rightarrow CaCO_3(s) + H_2O$$

Note that, if the partial pressure of $CO_2(g)$ were allowed to drop in such a gedanken experiment, an equilibrium would be achieved at $P_{CO_2} = 10^{-13.0}$ atmosphere (a small partial pressure!). In that situation the reaction would have a free energy change of zero and could proceed indifferently in either direction. We can then acknowledge the notable exception that such thermodynamically degenerate situations provide to our phase rule: in addition to the number of components minus one, there *can* be any number of species with fixed activities if those can be formed from the others according to a reaction with zero-free energy change. This is an unlikely occurrence in aquatic systems which have temperatures and pressures fixed independently of their chemistry. The determination of the temperature-pressure domains where such coexistence of phases is possible is of course an important aspect of high-temperature geochemistry.

The reasoning that we just applied to the Ca-CO$_3$ system can be applied to any other system in which we hypothesize as many species with fixed activities as we have components. By induction, we then have a demonstration of the phase rule as stated here. Note that the same reasoning can also be applied to any *subset* of solids and gases at fixed partial pressure in any given system. We thus have effectively demonstrated a much more powerful statement of the phase rule: *Species with fixed activities must be independent (and can always be chosen as components)*.

For example, according to this rule, two solids with identical formulae such as calcite and aragonite ($CaCO_3$) cannot coexist at equilibrium no matter how many components there are, unless the reaction $CaCO_3$ (aragonite) = $CaCO_3$ (calcite) has a zero-free energy change, in which case the solids are thermodynamically indistinguishable.

Similarly, a reaction that involves exclusively solids, gases, and water (no solute) cannot be at equilibrium under arbitrary fixed partial pressures of the gases. The ability to write such a chemical reaction implies *ipso facto* that the solids and the gases are not independent components. The various species thus cannot coexist at equilibrium, and the reaction has to proceed completely in one direction.

For example, silica, aluminum oxide (Gibbsite), and kaolinite cannot coexist at equilibrium

$$2SiO_2(s) + Al_2O_3 \cdot 3H_2O(s) \rightarrow Al_2O_3 \cdot 2SiO_2 \cdot 2H_2O + H_2O \qquad (79)$$
$$\text{silica} \qquad \text{gibbsite} \qquad \text{kaolinite}$$

and neither can calcite, kaolinite, and anorthite under a fixed (arbitrary) pressure of CO_2

$$CaO \cdot Al_2O_3 \cdot 2SiO_2 + CO_2(g) + 2H_2O$$
$$\text{anorthite}$$

$$\rightarrow Al_2O_3 \cdot 2SiO_2 \cdot 2H_2O + CaCO_3(s) \quad (80)$$
$$\text{kaolinite} \qquad \text{calcite}$$

Thermodynamically, these reactions have to proceed all the way in one direction—to the right under ambient P_{CO_2}.

The CO_2 of the atmosphere must be buffered by some mechanism, however. One way that P_{CO_2} may be fixed geochemically is in fact through an equilibrium between some calcium (or manganese) silicate and carbonate phases similar to Reaction 80. The effective constant of such a reaction must then match the existing P_{CO_2} of the atmosphere.

5. THE ACQUISITION OF ALKALINITY IN FRESHWATER

The story of how freshwaters acquire their alkalinity is essentially the story of how freshwaters acquire their major dissolved cations, Ca^{2+}, Na^+, K^+, and Mg^{2+}, from basic rocks. This is because most of the important noncarbonate anions (Cl^- and SO_4^{2-}) in freshwaters originate from the dissolution of chloride and sulfate salts (e.g., NaCl and $CaSO_4$) which are not acidic and contribute no net negative alkalinity. The major exception to this is the dissolution of sulfide minerals such as pyrite (FeS_2) whose global contribution to freshwaters is difficult to estimate.

Following Garrels and MacKenzie,[2] consider, for example, the average composition of North American rivers. Table 5.5 shows how the major ions can be roughly accounted for in five steps:

1 The atmospheric contribution is estimated on the basis of rainwater composition.

2 All remaining chloride is taken to be halite dissolution:

$$NaCl \rightarrow Na^+ + Cl^- \tag{81}$$

3 The sulfate is attributed to calcium sulfate (gypsum):

$$CaSO_4 \cdot 2H_2O \rightarrow Ca^{2+} + SO_4^{2-} + 2H_2O \tag{82}$$

4 Magnesium and the balance of Ca^{2+} are ascribed to carbonate rock:

$$CaCO_3 + CO_2(g) + H_2O \rightarrow Ca^{2+} + 2HCO_3^- \tag{83}$$

$$MgCO_3 + CO_2(g) + H_2O \rightarrow Mg^{2+} + 2HCO_3^- \tag{84}$$

5 Finally, the silicate and the monovalent cations are taken to originate from the weathering of feldspars. In order to obtain the low ratio of H_2SiO_3 to the remaining $Na^+ + K^+$ (approximately 1 compared to about 3 in the rock), we can hypothesize re-precipitation of some of the silicic acid:

$$K_2O \cdot Al_2O_3 \cdot 6SiO_2(s) + 2CO_2(g) + 2.4H_2O$$
$$\text{K-feldspar}$$

$$\rightarrow 2K^+ + 2HCO_3^- + Al_2O_3 \cdot 2SiO_2(s)$$
$$\text{kaolinite}$$

$$+ 2.6SiO_2(s) + 1.4H_2SiO_3(aq) \tag{85}$$

and

$$Na_2O \cdot Al_2O_3 \cdot 6SiO_2(s) + 2CO_2(g) + 2.4H_2O$$
$$\text{albite}$$

$$\rightarrow 2Na^+ + 2HCO_3^- + Al_2O_3 \cdot 2SiO_2(s)$$
$$\text{kaolinite}$$

$$+ 2.6SiO_2(s) + 1.4H_2SiO_3(aq) \tag{86}$$

This very simplified genealogy of average North American river water shows that some 80% of the alkalinity is attributable to carbonate rock dissolution (it makes no difference how the Mg^{2+} and Ca^{2+} are distributed between sulfate and carbonate) and 20% to the weathering of alumino-silicate rocks. More detailed calculations based on different estimations for the world average river water[1] (with a 50% higher alkalinity) show only 60% of the alkalinity to be

TABLE 5.5

Synthesis of Average North American Freshwater from Rock Types[a,b]

	Average Water	From Atmosphere[c]	From Halite	From Gypsum	From Limestone	From Feldspars
HCO_3^-	1.25	0.04			1.01	0.20
SO_4^{2-}	0.42	0.03		0.39		
Cl^-	0.23	0.10	0.13			
Ca^{2+}	1.05	0.04		0.39	0.62	
Mg^{2+}	0.42	0.03			0.39	
Na^+	0.39	0.09	0.13			0.17
K^+	0.04	0.01				0.03
H_2SiO_3	0.15	0.01				0.14

[a] Source: Adapted from Garrels and MacKenzie (1971).[2]
[b] Note: The figures given are in milliequivalent per liter; SiO_2 is given in millimolar.
[c] From rainwater analysis. Does not include the CO_2 utilized in the weathering of limestone and feldspar.

contributed by dissolution of carbonates, most of the remainder originating from the dissolution of calcium and magnesium silicates and alumino-silicates.

The major interest in the sort of genealogical analysis shown in Table 5.5 is not merely to correlate average water and average rock composition but rather to understand the differences among natural waters, to relate a particular water chemistry to the regional mineralogy. The next four examples illustrate on the basis of simple equilibrium models how such important solids as carbonates, feldspars, montmorillonites, and sulfides contribute to the alkalinity of freshwaters. While these examples are certainly instructive and the results of the calculations are within the range of observations, they should not be taken too literally. It should be remembered that the thermodynamic properties of some of the actual mineral phases are not easily defined and that equilibrium may be a poor approximation of reality, particularly in well-drained areas where the residence time of the water in the soil is short.

Example 11. Dissolution of $CaCO_3(s)$ at a Given P_{CO_2}

As the first and simplest example let us consider the dissolution of calcium carbonate under a fixed partial pressure of CO_2.

Recipe

   1 $P_{CO_2} = 10^{-2.5}$ at.   (Note the high value typical of groundwater systems.)

   2 $CaCO_3(s)$ is present in equilibrium with the water.

The choice of principal components is straightforward, leading to Tableau 5.1:

TABLEAU 5.1

	H^+	CO_2	$CaCO_3$	pK
H^+	1			
OH^-	-1			14.0
HCO_3^-	-1	1		7.8
$H_2CO_3^*$		1		1.5
CO_3^{2-}	-2	1		18.1
Ca^{2+}	2	-1	1	-9.8
CO_2		1		
$CaCO_3$			1	

$$TOTH = (H^+) - (OH^-) - (HCO_3^-) - 2(CO_3^{2-}) + 2(Ca^{2+}) = 0 \quad (87)$$

The major terms are (HCO_3^-) and (Ca^{2+}):

$$10^{-7.8}(H^+)^{-1}P_{CO_2} = 2 \times 10^{9.8}(H^+)^2(P_{CO_2})^{-1} \quad (88)$$

$$(H^+)^3 = 10^{-17.9}(P_{CO_2})^2 = 10^{-22.9} \quad (89)$$

therefore

$$pH = 7.63$$

$$(Ca^{2+}) = 10^{9.8} \times 10^{-15.27} \times 10^{2.5} = 10^{-2.97} = 1.07 \times 10^{-3}$$

$$(Alk) = (HCO_3^-) = 2(Ca^{2+}) = 2.14 \times 10^{-3}$$

The alkalinity of such a system is typical of freshwaters in calcareous regions (and about twice average river water). A general relationship between alkalinity and P_{CO_2} for this system can be obtained:

$$Alk = (HCO_3^-) = 10^{-7.8}P_{CO_2}(H^+)^{-1} \quad (90)$$

Introducing $(H^+) = 10^{-5.97}(P_{CO_2})^{2/3}$ from Equation 89:

$$Alk = 10^{-1.83}P_{CO_2}^{1/3} \quad (91)$$

The process by which the water acquires its alkalinity in a situation like that of Example 11 can be represented as a single reaction:

$$CO_2(g) + CaCO_3(s) + H_2O = 2HCO_3^- + Ca^{2+} \quad (92)$$

Half of the carbon thus dissolved into the water originates from the rock and half from the gas phase. If the effective partial pressure of CO_2 is increased, the amount of carbonate rock that is dissolved also increases. As a result precipitation-dissolution of carbonate rock is emphatically not a homeostatic mechanism for the total carbon in the atmosphere and the hydrosphere. For example, the $CO_2(g)$ produced by fossil fuel burning cannot ultimately find its way into limestone according to the low-temperature geochemical processes of the carbonate system. In fact the historical increase in P_{CO_2} should ultimately result in an intensified weathering of carbonate rock on the continents and/or in a decrease of net carbonate sedimentation on the ocean floor.

As illustrated in Table 5.5, sodium and potassium in freshwaters originate principally from the weathering of alumino-silicates. The reaction usually involves the formation of another solid (it is then known as "incongruent dissolution") and has the general formula:

$$(Na, K) Al \cdot silicate + H^+ \rightarrow Na^+, K^+ + H_2SiO_3 + weathered\ Al \cdot silicate \quad (93)$$

The stoichiometry is of course variable, involving more or less water, and the hydrogen ion is usually provided by the dissolution and dissociation of CO_2. Let us first consider the weathering of an igneous rock, albite (pure Na-feldspar), to the common clay, Na-montmorillonite.

Example 12.   Weathering of Albite to Na-Montmorillonite

Recipe

1   $P_{CO_2} = 10^{-2.5}$ at.
2   Albite: $Na_2O \cdot Al_2O_3 \cdot 6SiO_2(s)$.
3   Na-montmorillonite: $Na_2O \cdot 7Al_2O_3 \cdot 22SiO_2 \cdot nH_2O(s)$.

From Table 5.4 we can write the reactions:

$$\tfrac{7}{20}albite + \tfrac{3}{5}H^+ + H_2O = \tfrac{1}{20}Na\text{-mtte} + H_2SiO_3$$

$$+ \tfrac{3}{5}Na^+; \quad pK = -0.43 \quad (94)$$

Choosing H^+ and Na^+ to round up the set of components yields Tableau 5.2:

TABLEAU 5.2

	H^+	Albite	Na-mtte	CO_2	Na^+	pK
H^+	1					
OH^-	-1					
$H_2CO_3^*$				1		1.5
HCO_3^-	-1			1		7.8
CO_3^{2-}	-2			1		18.1
Na^+					1	
H_2SiO_3	$\tfrac{3}{5}$	$\tfrac{7}{20}$	$-\tfrac{1}{20}$		$-\tfrac{3}{5}$	-0.43
$HSiO_3^-$	$-\tfrac{2}{5}$	$\tfrac{7}{20}$	$-\tfrac{1}{20}$		$-\tfrac{3}{5}$	9.43
CO_2				1		
Albite		1				
Na-mtte			1			

$$TOTH = (H^+) - (OH^-) - (HCO_3^-) - 2(CO_3^{2-})$$

$$+ \tfrac{3}{5}(H_2SiO_3) - \tfrac{2}{5}(HSiO_3^-) = 0 \quad (95)$$

$$TOTNa = (Na^+) - \tfrac{3}{5}(H_2SiO_3) - \tfrac{3}{5}(HSiO_3^-) = 0 \quad (96)$$

The major terms are (Na^+), (H_2SiO_3), and (HCO_3^-):

$$\tfrac{3}{5}(H_2SiO_3) = (HCO_3^-) \tag{97}$$

$$\tfrac{3}{5}(H_2SiO_3) = (Na^+) \tag{98}$$

Introducing the mass law for (H_2SiO_3) in Equation 98 leads to

$$\tfrac{3}{5} \times 10^{0.43}(H^+)^{3/5}(Na^+)^{-3/5} = (Na^+) \tag{99}$$

therefore

$$(Na^+) = [\tfrac{3}{5} \times 10^{0.43}(H^+)^{3/5}]^{5/8} = 10^{0.13}(H^+)^{3/8} \tag{100}$$

$$\tfrac{3}{5}(H_2SiO_3) = 10^{0.13}(H^+)^{3/8} \tag{101}$$

Substituting Equation 101 into Equation 97 and expressing the mass law for (HCO_3^-) yields

$$10^{0.13}(H^+)^{3/8} = 10^{-7.8} \times P_{CO_2}(H^+)^{-1} \tag{102}$$

therefore

$$(H^+)^{11/8} = 10^{-10.43}$$

$$pH = 7.59$$

$$(Na^+) = 10^{0.13}(10^{-7.59})^{3/8} = 10^{-2.71} = 1.9 \times 10^{-3}\ M$$

$$(Alk) = (Na^+) = (HCO_3^-) = 1.9 \times 10^{-3}\ M$$

The alkalinity of this system is slightly more than that of the average freshwater. The general relationship between alkalinity and P_{CO_2} can be written

$$(Alk) = 10^{-2.03} \times P_{CO_2}^{3/11} \tag{103}$$

The weathering of alumino-silicate usually proceeds further than the formation of montmorillonites. The cations can be completely dissolved, leading to the formation of kaolinite, the most thermodynamically stable of the clays.

Example 13.   Weathering of Na-Montmorillonite to Kaolinite

Recipe
  1 $P_{CO_2} = 10^{-2.5}$ at.
  2 Na-montmorillonite: $Na_2O \cdot 7Al_2O_3 \cdot 22SiO_2 \cdot nH_2O(s)$.
  3 Kaolinite: $Al_2O_3 \cdot 2SiO_2 \cdot 2H_2O(s)$.

From Table 5.4 we can write the reaction

$$\tfrac{1}{8}\text{Na-montmorillonite} + \tfrac{1}{4}H^+ + \tfrac{15}{8}H_2O = \tfrac{7}{8}\text{kaolinite} + H_2SiO_3$$

$$+ \tfrac{1}{4}Na^+; \quad pK = 2.27 \tag{104}$$

Choosing H^+ and Na^+ as components in addition to the solids and gases leads to Tableau 5.3:

TABLEAU 5.3

	H^+	Na-mtte	Kaolinite	CO_2	Na^+	pK
H^+	1					
OH^-	-1					
$H_2CO_3^*$				1		1.5
HCO_3^-	-1			1		7.8
CO_3^{2-}	-2			1		18.1
Na^+					1	
H_2SiO_3	$\frac{1}{4}$	$\frac{1}{8}$	$-\frac{7}{8}$		$-\frac{1}{4}$	2.27
$HSiO_3^-$	$-\frac{3}{4}$	$\frac{1}{8}$	$-\frac{7}{8}$		$-\frac{1}{4}$	12.13
CO_2				1		
Na-mtte	1					
Kaolinite			1			

$$TOTH = (H^+) - (OH^-) - (HCO_3^-) - 2(CO_3^{2-})$$

$$+ \tfrac{1}{4}(H_2SiO_3) - \tfrac{3}{4}(HSiO_3^-) = 0 \tag{105}$$

$$TOTNa = (Na^+) - \tfrac{1}{4}(H_2SiO_3) - \tfrac{1}{4}(HSiO_3^-) = 0 \tag{106}$$

With the usual approximations

$$(HCO_3^-) = \tfrac{1}{4}(H_2SiO_3) = (Na^+) \tag{107}$$

First equating (Na^+) and $\tfrac{1}{4}(H_2SiO_3)$

$$(Na^+) = \tfrac{1}{4} \times 10^{-2.27}(H^+)^{1/4}(Na^+)^{-1/4} \tag{108}$$

therefore

$$(Na^+) = \left[10^{-2.87}(H^+)^{1/4}\right]^{4/5} = 10^{-2.30}(H^+)^{1/5} \tag{109}$$

Equating now to (HCO_3^-) expressed from its mass law

$$10^{-2.30}(H^+)^{1/5} = 10^{-7.8}P_{CO_2}(H^+)^{-1}$$

$$(H^+)^{6/5} = 10^{-8.0} \tag{110}$$

therefore

$$pH = 6.67$$

$$(Na^+) = 10^{-2.3}(10^{-6.67})^{1/5} = 10^{-3.63} = 2.3 \times 10^{-4}\ M$$

$$Alk = (HCO_3^-) = (Na^+) = 2.3 \times 10^{-4}\ M$$

Note that the alkalinity of such a system is even less sensitive to the partial pressure of CO_2 than in the other examples:

$$Alk = 10^{-3.22} \times P_{CO_2}^{1/6} \tag{111}$$

Many more examples of such weathering reactions could be presented. They could also be combined in the same model to obtain rather representative compositions of freshwater systems. Even if the absolute concentrations of ions do not correspond to an equilibrium for the weathering reactions, the ratios of major elements dissolved into the water are usually good indicators of the minerals from which they originate.

As a drastic departure from these examples of weathering of aluminosilicate rocks, let us examine the case of an acid mine drainage: the acquisition of negative alkalinity (i.e., mineral acidity) from sulfide rocks.

Example 14.   Dissolution and Oxidation of Pyrite

In acid mine drainage sites pyrite, $FeS_2(s)$, is dissolved and oxidized by the oxygen-rich groundwater. We may schematize the process in two steps:

$$FeS_2(s) + \tfrac{7}{2}O_2 + H_2O \rightarrow Fe^{2+} + 2SO_4^{2-} + 2H^+ \tag{112}$$

$$Fe^{2+} + \tfrac{1}{4}O_2 + \tfrac{5}{2}H_2O \rightarrow Fe(OH)_3(s) + 2H^+ \tag{113}$$

The second of these processes which depends on the oxidation of ferrous to ferric iron is kinetically hindered at low pH. For the purpose of illustration let us consider the groundwater in the mine and assume that only the first of these two reactions, the oxidative dissolution of pyrite, takes place. Although an equilibrium calculation for such a reaction is mathematically possible, it is chemically meaningless. As we shall see in Chapter 7, the value of the equilibrium constant is so large that the reaction effectively proceeds to completion in the presence of oxygen. To allow a quantitative treatment of the problem, let us assume that the reaction is limited by the total concentration of oxygen originally present in the water, say,

$$(O_2)_T = 3 \times 10^{-4} M$$

The total amount of $FeS_2(s)$ reacted is then

$$3 \times 10^{-4} \times \tfrac{2}{7} = 10^{-4.07} M$$

Since the products of Reaction 112 are the principal components at the CO_2 equivalence point, the two protons that are produced correspond to two negative alkalinity equivalents (see Chapter 4). In other words the oxidative dissolution of pyrite is effectively adding $FeSO_4$ and H_2SO_4 to the system:

$$\Delta Alk = -2 \times 10^{-4.07} = -10^{-3.77} \text{ eq liter}^{-1}$$

If we assume that the groundwater originally had a zero alkalinity (unpolluted rainwater), this gives us directly the mineral acidity of the water:

$$H\text{-}Acy = -Alk = 10^{-3.77} = 1.7 \times 10^{-4} \text{ eq liter}^{-1}$$

As this mine drainage water mixes with more alkaline waters, it will contribute more than its calculated negative alkalinity to the mixture. When the pH is high enough for the kinetics to become favorable, the alkalinity (and the pH) will be reduced again, due to iron oxidation and precipitation, according to Reaction 113.

6. THE CONTROL OF ALKALINITY IN THE OCEANS

The question of the alkalinity of seawater can be addressed on two different time and space scales. First, considering geological times—over which the oceans are well mixed—one may ask what geochemical processes control the concentration of the major ions in seawater and thus the alkalinity of the oceans. Second, on a much finer scale one may want to examine the relatively small spatial and temporal variations in seawater alkalinity and to understand the causes of such variations.

6.1   Seawater Alkalinity over Geological Times

If we compare the major ion composition of the average river to that of seawater (Table 5.6), it is quite striking that seawater is not simply concentrated river water. The relative concentrations of the major cations is proportionally much lower in seawater than in rivers compared, for example, to chloride concentrations. As noted by Sillén,[16] we are very close to the equivalence point of the acidimetric titration of the oceans. What then are the oceanic mechanisms that remove cations in seawater more effectively than they remove anions? What are the alkalinity sinks that maintain the alkalinity of the ocean at its present value?

Chemical oceanography texts of yesteryear, which relied on experimental information before the discovery of the thermal, chemical, and biological activity at ridge crests, had few known alkalinity sinks available to them.[17] They listed

TABLE 5.6

Major Ion Composition of Average River and Seawater[a]

	Average River (mM)	Surface Seawater (mM)
HCO_3^-	0.96	2.38
SO_4^{2-}	0.12	28.2
Cl^-	0.22	545.0
Ca^{2+}	0.37	10.2
Mg^{2+}	0.18	53.2
Na^+	0.26	468.0
K^+	0.07	10.2

[a] Source: From Holland (1978).[1]

only three kinds of sedimentary material that seem to be removed to ocean sediments at rates in any way comparable to river inputs of major ions:

1   The suspended load of rivers (physical weathering) that is deposited essentially unaltered at the bottom of the oceans (and is referred to as "detrital" material by oceanographers to confuse biologists).

2   Organic matter from the "soft parts" of sea living organisms.

3   Calcium carbonate (calcite and aragonite) and silica (opal) precipitated as the "hard parts" of plants and animals (and thus also referred to as "organic" sediments by oceanographers to confuse chemists).

Of these various sedimentary deposits only calcium carbonate represents an alkalinity sink. It was then widely believed that the other major cations, Mg^{2+}, Na^+, and K^+, were removed by some as yet unidentified "reverse weathering" reactions. The formation of authigenic alumino-silicates in ocean sediments was generally hypothesized as a major sink of cations and alkalinity although there was really no experimental evidence for it.

On the basis of a rough mass balance that takes into account the principal sinks now documented for the major ions in seawater, we can calculate that calcium carbonate precipitation may in fact be the only important alkalinity sink in the ocean.[18] As shown in Table 5.7, the other major ions appear to be simply exchanged for calcium in a variety of processes. (Note that Table 5.7, much like Table 5.5, is partly tautological. Some numbers are chosen specifically to balance out elemental fluxes, and electroneutrality guarantees that everything works out in the end.)

An input-output balance for the major elements in seawater starts with an estimate of riverine inputs (Table 5.7, Column 3). The chloride cycle is then closed by ascribing it entirely to aerosol formation and dissolution and—over geological times—to evaporite formation and weathering. The other elements are included in this cycle in proportions typical of sea salt aerosols and evaporite rocks. The ion exchange process taking place when suspended river material (detrital clays) contacts seawater is then estimated on the basis of laboratory experiments. Its net effect is to exchange Na^+, K^+, and Mg^{2+} for Ca^{2+}, roughly balancing out the sodium fluxes in the process. The major sink for magnesium and sulfate in the ocean appears to be the high-temperature equilibration of seawater with basalt at ridge crests. The hot water from the "vents" is effectively stripped of magnesium and sulfate. (The results for Na^+ and Cl^- are variable.) The total fluxes corresponding to this process which are estimated on the basis of an oceanic heat budget provide a surprisingly good balance for Mg^{2+} and SO_4^{2-}. Although its flux has not been estimated independently, potassium is known to react with basalt at low temperature, on the slope of the oceanic ridges, and to exchange for calcium. In Table 5.7 the K^+ flux due to hydrothermal activity is thus simply chosen for an exact mass balance. Note that so far the net effects of ion exchange and hydrothermal activity on alkalinity are pretty much negligible. When the calcium balance is completed by choosing an

TABLE 5.7
Input-Output Balance for Major Seawater Ions and Alkalinity[a]

	Concentration	Ocean Inventory	River Input	Atmospheric/ Evaporite Cycling	Ion Exchange	Hydrothermal Activity	Carbonate Deposition	Net
Cl^-	545	710	10.0	(−10.0)		?		.0
Na^+	468	608	11.8	−9.3	−1.9	?		0.6
Mg^{2+}	53	69	8.0	−0.5	−1.2	−7.8		−1.5
SO_4^{2-}	28	37	3.7	−0.5		−3.8		−0.6
K^+	10	13	3.2	−0.1	−0.4	(−2.7)		.0
Ca^{2+}	10	13	17.1	−0.1	2.6	5.1	(−24.7)	.0
Alk	2.4	3.1	47.8		0.5	−0.5	−49.4	−1.6
	$m$mol/kg	10^{18} mol				10^{12} mol/y		

[a] Source: From McDuff and Morel (1980).[18]

appropriate value for the calcium carbonate deposition, the alkalinity is also balanced *ipso facto*. This chosen rate of deposition of $CaCO_3$ is in reasonable agreement with estimates from isotope studies.

According to this simple model of ocean chemistry, the concentrations of chloride and sodium in seawater are thus controlled by a balance between the rate of evaporite formation and its weathering. This is mostly dependent on the chance geological formation of closed basins and is probably reasonably constant on a scale of millions of years. Sulfate, potassium, and magnesium are controlled by the balance of weathering and hydrothermal activities. Both of these processes may be dependent on tectonic activity, and their relative intensities may be less variable than their absolute intensities. Finally, calcium and carbonate are controlled by calcium carbonate precipitation, the partial pressure of CO_2 in the atmosphere (assuming P_{CO_2} to be controlled by other processes over geological time), and the constraint of charge balance. The resulting carbonate-pH model of the ocean then consists of three governing equations:

1 Charge balance. Neglecting weak acids and placing on the right-hand side the ions controlled by input-output processes:

$$2(Ca^{2+}) \cong (Cl^-) + 2(SO_4^{2-}) - (Na^+) - (K^+) - 2(Mg^{2+}) \qquad (114)$$

2 Solubility of $CaCO_3(s)$. Using an apparent constant that takes into account ionic strength effects, complex formation, and pressure and temperature effects (e.g., $pK_{a2} = 9.3$ for carbonate; the solubility constant is actually chosen to be equal to the reaction quotient corresponding to present-day ocean composition):

$$\frac{(Ca^{2+})(HCO_3^-)}{(H^+)} = K_1 = 10^{3.5} \qquad (115)$$

3 Solubility of $CO_2(g)$. Again using an apparent constant:

$$(HCO_3^-)(H^+) = K_2 P_{CO_2} = 10^{-7.4} P_{CO_2} (\cong 10^{-10.9} \ M \text{ now}) \qquad (116)$$

Algebraic manipulation of these three equations yields

$$pH \cong 5.45 - \tfrac{1}{2} \log(P_{CO_2}) - \tfrac{1}{2} \log(Ca^{2+}) = 8.2 \qquad (117)$$

$$Alk \cong (HCO_3^-) + 2(CO_3^{2-}) \qquad (118)$$

therefore

$$Alk \cong 10^{-1.95} \left[\frac{P_{CO_2}}{(Ca^{2+})} \right]^{1/2} + \frac{10^{-5.5}}{(Ca^{2+})} \cong 2.3 \times 10^{-3} \ M \qquad (119)$$

With such a model one can make a simple study of the long-term stability of the acid-base chemistry of the oceans with respect to possible historical or eventual geological events such as changes in geothermal activity, variations in weathering rate, or increases in atmospheric CO_2.

6.2   Local Variations

Although on a large time and space scale the ocean might be considered globally at equilibrium with $CaCO_3(s)$, measurements in the water column show that $CaCO_3(s)$ is in fact supersaturated in surface waters and undersaturated at depths (Figure 5.10). This is largely due to the concretive activities of plants and animals in the euphotic zone. Coccoliths and forams precipitate calcite and aragonite; diatoms and radiolarians precipitate opal (to much below its saturation). As both the hard and the soft parts of such organisms settle to deepwater and the organic matter is oxidized, the solids dissolve progressively. This is because both SiO_2 and $CaCO_3$ are undersaturated, as shown on Figure 5.10. (Recall from Chapter 2 that the solubility product of $CaCO_3$ increases markedly with depth.) Obviously, these processes must have an effect on the alkalinity of the water. Let us calculate in a simple example the magnitude of such an effect and the relative importance of the various processes.

Example 15.   Removal of Alkalinity Due to Biological Activity in Surface Water*

Consider a volume of Pacific deepwater upwelled into the euphotic zone of the productive equatorial region. The relevant initial conditions of the deepwater are

$$(C_T)^0 = 2.35 \times 10^{-3} \ M$$

$$Alk^0 = 2.45 \times 10^{-3} \text{ eq liter}^{-1}$$

$$(PO_4)_T^0 = 2.5 \times 10^{-6} \ M$$

$$(NO_3)_T^0 = 4.0 \times 10^{-5} \ M$$

$$(H_2SiO_3)_T^0 = 1.5 \times 10^{-4} \ M$$

$$(Ca)_T^0 = 1.030 \times 10^{-2} \ M$$

Consider now that the productivity of the water may be partly attributed to diatoms and partly to coccoliths in such a way that roughly 0.5 mol of SiO_2 and 0.6 mol of $CaCO_3$ are precipitated for every mole of carbon fixed photosynthetically. The overall reaction of formation for living matter can thus be written in this particular case (using the principal components at the CO_2 equivalence point):

$$H_2PO_4^- + 16NO_3^- + 166H_2CO_3^* + 50H_2SiO_3 + 60Ca^{2+}$$

$$\rightarrow \underset{\text{soft parts}}{C_{106}H_{263}O_{110}N_{16}P_1} + 50SiO_2 + \underset{\text{hard parts}}{60CaCO_3}$$

$$+ 103H^+ + 34H_2O + 138O_2 \qquad (120)$$

* The conditions for this example were provided by R. Collier.

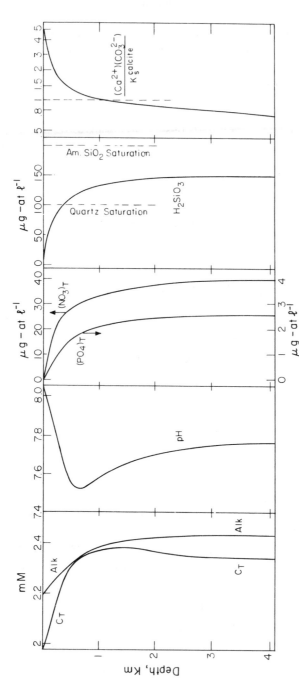

Figure 5.10 Typical vertical profiles for total carbonate, alkalinity, pH, phosphate, nitrate, and silicate concentrations, and calcium carbonate saturation in the (Pacific) ocean. These idealized profiles[17,19,20] show the depletion of the principal algal nutrients, N, P, and Si, in surface waters and the resulting alkalinity decrease and pH increase (Example 15). Note the pH minimum at the point where C_T and alkalinity are approximately equal. Calcite which is supersaturated in surface waters becomes undersaturated at depth due in part to the decrease in pH and to the effect of pressure on the solubility of $CaCO_3(s)$ (see Figure 2.3).

As the reaction proceeds to the right, the alkalinity of the water decreases by 103 equivalents per mole of carbon fixed (120 eq due to $CaCO_3$ precipitation; -17 eq due to photosynthetic nitrate and phosphate uptake). Considering that the productivity is limited by the nutrient supply, we obtain the change in composition (note that N and P are present in the desired 16:1 ratio):

$$\Delta(PO_4)_T = -2.5 \times 10^{-6} \ M$$

$$\Delta(NO_3)_T = -4.0 \times 10^{-5} \ M$$

$$\Delta C_T = -4.0 \times 10^{-5} \times \tfrac{166}{16} = -4.15 \times 10^{-4} \ M$$

$$\Delta Alk = -4.0 \times 10^{-5} \times \tfrac{103}{16} = -2.57 \times 10^{-4} \ M$$

$$\Delta(Ca)_T = -4.0 \times 10^{-5} \times \tfrac{60}{16} = -1.50 \times 10^{-4} \ M$$

$$\Delta(H_2SiO_3)_T = -4.0 \times 10^{-5} \times \tfrac{50}{16} = -1.25 \times 10^{-4} \ M$$

therefore

$$(PO_4)_T = 0$$

$$(NO_3)_T = 0$$

$$C_T = 1.94 \times 10^{-3} \ M$$

$$Alk = 2.19 \times 10^{-3} \ \text{eq liter}^{-1}$$

$$(Ca)_T = 1.015 \times 10^{-2} \ M$$

$$(H_2SiO_3)_T = 2.5 \times 10^{-5} \ M$$

This example illustrates a few salient points with regard to the effect of biological activity on alkalinity in the oceans. First it is important to underscore that most of the alkalinity changes are due to $CaCO_3$ precipitation (and re-dissolution) with only a minor effect due to NO_3^- uptake (and regeneration) and no effect at all because of SiO_2 precipitation. Alkalinity variations are thus highly dependent on the kind of organisms present in surface water, the principal parameter being the ratio of carbon precipitated as $CaCO_3$ to the carbon fixed photosynthetically. (Note that another factor in alkalinity variations implicit to this discussion—and always important in ocean chemistry—is the advection of water masses, since adjacent water layers may have different origins and exhibit widely different chemistry.)

All the changes in concentrations calculated in this example are measurable and rather typical of the differences between deep and surface Pacific water (original vs. final composition). A detailed explanation of profiles such as those of Figure 5.10 is complicated by the fact that suspended particulate nitrogen, organic carbon, and inorganic carbon are regenerated at different rates in the water column. For example, the organic carbon to nitrogen ratio in particles increases with depth (e.g., from 6 to 8, particularly in oligotrophic areas), demonstrating a faster regeneration of nitrogen than of carbon. However, the nitrogen released to the upper part of the water column, mostly as ammonium and urea, is reutilized rapidly by the biota with a zero net effect on alkalinity. Although the total net carbon fixation can thus be far in excess of the original nitrate concentration compared to the Redfield ratio, the alkalinity variations are still stoichiometrically linked to the changes in calcium and nitrate concentrations. In all cases a major factor in determining the shape of profiles such as those of Figure 5.10 is the kinetics of calcium carbonate dissolution in the water column, a subject of great complexity and of active research.

7. KINETICS OF PRECIPITATION AND DISSOLUTION

As noted previously, many of the precipitation-dissolution reactions that take place in aquatic systems are not rapid compared to the hydraulic residence time. A kinetic description of the processes would then be desirable. However, the physical, chemical, and biological processes that control the kinetics of solid precipitation and dissolution in natural waters are highly complex, intertwined, often poorly understood, and not readily amenable to theoretical mathematical descriptions. As a result one is led to use empirical rate laws that aggregate conveniently the elementary processes but have a limited domain of applicability.

Aside from the nucleation process (i.e., the initial formation of solid nuclei), the precipitation and the dissolution of a solid phase are reasonably symmetrical processes. As depicted in Figure 5.11, four steps can be distinguished in the attachment and detachment of solutes to and from a solid surface: (1) diffusion in the solution boundary layer adjacent to the surface, (2) adsorption-desorption reaction with the solid surface, (3) migration on the surface to or from a step edge (two-dimensional diffusion), and (4) migration along a step edge to or from a kink (one-dimensional diffusion). Any of these steps, alone or in combinations, may limit the kinetics of precipitation or dissolution. When Step 1 only is limiting, the kinetics are said to be diffusion controlled; when Steps 2, 3, and/or (4) only are limiting, the kinetics are said to be controlled by a surface reaction. Considering, for simplicity, that the precipitation-dissolution reaction involves only one solute with an equilibrium concentration, C_S, and a bulk concentration, C, Figure 5.12 illustrates the differences between these two extreme cases. If the overall reaction is entirely controlled by the kinetics of the surface reaction (i.e., the transport of solute to and from the surface is comparatively very fast), then the solute concentration is uniform in the solution and equal to C up to the solid-solution interface. Conversely, if transport is limiting, the solute concentration near the solid surface varies from its equilibrium value, C_S, at the surface, to the bulk concentration, C, far from the surface.

The initial formation of solid nuclei by precipitation in a saturated solution (homogeneous nucleation) is a very complicated process, usually involving the formation of polymeric species. Owing to their large surface to volume ratios, and hence their large surface energies (interfacial tension), small solids are

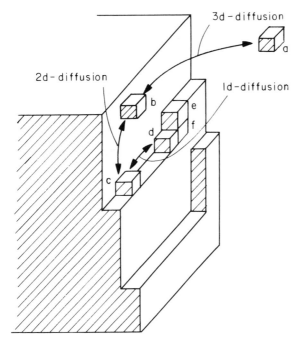

Figure 5.11 The various steps in the attachment and detachment of an ion or molecule to and from a solid lattice. During precipitation the ion or molecule becomes increasingly stable (lower free energy) as it becomes embedded deeper in the solid: $a < b < c < d < e < f$. Adapted from Nancollas and Reddy.[21]

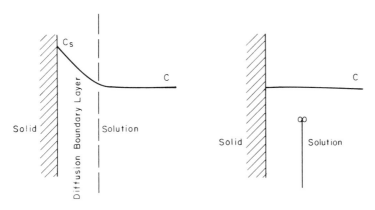

Figure 5.12 Control of dissolution kinetics by diffusion or surface reactions. If the kinetics are controlled by diffusion, there is a concentration gradient in a fluid layer adjacent to the solid; the thickness of that diffusion layer and the rate of dissolution is dependent on the mixing rate. At a sufficiently high mixing rate the diffusion gradient is effectively eliminated; the dissolution becomes controlled by a surface reaction independent of the mixing rate. A symmetrical situation could be depicted for the kinetics of precipitation.

inherently less stable than large solids. The solubility of finely dispersed solids ($< 1 \mu$m) is thus greater than the standard solubility of large crystals,[22] and an energy barrier has to be overcome before precipitation can commence. Experimentally, the kinetics of such homogeneous precipitation reactions are often difficult to reproduce, resulting in variable degrees of supersaturation before nucleation is observed. Fortunately, in natural waters there are usually plenty of suspended particles to serve as nuclei for precipitating substances (heterogeneous nucleation), and one can limit an elementary discussion of precipitation-dissolution kinetics in aquatic systems to that of crystal growth and dissolution. Lest we should forget, however, natural waters are not sterile, and, as noted in Example 15, organisms are responsible for carrying out important precipitation-dissolution reactions. The kinetics of these biologically controlled reactions are often only indirectly controlled by the chemical composition of the water. For example, the rate of SiO_2 or $CaCO_3$ precipitation may depend on the nitrate or phosphate concentrations that limit the growth of the precipitating organisms.

7.1 Diffusion-Controlled Precipitation and Dissolution

The case of the diffusion control of precipitation-dissolution reactions is formally very similar to the one that we have examined for the transfer of gases at the air-water interface (see Section 7 of Chapter 4). One typically considers simply the molecular diffusion of solutes across a boundary layer whose thickness is defined by the hydrodynamics of the system (turbulence and particle size). The possible reactions among solutes in the boundary layer have to be considered since the net transport of a particular solute to and from a solid surface may be effectively "enhanced" by reactions of other solutes, as demonstrated for the kinetics of CO_2 exchange at the air-water interface. For example, the transport of a total constituent such as C_T or $(S \cdot -II)_T$ may have to be considered rather than that of some individual species (e.g., CO_3^{2-} or S^{2-}) since the corresponding acid-base reactions are fast.

Usually one considers only the steady-state condition and does not account for the possible electrical forces on the transport of ions—although in some instances they could be important, as most solids in natural waters exhibit a sizable surface charge. The flux of a solute is then taken to be simply proportional to its concentration gradient, leading to an expression of the form:

$$\frac{dC}{dt} = E(C_S - C)$$

where dC/dt is the time rate of change of the concentration C in the bulk solution due to the precipitation-dissolution process (other processes may be taking place as well); C_S is the concentration of the solute at equilibrium with the solid; and E is an exchange coefficient that includes the hydrodynamic characteristics of the solid (size, surface area, and porosity in the case of compacted sediments[23,24]), the solid concentration, and the diffusion coefficient of the solute. Although theoretical expressions of E can be developed for simple situations, it is generally taken as an empirical parameter, with dimensions of inverse time. Some of the empiricisms can be resolved by providing an explicit functional dependency of E on the hydrodynamic regime. Although it may appear as an obvious theoretical improvement to include explicitly the surface area in the rate equation (rather than implicitly in E), it is of little help practically. This is because the effective reactive surface area of a solid suspension is particularly difficult to assess experimentally. None of the classical techniques for measuring surface area—gas adsorption, dye adsorption, particle size measurement—provide a value that is directly pertinent to precipitation-dissolution reactions.

An important situation where the solid-solution exchange is often limited by diffusional transport is the dissolution of an undersaturated solid in soil or sediments. If we consider as a limiting case the example of a rock dissolving in an infinite (or continuously renewed) water reservoir, the flux of solutes from the solid to the water should be constant since there is no concentration buildup in the solution—C is constant—and no depletion of reactive solid. However, in many instances approximating these conditions, it is observed that the solute fluxes actually decrease with time. This is usually attributed to the buildup of a surface phase due to reprecipitation of another (weathered) solid or to biological fouling. The new surface phase poses an increasing obstacle to diffusion and effectively decreases the dissolution rate of the solid. In other words, the exchange coefficient E decreases, usually linearly, with time, and the corresponding empirical functionality can be introduced in the rate equation.

7.2 Surface Reaction Control of Precipitation and Dissolution

It used to be thought that most dissolution reactions in natural waters were diffusion controlled. It is now known that the kinetics of dissolution of many sparingly soluble minerals—including such important phases as opal, silica, calcite, and aragonite in seawater, and feldspars in groundwaters—are in fact controlled by a surface reaction, as are most precipitation reactions. When this is the case, the kinetics of the overall precipitation or dissolution reaction are independent of the hydrodynamics of the system. For example, in laboratory experiments, the stirring rate, above some minimum value, has no effect on the rate of the reaction. Intermediate situations between pure surface reaction and diffusion controls, as illustrated in Figure 5.12, are of course encountered.

It is rather intuitive that molecules at the surface of a solid are less stable when they are in contact with more water and more stable when they are embedded deeper into the solid. For example, in the diagram of Figure 5.11 the surface molecules are increasingly stable in the order b, c, d, e, and f. The reactivity of a surface is thus highly dependent upon its geometry at the molecular level as well as on its chemical nature. For example, dissolution reactions are often dependent on the density of surface "kinks" since the molecules in these positions are surrounded by more water and are thus more reactive.[25] A very smooth surface, without "kinks" or "holes," is less susceptible to dissolution than one that is chemically similar but has a more rugged surface. The thermodynamics and the kinetics of new hole formation on a smooth surface is in fact rather similar to that of nuclei formation in a homogeneous solution. This problem of "hole nucleation" may well play a role in the preservation of unstable biogenic particles such as diatom frustules in the deep sea. Another reason may be the presence of adsorbed material, foreign ions or organic matter, which may adsorb preferentially at discontinuities on the surface and retard considerably the dissolution process.

Despite the apparent complexity of the precipitation-dissolution processes, reasonably simple kinetic functionalities are observed in controlled laboratory experiments. Most surface-controlled reactions can be described by an equation of the form:

$$\frac{dC}{dt} = -k(C - C_S)^\alpha \tag{122}$$

where C and C_S have their previous meaning, k is a rate constant (often expressed as $K \times S$, where S is the surface area of the solid), and α is the order of the reaction, ideally an integer ($\alpha = 0, 1, 2, 3$).*

When a solid is formed by the reaction of, or dissolves into, two ions, one of which is in great excess of the other, Equation 122 can be used by taking C to be the concentration of the less abundant ion. For example, the kinetics of formation of calcite in the presence of excess carbonate can be written:

$$\frac{d(Ca^{2+})}{dt} = -k[(Ca^{2+}) - (Ca^{2+})_s]^2 \tag{123}$$

where $(Ca^{2+})_s$ is the concentration of Ca^{2+} in equilibrium with the solid at the given carbonate concentration.

If the changes in concentrations of two reactants have to be considered, another form of Equation 122 is often used:

$$\frac{dC}{dt} = -k(\Omega - 1)^\alpha \tag{124}$$

where the saturation quotient Ω is the ratio of the ion product of the reactants to the equilibrium constant, for example, for calcite:

$$\Omega = \frac{\{Ca^{2+}\}\{CO_3^{2-}\}}{K_s} \tag{125}$$

Equation 124 reduces to 122 if the solid is formed from a single solute or one of the reactants is in large excess.

Many precipitation reactions are observed to obey second-order kinetics ($\alpha = 2$). This is, for example, the case for such important calcium solids as calcite ($CaCO_3$), gypsum ($CaSO_4 \cdot 2\ H_2O$), and brushite ($CaHPO_4 \cdot 2\ H_2O$). Note that, when the kinetics are first order ($\alpha = 1$), as is the case, for example, in the dis-

* For dissolution reactions of even order (e.g., $\alpha = 0$ or 2), k is negative.

solution of calcium carbonate and silica near seawater conditions,[25,26] the rate expression is formally similar to that of a diffusion-controlled reaction. Experimentally one may differentiate between the two by studying the effect of mixing rate on the overall kinetics.

Although new experimental data will one day allow us to describe mathematically the pertinent precipitation-dissolution kinetics in natural waters, there are still too many poorly understood factors to allow us to write generally applicable rate expressions. Major progress in this field is coming from the application of more sophisticated surface chemistry. Precipitation-dissolution rates are found to be a function of surface pH, surface charge, or the concentration of various adsorbed ions. By including the proper chemical description of the solid surface, we may become able to predict such dramatic effects as the inhibition of calcite dissolution in seawater by micromolar concentrations of phosphate. There are many theoretical and experimental obstacles to our understanding of chemical processes at the solid-water interface, however, and it is not a remarkably optimistic statement to link the progress of any field to progress in surface chemistry. (The same basic difficulty to describe thermodynamically the continuum between an ideal solid and an ideal solution phase is encountered in both fields. In one case a temporal variation is considered, in the other, a spatial one.)

To finish this discussion on a more upbeat tone, let us recall that many precipitation-dissolution reactions in natural waters are sufficiently fast or slow that we may often ignore their kinetics. For qualitative estimates, we should note that rates of dissolution and stabilities of solid phases are generally inversely correlated. In fact we may be able to refine considerably such a rule of thumb within particular classes of compounds, as illustrated by the remarkable linear free energy relationship observed for various silica phases in Figure 5.13.

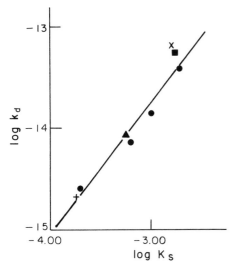

Figure 5.13 Linear free energy relationship between the rate of dissolution (k_d) and the solubility (K_s) for various silica phases. Different symbols correspond to different authors. From Wollast, 1974.[26]

REFERENCES

1   H. D. Holland, *The Chemistry of the Atmosphere and Oceans*, Wiley, New York, 1978.
2   R. M. Garrels and F. T. MacKenzie, *Evolution of Sedimentary Rocks*, Norton, New York, 1971.
3   H. C. Helgeson, J. M. Delany, H. W. Nesbitt, and D. K. Bind, *Am. J. Sci.*, **278**, (1978).
4   R. M. Smith and A. E. Martell, *Critical Stability Constants: Inorganic Complexes*, vol. 4, Plenum, New York, 1976.
5   C. F. Baes, Jr., and R. E. Mesmer, *The Hydrolysis of Cations*, Wiley, New York, 1976.
6   R. A. Robie and D. R. Waldbaum, *U.S. Geol. Survey Bull.*, **1259**, 1–256 (1968).
7   J. J. Morgan, in *Principles and Applications in Water Chemistry*, S. D. Faust and J. V. Hunter, Eds., Wiley, New York, 1967.
8   W. Stumm and J. J. Morgan, *Aquatic Chemistry*, 2nd ed., Wiley, New York, 1981, p. 546.
9   K. S. Pitzer, *J. Phys. Chem.*, **77**, 268–277 (1973).
10 K. S. Pitzer, in *Activity Coefficients in Electrolyte Solutions*, vol. 1, R. M. Pytkowicz, Ed., CRC Press, Boca Raton, Fl., 1979.
11 C. E. Harvie and J. H. Weare, *Geochim. Cosmochim. Acta*, **44**, 981–997 (1980).
12 C. E. Harvie, J. H. Weare, L. A. Hardie, and H. P. Engster, *Science*, **208**, 498–500 (1980).
13 H. P. Engster, C. E. Harvie, and J. H. Weare, *Geochim. Cosmochim. Acta*, **44**, 1335–1347 (1980).
14 J. W. Gibbs, *The Collected Works*, Yale Univ. Press, New Haven, 1948, p. 96.
15 L. G. Sillén, in *Oceanography*, M. Sears, ed., AAAS, Washington, D.C., 1961.
16 L. G. Sillén, *Science*, **165**, 1189–1197 (1967).
17 W. S. Broecker, *Chemical Oceanography*, Harcourt, Brace, Jovanovich, New York, 1980.
18 R. E. McDuff and F. M. M. Morel, *Env. Sci. Technol.*, **14**, 1182–1186 (1980).
19 J. P. Riley and G. Skirrow, *Chemical Oceanography*, 2nd ed., Academic, London, 1975.
20 H. V. Sverdrup, M. W. Johnson, and R. H. Fleming, *The Oceans*, Prentice-Hall, Englewood Cliffs, N.J., 1942.
21 G. H. Nancollas and M. M. Reddy, in *Aqueous Environmental Chemistry of Metals*, A. J. Rubin, Ed., vol. 5, Wiley, New York, 1974.
22 W. Stumm and J. J. Morgan, *Aquatic Chemistry*, 2nd ed., Wiley, New York, 1981, pp. 295–299.
23 R. A. Berner, *Principles of Chemical Sedimentology*, McGraw-Hill, New York, 1971.
24 A. F. White and H. C. Claassen, in *Chemical Modeling in Aqueous Systems*, E. A. Jenne, Ed., ACS Symposium Series 93, American Chemical Society, Washington, D.C., 1979.
25 R. A. Berner and J. W. Morse, *Am. J. Sci.*, **274**, 108–134 (1974).
26 R. Wollast, in *The Sea*, vol. 5, E. D. Goldberg, Ed., Wiley, New York, 1974.

PROBLEMS

5.1 Given $P_{CO_2} = 10^{-3.5}$ at, $P_{H_2S} = 10^{-6}$ at, $Cd_T = 10^{-8}$ M, what is the principal stable form of cadmium at pH $= 7$? In what pH range(s) do any two or three solids coexist? (Use solubility products from Table 6.1.)

5.2 Calculate the free cadmium ion activity (Cd^{2+}), given $(Ca^{2+}) = 10^{-2}$ M, pH $= 8.5$, $Cd_T = 10^{-5}$ M, and assuming precipitation of carbonates. (Use solubility products from Table 6.1.)

5.3 What is the pH of a solution saturated with $CaCO_3(s)$ containing $Ca_T = 10^{-4}$ M and $C_T = 10^{-2}$ M in solution?

5.4 At what Fe(II) concentration would $FeCO_3(s)$ precipitate in groundwater at pH $= 7.5$, at equilibrium with $P_{CO_2} = 10^{-1.5}$ at?

5.5 The predominant forms of heavy metals (Zn, Pb, Fe, Cu, Ni, etc.) in oxic sediments are the carbonates and the oxide or hydroxide forms. An important consideration in modeling the sediment is what is the stable chemical form of the metal in question. Consider the reaction:

$$Pb(OH)_2(s) + CO_2(aq) = PbCO_3(s) + H_2O$$

ΔG^0 for the reaction is $= 5.98$ kcal mole^{-1}.

a. At what partial pressure of CO_2 will $Pb(OH)_2(s)$ and $PbCO_3(s)$ be at equilibrium?

b. If the system is at equilibrium with the atmosphere, $P_{CO_2} = 10^{-3.5}$ at, calculate ΔG for the reaction. What is the stable form of Pb in the sediments?

5.6 Given $(CO_3)_T = 10^{-2}$ M, $(S\cdot-II)_T = 10^{-7}$ M, $Cu_T = 10^{-8}$ M, what are the stable forms of copper as a function of pH? Make an appropriate diagram. Do some of the solids coexist at any pH?

5.7 a. Draw a log C-pH diagram for aluminum species, first, at equilibrium with $Al(OH)_3(s)$ and, second, at equilibrium with kaolinite and am$\cdot SiO_2(s)$.

b. For the recipe $(AlCl_3)_T = 10^{-3}$ M, what are the pH and the composition of the system, assuming equilibrium with $Al(OH)_3(s)$? Assuming equilibrium with kaolinite and am$\cdot SiO_2(s)$?

c. In each case for what concentrations (minimum and/or maximum) is an addition of $(AlCl_3)_T$ entirely soluble?

5.8 Following the algebraic and graphical method given in the chapter, study the solubility of Fe(II) in a system consisting of

	Case 1	Case 2	Case 3
$(Fe\cdot II)_T =$	10^{-4}	10^{-4}	10^{-4}
$(CO_3)_T =$	10^{-3}	10^{-5}	$10^{-3.8}$

5.9 a. Calculate the composition of a groundwater in equilibrium with $P_{CO_2} = 10^{-1.5}$ at, $CaCO_3$, albite, and kaolinite.

b. As a function of C_T (total soluble carbon), calculate the composition of this water as it flows out (no more rock to dissolve but precipitation can occur) and equilibrates with the atmosphere ($P_{CO_2} = 10^{-3.5}$ at).

5.10 Using seawater apparent constants (see Chapter 6), consider seawater to be at equilibrium with $CaCO_3(s)$ and the calcium concentration to be fixed (independently) at $(Ca^{2+}) = 10^{-2}$ M.

a. Derive a plot of pH versus C_T.

b. Does photosynthesis, independent of hard part formation, tend to precipitate or dissolve $CaCO_3(s)$? (Must Ca^{2+} be added or withdrawn from the system to maintain equilibrium?)

c. How does the alkalinity of the water vary with C_T?

d. Discuss the issue of the stability (homeostasis) of the carbon system with respect to $(CO_2)_T$ changes.

5.11   Consider a system made up of $Al(OH)_3$ and H_2SiO_3 in distilled water (no carbonate), total concentrations $TOTAl$ and $TOTSi$, and dissolved concentrations Al_T and Si_T.

a. Supposing the pH to be buffered at 7.0, develop a $TOTAl$ versus $TOTSi$ phase diagram. Use a range of 10^{-10} to 1 M for both components. Indicate clearly the regions of coexistence of solids.

b. Supposing the pH to be unbuffered, calculate the pH as a function of Al_T and Si_T. How is the phase diagram modified?

c. What is the equilibrium composition of the solution when $TOTAl = 0.10$ M and $TOTSi = 0.11$ M, 0.101 M, 0.10 M, and 0.09 M? At high concentrations (say $TOTAl \cong TOTSi \cong 0.1$ M, how much in excess of $TOTAl$ should $TOTSi$ be, or vice versa, for $SiO_2(s)$ or $Al(OH)_3(s)$ to precipitate?

d. For each representative total concentrations, indicate on the diagram the final solution composition (i.e., draw arrows from $TOTAl$, $TOTSi$ to Al_T, Si_T coordinates).

5.12   This problem deals with the characteristics of rain and groundwater (resulting from equilibrium of rain with rock minerals) in polluted and unpolluted regions. Although the questions are obviously related, most of them can be answered independently.

a. What is the pH of the rain in an unpolluted region if it can be considered simply as distilled water in equilibrium with the atmosphere ($P_{CO_2} = 10^{-3.5}$ at)? What is its buffer capacity?

b. In this case, what are the pH and alkalinity of the groundwater ($P_{CO_2} = 10^{-1.5}$ at) in a calcite region? In a dolomite region? In an albite region?

c. In a polluted region the rain is actually acidic, pH = 3.7, due to roughly equimolar concentrations of nitric and sulfuric acids. What is the mineral acidity of the polluted rainwater? What is its buffer capacity?

d. In the case of acidic precipitation, what are the pH and the alkalinity of the groundwater ($P_{CO_2} = 10^{-1.5}$ at) in a calcite region? In a dolomite region? In an albite region?

CHAPTER SIX

COMPLEXATION

In Chapters 4 and 5 we examined chemical processes that play a major controlling role in the geochemical cycles of elements. The dissolution and dissociation of weakly acidic gases and the weathering and sedimentation of rocks control the gross chemical composition of natural waters. Complex formation—defined here loosely as the reaction of two soluble species to form a third one – plays a minor role by comparison. In principle, the total soluble concentration of elements can be increased by complexation, and this is certainly important for some elements in some natural waters. However, in many aquatic systems, particularly in the oceans, most elements appear undersaturated rather than supersaturated with respect to solid phases. Although complexation may be important over relatively short time and space scales, it is not a dominant process in the global cycle of elements over geological times.

The subject of aquatic complexation is in a paradoxical state. We know a great deal about coordination chemistry, and we can make many chemical models that predict the existence of complexes in natural waters. Yet it is so difficult to analyze for individual chemical species under the conditions prevailing in aquatic systems where the concentrations are very low and the constituents very many, that we usually cannot convincingly demonstrate the existence of these complexes. The subject of coordination of trace elements in natural waters is probably the greatest remaining challenge to analytical chemists; the objective is to demonstrate and quantify the existence of fractions of chemical constituents as picomolar concentrations of perhaps ephemeral species.

The question of complexation is ultimately that of chemical speciation: What are the principal soluble species of the major and minor components of natural waters? What are the concentrations or the activities of free ions in comparison to the total (analytical) concentrations of chemical constituents? Beside irrepressible curiosity and chemical romanticism, a major motivation for asking these questions arises from our interest in the aquatic biota. Chemical constituents of natural waters affect the biota as essential nutrients and potential toxicants. It is now firmly established that these interactions are directly dependent on the chemical speciation of the constituents. For example, the toxicity of many trace metals to planktonic microorganisms is determined by their free ion activities, not their total soluble concentrations. This has been shown for the effects of Cu^{2+}, Cd^{2+}, Zn^{2+}, Ni^{2+}, Hg^{2+}, and Pb^{2+} on bacteria, phytoplankton, and zooplankton (and to some degree on fish). The availability of necessary metals such as Fe, Mn, Zn, and Cu to plants is also determined by free metal ion activities. Metal toxicity and limitation are not just matters of academic exercises or pollution control; they are thought to be natural controlling environmental factors for aquatic ecosystems. We need to understand the speciation of elements in water if we want to understand the relationship of aquatic organisms with their external milieu. Although the role of complexation in natural waters may be more subtle than geochemists typically care about or than analysts can presently measure, it is a dominant aspect of aquatic chemistry for biologists.

In keeping with the thermodynamic view that we have taken so far, we do not discuss in this chapter any of the fundamental theories of coordination chemistry. Questions relating to the nature of coordination bonds or comparisons among the coordinative properties of elements and compounds are addressed minimally and only occasionally. For the most part we consider that chemical species are totally described by their stoichiometry and their free energies, and do not rely on images of molecular structures or electron orbitals.

Chemical complexes in natural waters can be conveniently classified into three groups: ion pairs of major constituents, inorganic complexes of trace elements, and organic complexes. These are discussed briefly at the beginning of the chapter where necessary definitions are provided. From simple mole balance considerations the speciation of abundant aquatic constituents such as Ca^{2+}, Na^+, Cl^-, and HCO_3^- cannot be affected by complexation with those present in trace amounts. Although the nature of the coordination processes may not be very different, the topics of inorganic complexation of major and of trace constituents are thus effectively separate and are presented in consecutive sections. Owing to our relative ignorance of the nature and properties of the dissolved organic matter in natural waters, the question of organic complexation must be addressed largely at a theoretical level. To make the presentation more tangible, the scant available data are thus generalized liberally and speculatively, and the discussion is organized around a few typical organic compounds. Finally, our bias toward biologically relevant processes leads us to a brief concluding discussion of biological interactions, following some considerations of chemical homeostasis ("chemostasis") and kinetics.

1. AQUEOUS COMPLEXES

Consider a metal ion such as Cu^{2+} in water. Although we commonly talk of the "free cupric ion" in solution, this is really a misnomer, for the metal ion is actually associated with its surrounding water molecules. Inorganic chemists distinguish in fact four solvent regions around a metal ion (see Figure 6.1):

1   A primary solvation shell where the water molecules are considered chemically bound to the ion. In the case of copper (and many other metals) there are six such water molecules, leading to the more refined chemical symbolism $Cu(H_2O)_6^{2+}$ instead of Cu^{2+} for the hydrated cupric ion.

2   A secondary solvation shell where the water molecules are ordered by the electrostatic influence of the ion. The volume of this shell increases with the charge of the ion and is inversely related to its size.

3   A transition region separating the hydrated metal ion from the bulk solution. In this region the water molecules are considered less ordered than in either the solvation shell or the bulk solution.

4   The bulk solution where the presence of the metal ion is not felt.

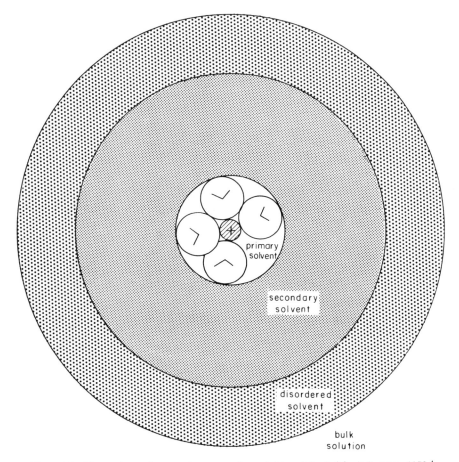

Figure 6.1   The various solvent regions around a metal ion. Adapted from Burgess, 1978.[1]

From a thermodynamic view the energy of hydration of an ion is critical to its very existence. Ionization of a metal or separation of cations and anions from a crystal lattice are energetically very unfavorable processes. For example, it takes some 645 kcal mol^{-1} to ionize Cu to Cu^{2+} in the gas phase. The presence of the cupric ion in solution is thus made possible only by a considerable energy of solvation which exceeds the unfavorable ionization energy. Of course the energetics of the metal-solvent interactions do not normally concern us when we study the thermodynamics of complex formation in aqueous solution and consider only the differences among the free energies of aquated species. However, the large energy of solvation is important for the kinetics of complex formation which are often controlled by the rate of dissociation of the co-ordinated water molecules, an energetically unfavorable process. The point here is that free metal ions in solutions are in fact aquo complexes; the water itself is a *ligand* that binds metals, and every complexation reaction in water is effectively a *ligand-exchange reaction*.

Other ligands that can replace water molecules around the *central atom* (the metal ion) are chemical species that have a free pair of electrons to share

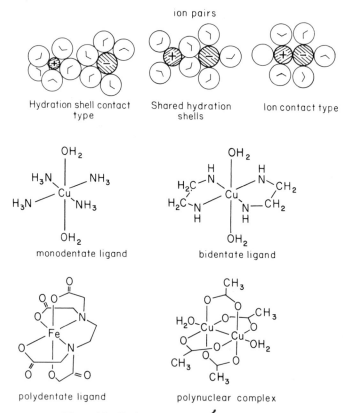

Figure 6.2   Various types of aqueous complexes

with the metal. These include simple anions such as the halides Cl^-, F^-, Br^-, and I^-, or more complex inorganic compounds such as NO_3^-, CO_3^{2-}, SO_4^{2-}, NH_3, S^{2-}, PO_4^{3-}, SO_3^{2-}, and CN^-, as well as a great variety of organic molecules with suitable functional groups usually containing oxygen, nitrogen, or sulfur atoms as purveyors of electron pairs (e.g., $R—COO^-$, $R—OH$; $R—NH_3$; $R—SH$). For the sake of generality it is convenient to consider H^+ as a metal and OH^- as a ligand, thus including all acid-base reactions as a subset of coordination reactions.

The reaction of a metal with a ligand can be of an electrostatic or covalent nature, or both. When it is primarily electrostatic and the reactants retain some hydration water between them, the product is called an *ion pair* or an *outer sphere complex*. In natural waters this type of interaction is particularly important among the major ions of high ionic strength systems such as seawater. When the reaction of a metal with a ligand involves coordination at several positions, one speaks of *chelation* (in Greek, *chelos* = crab—has *two* binding claws). Such a reaction requires the combination of a metal with a *coordination number* greater than one (i.e., more than one site for coordination), and a *multidentate ligand*, an organic compound with several reactive functional groups. These organic compounds are called *chelators* or *chelating agents*. Among metals only H^+ has a coordination number of one; most metals of interest in aquatic chemistry have a coordination number of six and thus form octahedral complexes. In natural waters *polynuclear complexes*, those involving more than one metal, are probably rare except for polymeric hydroxide species. In this chapter we use the word complex somewhat loosely to cover all soluble species resulting from the metal-ligand combinations illustrated in Figure 6.2, including ion pairs and chelates.

Metal-ligand interactions are a particularly convenient framework within which to organize the reactions of interest in natural waters, namely, acid-base reactions, complexation, and solid formation. Table 6.1, for example, presents in a compact way the equilibrium constants and the stoichiometric coefficients for reactions among some 20 metals and 31 ligands. The first 15 ligands represent practically all reactive inorganic constituents commonly encountered in natural waters; they are roughly arranged in order of decreasing abundance. The other 16 ligands are chosen as representative compounds containing the types of functionalities present in aquatic organic matter, some of them being actually present at trace concentrations in natural waters. The metals are also roughly organized in order of decreasing abundance, with the list of trace metals being biased toward the more reactive elements.

The log of equilibrium constants given in Table 6.1 corresponds to a general complexation or precipitation reaction of the type

$$m\,M + l\,L + h\,H = M_m L_l H_h$$

$$\beta_{mlh} = \frac{(M_m L_l H_h)}{(M)^m (L)^l (H)^h}$$

TABLE 6.1 Stability Constants for Formation of Complexes and Solids from Metals and Ligands[a]

	OH^-	CO_3^{2-}	SO_4^{2-}	Cl^-	Br^-	F^-	NH_3	$B(OH)_4^-$
H^+	HL·w 14.00	HL 10.33 H₂L 16.68 H₂L·g 18.14	HL 1.99	—	—	HL 3.2	HL 9.24 L·g −1.8	HL 9.24 HL₃ 10.4 H₂L₃ 20.4 H₂L₄ 21.0 H₄L₅ 38.8
Na^+	—	NaL 1.27 NaHL 10.08	NaL 1.06	—				
K^+	—	—	KL 0.96	—				
Ca^{2+}	CaL 1.15 CaL₂·s 5.19	CaL 3.2 CaHL 11.59 CaL·s 8.22 CaL·s 8.35	CaL 2.31 CaL·s 4.62	—		CaL 1.1 CaL₂·s 10.4	—	
Mg^{2+}	MgL 2.56 Mg₄L₄ 16.28 MgL₂·s 11.16	MgL 3.4 MgHL 11.49 MgL·s 4.54 MgL·s 7.45	MgL 2.36	—	—	MgL 1.8 MgL₂·s 8.2	—	
Sr^{2+}	—	SrL·s 9.0	SrL 2.6 SrL·s 6.5			SrL₂·s 8.5		
Ba^{2+}	—	BaL 2.8 BaL·s 8.3	BaL 2.7 BaL·s 10.0	—		BaL₂·s 5.8		
Cr^{3+}	CrL 10.0 CrL₂ 18.3 CrL₃ 24.0 CrL₄ 28.6 Cr₃L₄ 47.8 CrL₃·s 30.0		CrL 3.0	CrL 0.1		CrL 5.2 CrL₂ 9.2 CrL₃ 12.0		
Al^{3+}	AlL 9.0 AlL₂ 18.7 AlL₃ 27.0 AlL₄ 33.0 Al₃L₄ 42.1 AlL₃·s 33.5					AlL 7.0 AlL₂ 12.6 AlL₃ 16.7 AlL₄ 19.1		
Fe^{3+}	FeL 11.8 FeL₂ 22.3 FeL₄ 34.4 Fe₂L₂ 25.0 FeL₃·s 42.7 FeL₃·s 39.5		FeL 4.0 FeL₂ 5.4	FeL 1.5 FeL₂ 2.1	FeL 0.6	FeL 6.0 FeL₂ 10.6 FeL₃ 13.7		
Mn^{2+}	MnL 3.4 MnL₂ 5.8 MnL₃ 7.2 MnL₄ 15.7 MnL₂·s 12.8	MnHL 12.1 MnL·s 9.3	MnL 2.3	MnL 0.6		MnL 1.9	MnL 1.0 MnL₂ 1.5	

Ion								
Fe²⁺	FeL 4.5 FeL₂ 7.4 FeL₃ 11.0 FeL₂·s 15.1	FeL·s 10.7	FeL 2.2			FeL 1.4		
Co²⁺	CoL 4.3 CoL₂ 5.1 CoL₃ 10.5 CoL₂·s 15.7	CoL·s 10.0	CoL 2.4	CoL 0.5		CoL 1.0	CoL 2.0 CoL₂ 3.5 CoL₃ 4.4 CoL₄ 5.0	
Ni²⁺	NiL 4.1 NiL₂ 9.0 NiL₃ 12.0 NiL₂·s 17.2	NiL·s 6.9	NiL 2.3	NiL 0.6		NiL 1.1	NiL 2.7 NiL₂ 4.9 NiL₃ 6.6 NiL₄ 7.7 NiL₅ 8.3	
Cu²⁺	CuL 6.5 CuL₂ 11.8 CuL₃ 16.4 CuL₄ 10.4 Cu₂L₂·s 19.3 CuL₂·s 20.4	CuL 6.7 CuL₂ 10.2 CuL₃·s 9.6 Cu₃(OH)₂L·s 33.8 Cu₃(OH)₂L₂·s 46.0	CuL 2.4 Cu₄(OH)₆L·s 68.6	CuL 0.5		CuL 1.5	CuL 4.0 CuL₂ 7.5 CuL₃ 10.3 CuL₄ 11.8	
Zn²⁺	ZnL 5.0 ZnL₂ 11.1 ZnL₃ 13.6 ZnL₄ 14.8 ZnL₂·s 15.5 ZnL₂·s 16.8	ZnL·s 10.0	ZnL 2.1 ZnL₂ 3.1	ZnL 0.4 ZnL₂ 0.0 ZnL₃ 0.5 Zn₂(OH)₃L·s 26.8		ZnL 1.2	ZnL 2.2 ZnL₂ 4.5 ZnL₃ 6.9 ZnL₄ 8.9	
Pb²⁺	PbL 6.3 PbL₂ 10.9 PbL₃ 13.9 PbL₂·s 15.3	PbL·s 13.1	PbL 2.8 PbL·s 7.8	PbL 1.6 PbL₂ 1.8 PbL₃ 1.7 PbL₄ 1.4 PbL₂·s 4.8	PbL 1.8 PbL₂ 2.6 PbL₃ 3.0 PbL₃·s 5.7	PbL 2.0 PbL₂ 3.4 PbL₂·s 7.4		
Hg²⁺	HgL 10.6 HgL₂ 21.8 HgL₃ 20.9 HgL₂·s 25.4	HgL·s 16.1	HgL 2.5 HgL₂ 3.6	HgL 7.2 HgL₂ 14.0 HgL₃ 15.1 HgL₄ 15.4 HgOHL 18.1	HgL 9.6 HgL₂ 18.0 HgL₃ 20.3 HgL₄ 21.6 HgL₂·s 19.8	HgL 1.6	HgL 8.8 HgL₂ 17.4 HgL₃ 18.4 HgL₄ 19.1	
Cd²⁺	CdL 3.9 CdL₂ 7.6 CdL₂·s 14.3	CdL·s 13.7	CdL 2.3 CdL₂ 3.2 CdL₃ 2.7	CdL 2.0 CdL₂ 2.6 CdL₃ 2.4 CdL₄ 1.7	CdL 2.1 CdL₂ 3.0	CdL 1.0 CdL₂ 1.4	CdL 2.6 CdL₂ 4.6 CdL₃ 5.9 CdL₄ 6.7	
Ag⁺	AgL 2.0 AgL₂ 4.0 AgL·s 7.7	Ag₂L·s 11.1	AgL 1.3 Ag₂L·s 4.8	AgL 3.3 AgL₂ 5.3 AgL₃ 6.4 AgL·s 9.7	AgL 4.7 AgL₂ 6.9 AgL₃ 8.7 AgL₄ 9.0 AgL·s 12.3	AgL 0.4	AgL 3.3 AgL₂ 7.2	AgL 0.6 AgHL₂ 22.9

(continued)

TABLE 6.1 (Continued)

	SiO_3^{2-}	S^{2-}	$S_2O_3^{2-}$	PO_4^{3-}	$P_2O_7^{4-}$	$P_3O_{10}^{5-}$	CN^-
H^+	HL 13.1 H₂L 23.0 H₂L₂ 26.6 H₄L₄ 55.9 H₆L₄ 78.2 H₂L·s 2.7	HL 13.9 H₂L 20.9 H₂L·g 21.9	HL 1.6 H₂L 2.2	HL 12.35 H₂L 19.55 H₃L 21.70	HL 9.4 H₂L 16.1 H₃L 18.3 H₄L 19.7	HL 9.3 H₂L 18.8 H₃L 21.3 H₄L 22.3	HL 9.2
Na^+			NaL 0.5	NaHL 13.5	NaL 2.3 Na₂L 4.2 NaHL 10.8	NaL 2.7 NaHL 11.6	
K^+			KL 1.0	KHL 13.4	KL 2.1	KL 2.8	
Ca^{2+}	CaL 4.2 CaHL 14.1 CaH₂L₂ 29.9		CaL 2.0	CaL 6.5 CaHL 15.1 CaH₂L 21.0 CaHL·s 19.0	CaL 6.8 CaHL 13.4 CaOHL 8.9 Ca₂L·s 14.7	CaL 8.1 CaHL 14.1 CaOHL 10.4	
Mg^{2+}	MgL 5.3 MgHL 14.3 MgH₂L₂ 30.8		MgL 1.8	MgL 4.8 MgHL 15.3 MgH₂L 20.8 Mg₃L₂·s 25.2 MgHL·s 18.2	MgL 7.2 MgHL 14.1 MgOHL 9.3	MgL 8.6 MgHL 14.5 MgOHL 11.0	
Sr^{2+}			SrL 2.0	SrL 5.5 SrHL 14.5 SrH₂L 20.3 SrHL·s 19.3	SrL 5.4 SrOHL 7.7 Sr₂L·s 12.9	SrL 7.2 SrHL 13.6 SrOHL 9.3	
Ba^{2+}			BaL 2.3 BaL·s 4.8	BaHL·s 19.8		BaL 6.3 BaHL 12.9 Ba₂L·s 16.1	
Cr^{3+}							
Al^{3+}							
Fe^{3+}	FeHL 22.7		FeL 3.3	FeHL 22.5 FeH₂L 23.2 FeL·s 26.4			FeL₆ 43.6
Mn^{2+}		MnL·s 10.5 13.5	MnL 2.0			MnL 9.9 MnHL 14.8	

Ion						
Fe²⁺	FeL·s 18.1		FeHL 16.0 FeH₂L 22.3 Fe₃L₂·s 36.0			FeL₆ 35.4
Co²⁺	CoL·s 21.3 CoL·s 25.6	CoL 2.1	CoHL 15.5	CoL 7.9 CoHL 14.1	CoL 9.7 CoHL 14.8	CoL CoHL₅
Ni²⁺	NiL·s 19.4 NiL·s 24.9 NiL·s 26.6	NiL 2.1	NiHL 15.4	NiL 7.7 NiHL 14.4	NiL 9.5 NiHL 14.7	NiL 7.7 NiL₄ 30.9 NiHL₄ 36.7 NiH₂L₄ 41.4 NiH₃L₄ 44.0
Cu²⁺	CuL·s 36.1		CuHL 16.5 CuH₂L 21.3	CuL 9.8 CuHL 15.5 CuL₂ 12.5 CuH₂L 19.2	CuL 11.1 CuHL 15.5	CuL₂ 16.3 CuL₃ 21.6 CuL₄ 23.1
Zn²⁺	ZnL 16.6 ZnL·s 24.7	ZnL 2.4 ZnL₂ 2.5 ZnL₃ 3.3 Zn₂L₂ 7.0	ZnHL 15.7 ZnH₂L 21.2 Zn₃L₂ 35.3	ZnL 8.7 ZnL₂ 11.0 ZnOHL 13.1	ZnL 10.3 ZnHL 14.9 ZnOHL 13.6	ZnL 5.7 ZnL₂ 11.1 ZnL₃ 16.1 ZnL₄ 19.6 ZnL·s 15.9
Pb²⁺	PbL·s 27.5	PbL 3.0 PbL₂ 5.5 PbL₃ 6.2 PbL₄ 7.3	PbHL 15.5 PbH₂L 21.1 Pb₃L₂·s 43.5 PbHL·s 23.8	PbL 9.5 PbL₂ 10.2		
Hg²⁺	HgL 7.9 HgL₂ 14.3 HgOHL 18.5 HgL·s 52.7 HgL·s 53.3	HgL₂ 29.2 HgL₃ 30.6		HgOHL 18.6		HgL 17.0 HgL₂ 32.8 HgL₃ 36.3 HgL₄ 39.0 HgOHL 29.6
Cd²⁺	CdL 19.5 CdHL 22.1 CdH₂L₂ 43.2 CdH₃L₃ 59.0 CdH₄L₄ 75.1 CdL·s 27.0	CdL 3.9 CdL₂ 6.3 CdL₃ 6.4 CdL₄ 8.2 Cd₂L₂ 13.0		CdL 8.7 CdOHL 11.8	CdL 9.8 CdHL 14.6 CdOHL 12.6	CdL 6.0 CdL₂ 11.1 CdL₃ 15.7 CdL₄ 17.9
Ag⁺	AgL 19.2 AgHL 27.7 AgHL₂ 35.8 AgH₂L₄ 26.3 AgH₂L₂ 45.7 Ag₂L·s 50.1	AgL 8.8 AgL₂ 13.7 AgL₃ 14.2 Ag₂L₄ 26.3 Ag₃L₅ 39.8 Ag₆L₈ 78.6	Ag₃L·s 17.6			AgL₂ 20.5 AgL₃ 21.4 AgOHL 13.2 AgL·s 15.7

(continued)

TABLE 6.1 (Continued)

	Ethylenediamine	NTA	EDTA	CDTA	IDA	Picolinate	Cysteine	Desferri-ferrioxamine B
H^+	HL 9.93 H_2L 16.78	HL 10.33 H_2L 13.27 H_3L 14.92 H_4L 16.02	HL 11.12 H_2L 20.01 H_3L 21.04 H_4L 23.76 H_5L 24.76	HL 13.28 H_2L 22.63 H_3L 23.98 H_4L 26.62 H_5L 28.34	HL 9.73 H_2L 12.63 H_3L 14.51	HL 5.39 H_2L 6.40	HL 10.77 H_2L 19.13 H_3L 20.84	HL 10.1 H_2L 19.4 H_3L 27.8
Na^+		NaL 1.9	NaL 2.5		NaL 0.8			
K^+			KL 1.7					
Ca^{2+}		CaL 7.6	CaL 12.4 CaHL 16.0	CaL 15.0	CaL 3.5	CaL 2.2 CaL_2 3.8		CaL 3.5
Mg^{2+}	MgL 0.4	MgL 6.5	MgL 10.6 MgHL 15.1	MgL 12.8	MgL 3.8	MgL 2.6 MgL_2 4.0		MgL 5.2
Sr^{2+}		SrL 6.3	SrL 10.5 SrHL 14.9	SrL 12.4	SrL 3.1	SrL 1.8 SrL_2 3.0		SrL 3.1
Ba^{2+}		BaL 5.9	BaL 9.6 BaHL 14.6	BaL 10.5 BaHL 17.8	Ba 2.5	BaL 1.6		
Cr^{3+}			CrL 26.0 CrHL 28.2 CrOHL 32.2		CrL 12.2 CrL 23.2			
Al^{3+}		AlL 13.4 AlOHL 22.1	AlL 18.9 AlHL 21.6 AlOHL 26.6 $Al(OH)_2L$ 30.0	AlL 22.1 AlHL 24.3 AlOHL 28.1	AlL 9.9 AlL_2 17.5			
Fe^{3+}		FeL 17.9 FeL_2 26.3	FeL 27.7 FeHL 29.2 FeOHL 33.8 $Fe(OH)_2L$ 37.7	FeL 32.6 FeOHL 36.5	FeL 12.5	FeL_2 13.9 $FeOHL_2$ 24.9		FeL 31.9 FeHL 32.6

Mn⁺	MnL 2.8 MnL₂ 4.9 MnL₃ 5.8	MnL 8.7 MnL₂ 11.6	MnL 15.6 MnHL 19.1	MnL 19.2 MnHL 22.4		MnL 4.0 MnL₂ 7.1 MnL₃ 8.8	MnL 5.6	
Fe²⁺	FeL 4.3 FeL₂ 7.7 FeL₃ 9.7	FeL 9.6 FeL₂ 13.6 FeOHL 12.6	FeL 16.1 FeHL 19.3 FeOHL 20.4 Fe(OH)₂L 23.7	FeL 20.8 FeHL 23.9	FeL 6.7 FeL₂ 11.0	FeL 5.3 FeL₂ 9.7 FeL₃ 13.0		FeHL 17.7 HeH₂L 23.1
Co²⁺	CoL 6.0 CoL₂ 10.8 CoL₃ 14.1	CoL 11.7 CoL₂ 15.0 CoOHL 14.5	CoL 18.1 CoHL 21.5	CoL 21.4 CoHL 24.7	CoL 7.9 CoL₂ 13.2	CoL 6.4 CoL₂ 11.3 CoL₃ 15.8		CoL 11.2 CoHL 18.0 CoHL 23.6
Ni²⁺	NiL 7.4 NiL₂ 13.6 NiL₃ 17.9	NiL 12.8 NiL₂ 17.0 NiOHL 15.5	NiL 20.4 NiHL 24.0 NiOHL 21.8	NiL 22.1 NiHL 25.4	NiL 9.1 NiL₂ 15.2	NiL 7.2 NiL₂ 13.2 NiL₃ 17.9	NiL 10.7 NiL₂ 20.9	NiL 11.8 NiHL 18.3 NiH₂L 23.8
Cu²⁺	CuL 10.5 CuL₂ 19.6 CuOHL 11.8	CuL 14.2 CuL₂ 18.1 CuOHL 18.6	CuL 20.5 CuHL 23.9 CuOHL 22.6	CuL - 23.7 CuHL 27.3	CuL 11.5 CuL₂ 17.6	CuL 8.4 CuL₂ 15.6	Cu(II)→Cu(I)	CuL 15.0 CuHL 24.1 CuH₂L 27.0
Zn²⁺	ZnL 5.7 ZnL₂ 10.6 ZnL₃ 13.9	ZnL 12.0 ZnL₂ 14.9 ZnOHL 15.5	ZnL 18.3 ZnHL 21.7 ZnOHL 23.4	ZnL 21.1 ZnHL 24.4	ZnL 8.2 ZnL₂ 13.5	ZnL 5.7 ZnL₂ 10.3 ZnL₃ 13.6	ZnL 10.1 ZnL₂ 19.1 ZnHL 16.4	ZnL 11.0 ZnHL 17.5 ZnH₂L 22.9
Pb²⁺	PbL 7.0 PbL₂ 8.5	PbL 12.6	PbL 19.8 PbHL 23.0	PbL 22.1 PbHL 25.3	PbL 8.3	PbL 5.0 PbL₂ 8.6	PbL 12.5	
Hg²⁺	HgL 14.3 HgL₂ 23.2 HgOHL 24.2 HgHL₂ 28.0	HgL 15.9	HgL 23.5 HgHL 27.0 HgOHL 27.7	HgL 26.8 HgHL 30.3 HgOHL 29.7	HgL 12.6	HgL 8.1 HgL₂ 16.2	HgL 15.3	
Cd²⁺	CdL 5.4 CdL₂ 9.9 CdL₃ 12.7	CdL 11.1 CdL₂ 15.1 CdOHL 13.4	CdL 18.2 CdHL 21.5	CdL 21.7 CdHL 25.1	CdL 6.6 CdL₂ 11.1	CdL 5.0 CdL₂ 8.8 CdL₃ 11.4		CdL 8.8 CdL 8.2 CdH₂L 22.7
Ag⁺	AgL 4.7 AgL₂ 7.7 AgHL 11.9	AgL 5.8	AgL 8.2 AgHL 14.9	AgL 9.9		AgL 3.6 AgL₂ 6.1		

(continued)

247

TABLE 6.1 (Continued)

	Glycine	Glutamate	Acetate	Glycolate	Citrate	Malonate	Salicylate	Phthalate
H^+	HL 9.78 H_2L 12.13	HL 9.95 H_2L 14.47 H_3L 16.70	HL 4.76	HL 3.83	HL 6.40 H_2L 11.16 H_3L 14.29	HL 5.70 H_2L 8.55	HL 13.74 H_2L 16.71	HL 5.41 H_2L 8.36
Na^+					NaL 1.4	NaL 0.7		NaL 0.7
K^+					KL 1.3			
Ca^{2+}	CaL 1.4	CaL 2.1	CaL 1.2	CaL 1.6	CaL 4.7 CaHL 9.5 CaH_2L 12.3	CaL 2.4 CaHL 6.6	CaL 0.4	CaL 2.4
Mg^{2+}	MgL 2.7	MgL 2.8	MgL 1.3	MgL 1.3	MgL 4.7 MgHL 9.2	MgL 2.9 MgHL 7.1		
Sr^{2+}	SrL 0.9	SrL 2.3	SrL 1.1	SrL 1.2	SrL 4.1	SrL 2.1 SrHL 6.5		
Ba^{2+}	BaL 0.8	BaL 2.2	BaL 1.1	BaL 1.1	BaL 4.1 BaHL 9.0 BaH_2L 12.4	BaL 2.1	BaL 0.2	BaL 2.3
Cr^{3+}			CrL 5.4 CrL_2 8.4 CrL_3 11.2			CrL 9.6		
Al^{3+}			AlL 2.4				AlL 14.2 AlL_2 24.0 AlL_3 32.8	AlL 5.0 AlL_2 8.7
Fe^{3+}	FeL 10.8	FeL 13.8	FeL 4.0 FeL_2 7.6 FeL_3 9.6	FeL 3.7 FeOHL 16.7 $FeOHL_2$ 19.4 $FeOHL_3$ 20.9	FeL 13.5 $Fe_2(OH)_2L_2$ 52.7	FeL 9.3	FeL 17.6 FeL_2 28.6 FeL_3 37.2	
Mn^{2+}	MnL 3.2		MnL 1.4	MnL 1.6	MnL 5.5 MnHL 9.4	MnL 3.3	MnL 6.8 MnL_2 10.7	MnL 2.7
Fe^{2+}	FeL 4.3	FeL 4.6	FeL 1.4	FeL 1.9	FeL 5.7 FeHL 9.9		FeL 7.4 FeL_2 12.1	
Co^{2+}	CoL 5.1 CoL_2 9.0 CoL_3 11.6	CoL 5.4 CoL_2 8.7	CoL 1.5	CoL 2.0 CoL_2 3.0	CoL 6.3 CoHL 10.3 CoH_2L 12.9	CoL 3.7 CoL_2 5.1 CoHL 7.0	CoL 7.5 CoL_2 12.3	CoL 2.8 CoHL 7.2

248

	A	B	C	D	E	F	G	H
Ni²⁺	NiL 6.2, NiL₂ 11.1, NiL₃ 14.2	NiL 6.5, NiL₂ 10.6	NiL 1.4	NiL 2.3, NiL₂ 3.4, NiL₃ 3.7	NiL 4.1, NiL₂ 5.8, NiHL 7.2	NiL 6.7, NiHL 10.6, NiH₂L 13.4	NiL 7.8, NiL₂ 12.6	NiL 3.0, NiHL 7.6
Cu²⁺	CuL 8.6, CuL₂ 15.6	CuL 8.8, CuL₂ 15.0	CuL 2.2, CuL₂ 3.6	CuL 2.9, CuL₂ 4.7, CuL₃ 4.7	CuL 5.7, CuL₂ 8.2, CuHL 8.3	CuL 7.2, CuHL 10.7, CuH₂L 13.9, CuOHL 16.4, Cu₂L₂ 16.3	CuL 11.5, CuL₂ 19.3	CuL 4.0, CuHL 7.7
Zn²⁺	ZnL 5.4, ZnL₂ 9.8, ZnL₃ 12.3	ZnL 5.8, ZnL₂ 9.5, ZnL₃ 9.8	ZnL 1.6, ZnL₂ 1.8	ZnL 2.4, ZnL₂ 3.6, ZnL₃ 3.9	ZnL 3.8, ZnL₂ 5.4, ZnHL 7.1	ZnL 6.3, ZnL₂ 8.6, ZnHL 13.5, ZnH₂L 12.9	ZnL 7.5	ZnL 2.9, ZnL₂ 4.2
Pb²⁺	PbL 5.5, PbL₂ 8.9		PbL 2.7, PbL₂ 4.1	PbL 2.5, PbL₂ 3.7, PbL₃ 3.6	Pb 4.0, PbL₂ 4.5	PbL 5.4, PbL₂ 8.1, PbHL 13.5, PbH₂L 13.1		
Hg²⁺	HgL 10.9, HgL₂ 20.1		HgL 6.1, HgL₂ 10.1, HgL₃ 14.1, HgL₄ 17.6			HgL 12.2		
Cd²⁺	CdL 4.7, CdL₂ 8.4, CdL₃ 10.7	CdL 4.8	CdL 1.9, CdL₂ 3.2	CdL 1.9, CdL₂ 2.7	CdL 3.2, CdL₂ 4.0, CdHL 6.9	CdL 5.1, CdL₂ 7.2, CdHL 12.7, CdH₂L 12.6	CdL 6.4	CdL 3.4
Ag⁺	AgL 3.5, AgL₂ 6.9		AgL 0.7, AgL₂ 0.6	AgL 0.4, AgL₂ 0.5				

[a] Note: Constants are given as logarithms of the overall formation constants, β, for complexes and as logarithms of the overall precipitation constants for solids, at zero ionic strength and 25°C [From Smith and Martell 1975, 1976 and Martell and Smith 1974, 1977[2]. Exceptions are major ion interaction constants (Na^+, K^+, Ca^{2+}, Mg^{2+}, CO_3^{2-}, SO_4^{2-}, Cl^-) taken from Whitfield[3], hydrolysis (OH^-) and some carbonate (CO_3^{2-}) formation constants taken from Baes and Mesmer[4]; Cu^{2+}-CO_3^{2-} complexes constants taken from Sunda and Hanson[5]; the ZnS(aq) constant recalculated from the data of Sainte Marie et al.[6]; the $MnCO_3$(s) constant taken from Morgan.[7]] When necessary, constants have been extrapolated to $I = 0$ using the Davies equation (applied to all ions including H^+ and tri- and tetravalent ions), leading to the following values of (log of) activity coefficients:

I \ z	1	2	3	4
0.1 M, 2 M	0.11	0.44	0.99	1.76
0.3 M	0.13	0.52	1.17	2.08
0.5 M	0.15	0.60	1.35	2.40
1.0 M	0.14	0.56	1.26	2.24
3.0 M	0.07	0.28	0.63	1.12
4.0 M	0.03	0.12	0.27	0.48

For any given complex or solid several different reactions of formation or dissociation can be written so that it is necessary to specify the reaction considered when giving an equilibrium constant. This is achieved implicitly by standardizing the notation for the constants themselves as shown in Table 6.2.

An important caveat must accompany any compilation of thermodynamic data such as that of Table 6.1. While the choice of constants has been made with some care, there might still be some glaring errors or omissions. The original references have been examined only in a few instances, and previous compilations (particularly that of Smith and Martell and Martell and Smith[2]) have been relied upon extensively. A major issue is that of data consistency: for example, a complex formation constant reported by one author may have been calculated on the basis of a solubility constant that is not the same as that chosen in the compilation. For precise calculations it is essential to examine the original literature and crosscheck the methods for estimating the constants from experimental data. Note also that the absence of a reported constant for a metal-ligand combination in Table 6.1 does not imply that no important complex may form. It is in fact the value of such a table to make the missing information particularly visible. [The price paid for such advantage is that mixed complexes (e.g., HgClGly) cannot be accommodated in the format of the table.]

Some simple generalities about coordination chemistry can be deduced by inspection of Table 6.1. Consider first the sulfate column. The similarities among stability constants for metals of like charge is striking (e.g., $\log K \cong 2.2 - 2.8$ for divalent metals ions). It is a reflection of the principally electrostatic binding of the sulfate ion pair complexes. Similarly low equilibrium constants for complex formation correlated with ionic charge are also observed for carbonate (and bicarbonate) and halides.

A dominant feature of Table 6.1 is the high degree of similarity among organic ligands in their relative affinities for various metals. The absolute values of the stability constants may be different from one ligand to another, but their relative values from metal to metal are highly correlated. A good example is the complex stability sequence of the "transition"metals, $Mn^{2+} < Fe^{2+} < Co^{2+} < Ni^{2+} < Cu^{2+} > Zn^{2+}$, well-known as the Irving-Williams series.[9] On the basis of such empirical observations the metals can be organized into various groups exhibiting similar coordinative properties. Table 6.1 has been organized to highlight some of these correlations; the grouping of the metals reflects their coordination properties which can be explained by the configurations of their electron shells.[10]

It is apparent that trivalent metals are typically more reactive than divalent ones, but this is largely offset by the greater insolubility of their corresponding oxides and hydroxides. For example, many ligands have a relatively high affinity for Fe^{3+}, but the very high stability of the ferric hydroxides, including the solid $Fe(OH)_3$, keeps the free ferric ion activity so low that the extent of Fe complexation is limited. As a result Cu^{2+} and Hg^{2+}, the most reactive of the divalent metals, are probably most apt to form complexes in natural waters. Copper complexation is the major focus of experimental studies of trace metal

TABLE 6.2

Formulation of Stability Constants[a,b]

Mononuclear Complexes
Addition of ligand

$$M \xrightarrow{\frac{L}{K_1}} ML \xrightarrow{\frac{L}{K_2}} ML_2 \cdots \xrightarrow{\frac{L}{K_i}} ML_i \cdots \xrightarrow{\frac{L}{K_n}} ML_n$$

$$\xrightarrow{\quad\quad \beta_2 \quad\quad}$$
$$\xrightarrow{\quad\quad\quad \beta_i \quad\quad\quad}$$
$$\xrightarrow{\quad\quad\quad\quad \beta_n \quad\quad\quad\quad}$$

$$K_i = \frac{[ML_i]}{[ML_{(i-1)}][L]}$$

$$\beta_i = \frac{[ML_i]}{[M][L]^i}$$

Addition of protonated ligands

$$M \xrightarrow{\frac{HL}{*K_1}} ML \xrightarrow{\frac{HL}{*K_2}} ML_2 \cdots \xrightarrow{\frac{HL}{*K_i}} ML_i \cdots \xrightarrow{HL} ML_n$$

$$\xrightarrow{\quad\quad *\beta_2 \quad\quad}$$
$$\xrightarrow{\quad\quad\quad *\beta_i \quad\quad\quad}$$
$$\xrightarrow{\quad\quad\quad\quad *\beta_n \quad\quad\quad\quad}$$

$$*K_i = \frac{[ML_i][H^+]}{[ML_{(i-1)}][HL]}$$

$$*\beta_i = \frac{[ML_i][H^+]^i}{[M][HL]^i}$$

Polynuclear Complexes
In β_{nm} and $*\beta_{nm}$ the subscripts n and m denote the composition of the complex $M_m L_n$ formed.
[If $m = 1$, the second subscript $(= 1)$ is omitted.]

$$\beta_{nm} = \frac{[M_m L_n]}{[M]^m [L]^n}$$

$$*\beta_{nm} = \frac{[M_m L_n][H^+]^n}{[M]^m [HL]^n}$$

---

[a] Source: Adapted from Stumm and Morgan (1981).[8]
[b] Note: The same notation as that used in L. G. Sillén and A. E. Martell, *Stability Constants of Metal-Ion Complexes*, Special Publications, Nos. 17 and 25, Chemical Society, London, 1964 and 1971, is used. (In the text the notation β_1 or K_1 is used indifferently to indicate the constants of formation of $M_1 L_1$ complexes. β' or K' indicate mixed acidity constants expressed as a function of the activity of H^+ and of the concentrations of other reactants and products.)

speciation in aquatic systems and will serve as one of our principal examples throughout this chapter.

2. ION ASSOCIATION AMONG MAJOR AQUATIC CONSTITUENTS

If two constituents are present in widely different concentrations, say, by a factor of a 100 or more, the less abundant constituent can affect the activity of the other only negligibly through complex formation. To study the complexation of the major constituents of natural waters, it is then sufficient to consider a model composed of the components that account for 99% or so of the dissolved solids. For most natural waters, including seawater, these components are the metals H^+, Na^+, Ca^{2+}, Mg^{2+}, and K^+ and the ligands OH^-, Cl^-, SO_4^{2-}, and CO_3^{2-}. (HCO_3^- is included implicitly in the list of ligands as a "complex" of H^+ and CO_3^{2-}.) Our objective is to describe quantitatively to what degree these eight major aquatic constituents are bound to each other as complexes. For the purpose of comparison let us consider a freshwater and a seawater model, both at $pH = 8.1$.

Example 1. Ion Association in Freshwater at $pH = 8.1$*

$$TOTNa = 2.8 \times 10^{-4} \, M = 10^{-3.55}$$

$$TOTCa = 3.7 \times 10^{-4} \, M = 10^{-3.43}$$

$$TOTMg = 1.6 \times 10^{-4} \, M = 10^{-3.80}$$

$$TOTK = 6.0 \times 10^{-5} \, M = 10^{-4.22}$$

$$TOTCl = 2 \times 10^{-4} \, M = 10^{-3.70}$$

$$TOTSO_4 = 1 \times 10^{-4} \, M = 10^{-4.00}$$

$$TOTCO_3 = 1.0 \times 10^{-3} \, M = 10^{-3.00}$$

Example 2. Ion Association in Seawater at $pH = 8.1$*

$$TOTNa = 4.68 \times 10^{-1} \, M = 10^{-0.33}$$

$$TOTCa = 1.02 \times 10^{-2} \, M = 10^{-1.99}$$

$$TOTMg = 5.32 \times 10^{-2} \, M = 10^{-1.27}$$

$$TOTK = 1.02 \times 10^{-2} \, M = 10^{-1.99}$$

$$TOTCl = 2 \times 10^{-1} \, M = 10^{-0.26}$$

$$TOTSO_4 = 2.82 \times 10^{-2} \, M = 10^{-1.55}$$

$$TOTCO_3 = 2.38 \times 10^{-3} \, M = 10^{-2.62}$$

* Total concentrations from Holland (1978).[11]

The upper left corner of Table 6.1 provides the necessary list of species and thermodynamic constants, and the equilibrium problems can be formulated with appropriate mole balance and mass law equations. The complexation effects that we wish to study are reasonably subtle, however, since the formation constants of the complexes are only on the order of 10^0 to 10^2. In order to have any sort of precision in the seawater model, it is then imperative that proper ionic strength corrections be made. In this spirit of precision we choose from the literature, activity coefficients that are considered most appropriate for seawater[3] (see Table 6.3) rather than apply a general empirical expression such as the Davies equation. Concentration equilibrium constants are then readily calculated, and choosing H^+, Ca^{2+}, Mg^{2+}, K^+, Cl^-, SO_4^{2-}, and HCO_3^- as components yields Tableau 6.1 in which the constants are expressed for the formation of the species from the components.

Equilibrium calculations corresponding to our two examples are a bit tedious. Tables 6.4 and 6.5 show an iterative solution scheme particularly convenient with hand calculators: (1) initially the calculation is started by assuming the principal component concentrations to be equal to the total concentrations, (2) each of the species is then calculated from the corresponding mass law, and (3) the sum of species in each mole balance equation is compared to the imposed total concentration and the original component concentration is increased or decreased proportionally for the next iteration. For the freshwater model system the calculation converges in two iterations, while it takes

TABLE 6.3

Free Single Ion Activity Coefficients Used in the Ion Association Model[a]

Ion	γ	$-\log \gamma$
H^+	0.95	0.02
Na^+	0.71	0.15
K^+	0.63	0.20
All other $+1$ ions	0.68	0.17
Ca^{2+}	0.26	0.59
Mg^{2+}	0.29	0.54
OH^-	0.65	0.19
Cl^-	0.63	0.20
All other -1 ions	0.68	0.17
SO_4^{2-}	0.17	0.77
CO_3^{2-}	0.20	0.70
Uncharged species	1.13	-0.05

[a] Source: After Whitfield (1974).[3]

four iterations to obtain the solution of the seawater example. The test of convergence is of course the satisfaction of the mole balance equations.

In the freshwater example the complexes are of little importance for the speciation of the major constituents. The most significant complexation is that of sulfate which is calculated to be 10% bound to calcium and magnesium. In the seawater example, on the other hand, the effect of association among major ions is quite important. A major fraction of both sulfate (60%) and carbonate (34%) is complexed by the metals, and some 10% of the calcium and magnesium bound to the ligands. The results of Example 2 are by and large comparable to those of the historical Garrels and Thompson[12] model for seawater, and of course are similar to those of Whitfield[3] since most of the same constants and activity coefficients have been selected. The differences that exist among these various calculations point out the difficulty in estimating stability constants and activity coefficients in a system as complex as seawater. In all these traditional models of ion interactions in seawater, chloride complexes are considered unimportant. There is now evidence, however, that species such as NaCl, KCl,

TABLEAU 6.1

	H^+	Na^+	K^+	Ca^{2+}	Mg^{2+}	HCO_3^-	SO_4^{2-}	Cl^-	Freshwater	Seawater
H^+	1									+0.02
OH^-	−1								−14.0	−13.81
CO_3^{2-}	−1					1			−10.33	−9.80
HCO_3^-						1				
H_2CO_3	1					1			+6.35	+6.13
SO_4^{2-}							1			
HSO_4^-	1						1		+1.99	+1.39
Cl^-								1		
Na^+		1								
$NaCO_3^-$	−1	1				1			−9.06	−9.21
$NaHCO_3$	1	1				1			−0.25	−0.62
$NaSO_4^-$	1	1					1		+1.06	+0.31
K^+			1							
KSO_4^-			1				1		+0.96	+0.16
Ca^{2+}				1						
$CaOH^+$	−1			1					−12.85	−13.27
$CaCO_3$	−1			1		1			−7.13	−7.94
$CaHCO_3^+$				1		1			+1.26	+0.67
$CaSO_4$				1			1		+2.31	+0.90
Mg^{2+}					1					
$MgOH^+$	−1				1				−11.44	−11.81
$MgCO_3$	−1				1	1			−6.93	−7.69
$MgHCO_3^+$					1	1			+1.16	+0.62
$MgSO_4$					1		1		+2.36	+1.00

TABLE 6.4

Calculation of Major Ion Interactions in a Freshwater System

	1st Iteration	2nd Iteration	% Metal	% Ligand
H^+	8.1	8.1		
Na^+	3.55	3.55	100	
K^+	4.22	4.22	100	
Ca^{2+}	3.43	3.45	95	
Mg^{2+}	3.80	3.82	95	
HCO_3^-	3.00	3.02		96
SO_4^{2-}	4.00	4.05		90
Cl^-	3.70	3.70		100
OH^-	5.9	5.9		
CO_3^{2-}	5.23	5.25		1
H_2CO_3	4.75	4.77		2
HSO_4^-	10.11	10.16		
$NaCO_3^-$	7.51	7.53		
$NaHCO_3$	6.80	6.82		
$NaSO_4^-$	6.49	6.54		
KSO_4^-	7.26	7.31		
$CaOH^+$	8.18	8.20		
$CaCO_3$	5.46	5.50 ⎱	3	1
$CaHCO_3^+$	5.17	5.21 ⎰		
$CaSO_4$	5.12	5.19	2	6
$MgOH^+$	7.14	7.16		
$MgCO_3$	5.63	5.67 ⎱	3	
$MgHCO_3^+$	5.64	5.68 ⎰		
$MgSO_4$	5.44	5.51	2	3
ΣNa	$10^{-3.55}$	$10^{-3.55}$		
ΣK	$10^{-4.22}$	$10^{-4.22}$		
ΣCa	$10^{-3.41}$	$10^{-3.43}$		
ΣMg	$10^{-3.78}$	$10^{-3.80}$		
ΣHCO_3	$10^{-2.98}$	$10^{-3.00}$		
ΣSO_4	$10^{-3.95}$	$10^{-4.00}$		
ΣCl	$10^{-3.7}$	$10^{-3.7}$		

TABLE 6.5

Calculation of Major Ion Interactions in Seawater

	1st Iteration	2nd Iteration	3rd Iteration	4th Iteration	% Metal	% Ligand
H^+	8.1	8.1	8.1	8.1		
Na^+	0.33	0.35	0.34	0.34	98	
K^+	1.99	2.01	2.00	2.00	98	
Ca^{2+}	1.99	2.08	2.03	2.03	91	
Mg^{2+}	1.27	1.38	1.32	1.32	89	
HCO_3^-	2.62	2.83	2.80	2.81		64
SO_4^{2-}	1.55	1.96	1.93	1.95		40
Cl^-	0.26	0.26	0.26	0.26		100
OH^-	5.71	5.71	5.71			
CO_3^{2-}	4.32	4.53	4.50	4.51		1
H_2CO_3	4.59	4.80	4.77	4.78		1
HSO_4^-	8.26	8.67	8.64	8.66		
$NaCO_3^-$	4.06	4.29	4.25	4.26		2
$NaHCO_3$	3.57	3.80	3.76	3.77		7
$NaSO_4^-$	1.57	2.00	1.96	1.98	2	37
KSO_4^-	3.38	3.81	3.78	3.80	2	1
$CaOH^+$	7.16	7.25	7.20	7.20		
$CaCO_3$	4.45	4.75	4.67	4.68		1
$CaHCO_3^+$	3.94	4.24	4.16	4.17	1	3
$CaSO_4$	2.64	3.14	3.06	3.08	8	3
$MgOH^+$	4.98	5.09	5.03	5.03		
$MgCO_3$	3.48	3.80	3.71	3.72		8
$MgHCO_3^+$	3.27	3.59	3.50	3.51	1	13
$MgSO_4$	1.82	2.34	2.25	2.27	10	19
ΣNa	$10^{-0.31}$	$10^{-0.34}$	$10^{-0.33}$	$10^{-0.33}$		
ΣK	$10^{-1.97}$	$10^{-2.00}$	$10^{-1.99}$	$10^{-1.99}$		
ΣCa	$10^{-1.90}$	$10^{-2.04}$	$10^{-1.99}$	$10^{-1.99}$		
ΣMg	$10^{-1.16}$	$10^{-1.33}$	$10^{-1.27}$	$10^{-1.27}$		
ΣHCO_3	$10^{-2.41}$	$10^{-2.65}$	$10^{-2.61}$	$10^{-2.62}$		
ΣSO_4	$10^{-1.14}$	$10^{-1.58}$	$10^{-1.53}$	$10^{-1.55}$		
ΣCl	$10^{-0.26}$	$10^{-0.26}$	$10^{-0.26}$	$10^{-0.26}$		

$MgCl^+$, and $CaCl^+$ may represent a significant fraction (13, 17, 43, 47%, respectively) of the total metal concentrations.[13] The recalculation of the ion-pairing model with the additional chloride constants is straightforward. However, the extension of these results to trace elements (see Section 3) would require a reinterpretation of the original experimental coordination data with equilibrium constants that are consistent with the new ion-pairing model.

As discussed in Chapter 2, interactions among ions in complex systems span the whole spectrum from unspecific long-range electrostatic interactions (of the type accounted for by the Debye-Hückel theory) to specific complex formation. In the *ion-association model* of seawater (Example 2), the mutual interactions among major ions are divided into these two types, utilizing simultaneously single ion activity coefficients and formation constants for the ion pair complexes. Such division between ideal and nonideal interactions among ions is largely arbitrary, particularly for ion pairs. As is the case for nonideal effects, the major part of the energy for ion pair formation is electrostatic and may be considered "long range" since the primary solvation shell is thought to be intact. Still we have chosen to consider that these long-range electrostatic interactions and not the others resulted in the formation of independent chemical entities. To paraphrase Horne,[14] we have chosen to bless some liaisons into marriages and considered the others to be outside the ideality laws.

A different approach is clearly possible: all electrostatic interactions among ions, including what we have taken to be ion pair formation, can be considered as nonideal interactions. In this approach a *total activity coefficient*, γ_i^T, is defined for each ion as the ratio of the free ion activity to the *total* ion concentration:

$$\gamma_i^T = \frac{\{C_i\}}{(C_i)_T} \tag{1}$$

For example, according to Example 2, the total activity coefficients for the major seawater ions are calculated to be

$$\gamma_{Na}^T = \frac{\{Na^+\}}{Na_T} = \gamma_{Na} \frac{(Na^+)}{Na_T} = 0.71 \times 0.98 = 0.70 \tag{2}$$

$$\gamma_K^T = \frac{\{K^+\}}{K_T} = \gamma_K \frac{(K^+)}{K_T} = 0.63 \times 0.98 = 0.62 \tag{3}$$

$$\gamma_{Ca}^T = \frac{\{Ca^{2+}\}}{Ca_T} = \gamma_{Ca} \frac{(Ca^{2+})}{Ca_T} = 0.26 \times 0.91 = 0.24 \tag{4}$$

$$\gamma_{Mg}^T = \frac{\{Mg^{2+}\}}{Mg_T} = \gamma_{Mg} \frac{(Mg^{2+})}{Mg_T} = 0.29 \times 0.89 = 0.26 \tag{5}$$

$$\gamma_{SO_4}^T = \frac{\{SO_4^{2-}\}}{(SO_4)_T} = \gamma_{SO_4} \frac{(SO_4^{2-})}{(SO_4)_T} = 0.17 \times 0.4 = 0.068 \tag{6}$$

$$\gamma_{Cl}^T = \frac{\{Cl^-\}}{Cl_T} = \gamma_{Cl} \frac{(Cl^-)}{Cl_T} = 0.63 \times 1.0 = 0.63 \tag{7}$$

For the carbonate system the coefficients are defined for each acid-base species:

$$\gamma_{H_2CO_3}^T = \frac{\{H_2CO_3^*\}}{(H_2CO_3)_T} = \gamma_{H_2CO_3}\frac{(H_2CO_3^*)}{(H_2CO_3^*)} = 1.13 \tag{8}$$

$$\gamma_{HCO_3}^T = \frac{\{HCO_3^-\}}{(HCO_3)_T}$$

$$= \gamma_{HCO_3}\frac{(HCO_3^-)}{(HCO_3^-) + (NaHCO_3) + (CaHCO_3^+) + (MgHCO_3^+)}$$

$$= 0.68 \times 0.74 = 0.50 \tag{9}$$

$$\gamma_{CO_3}^T = \frac{\{CO_3^{2-}\}}{(CO_3)_T} = \gamma_{CO_3}\frac{(CO_3^{2-})}{(CO_3^{2-}) + (NaCO_3^-) + (CaCO_3) + (MgCO_3)}$$

$$= 0.20 \times 0.10 = 0.020 \tag{10}$$

Such total activity coefficients are of course dependent on the ionic composition of the system, and they have to be defined anew for each different system. It is important to realize that total activity coefficients rather than ion pair formation constants are the quantities most directly amenable to experimental determination. Our calculation of total activity coefficients on the basis of the ion pair model is essentially a reversal of the process by which the parameters of the ion pair model are estimated in the first place.

In the same way that concentration equilibrium constants can be defined for any ionic strength by including activity coefficients into thermodynamic constants, *apparent equilibrium constants* can be defined for a particular solution by incorporating the total activity coefficients. For example, for the carbonate system in seawater, the first and second apparent mixed acidity constants are defined as

$$(K_1^{app})' = \frac{\{H^+\}(HCO_3)_T}{(H_2CO_3)_T} = K_1\frac{\gamma_{H_2CO_3}^T}{\gamma_{HCO_3}^T} = 10^{-6.00} \tag{11}$$

$$(K_2^{app})' = \frac{\{H^+\}(CO_3)_T}{(HCO_3)_T} = K_2\frac{\gamma_{HCO_3}^T}{\gamma_{CO_3}^T} = 10^{-8.93} \tag{12}$$

[These apparent constants vary somewhat depending on the chosen ion pair formation model. For example, Stumm and Morgan[15] report: $(pK_1^{app})' = 6.035$; $(pK_2^{app})' = 9.09$.] We obtain the apparent solubility product of calcite in the same way

$$K_s^{app} = (Ca^{2+})_T(CO_3^{2-})_T = \frac{K_s}{\gamma_{Ca}^T\gamma_{CO_3}^T} = 10^{-6.03} \tag{13}$$

Much of the interest in calculating the degree of major ion association in seawater centers on its effects on the carbonate system and particularly calcium carbonate solubility. (Note that $CaCO_3(s)$ has not been considered in Example 2, thus allowing for calcite supersaturation in the seawater.)

In reference to the ion pair model of Example 2, apparent acidity and solubility constants effectively lump together the various ion pairs for each carbonate species, not unlike the way that H_2CO_3 and $CO_2(aq)$ are lumped together to define $H_2CO_3^*$. This is a very convenient convention for studying carbonate chemistry in seawater: by simple redefinition of the acidity constants, $pK_{a1} = 6.0$, $pK_{a2} = 8.9$, we can consider seawater as a typical CO_2, C_T, Alk, pH system, ignoring the complications of major ion interactions. Note that calculations with apparent constants for the carbonate system in seawater and complete ion-pairing calculations are equivalent only because the major cations are little affected by ion pairing. Were it otherwise, the free concentrations of the metals and hence their effects on the carbonate speciation would be affected by the pH through carbonate complexation and the apparent constants would vary with pH.

It has to be realized that the two approaches, total activity coefficients or ion pair complexes, provide equally valid thermodynamic representations of ion interactions in water. Whether or not we wish to consider major ion interactions as actual complex formation is more a matter of philosophy and terminology than a matter of fact. In the end it is really a matter of practicality. If one is considering systems with highly variable ionic strength and composition, then it seems more convenient to use a universal description of ion association according to the ion-pairing approach. (For concentrated brines the ion-pairing model is insufficient, so other ion interaction models including first- and second-order effects among all major ions must be considered; see Chapter 2.) If, on the other hand, one is dealing exclusively with seawater composition, then the total activity coefficient approach becomes very efficient. The whole major seawater ion model of Tableau 6.1 then can be reduced to its first 11 lines, eliminating all ion pairs but adding the proper total activity coefficients in the last column.

3. INORGANIC COMPLEXATION OF TRACE ELEMENTS

Unlike the question of interactions among major ions, where everything depends on everything else, the question of the inorganic speciation of minor constituents of natural waters can be treated one element at a time, independently of the other constituents. This is because the important complexes are formed with constituents in large excess whose free concentrations (activities) are unaffected by complexation with a trace element.

To focus this discussion, let us consider a divalent trace metal M^{2+}. The inorganic ligands that may form important complexes with this metal in natural waters are relatively few: OH^-, Cl^-, SO_4^{2-}, CO_3^{2-}, S^{2-} (HCO_3^- and HS^- are included implicitly). Other reactive inorganic ligands (e.g., F^-, Br^-, NH_3, PO_4^{3-}, CN^-) are usually present in concentrations too low to produce any significant metal binding.

Consider the following complex formation reactions with ligands A^- and B^-:

$$M^{2+} + mA^- = MA_m^{(m-2)-}; \beta_m \tag{14}$$

$$M^{2+} + nB^- = MB_n^{(n-2)-}; \beta_n \tag{15}$$

Introducing the mass laws of Reactions 14 and 15 into the mole balance equation for M^{2+} allows the separation of the metal and ligand concentrations into the two factors of a product:

$$TOTM = (M^{2+}) + (MA_m^{(m-2)-}) + (MB_n^{(n-2)-}) + \cdots \tag{16}$$

$$TOTM = (M^{2+})[1 + \beta_m(A^-)^m + \beta_n(B^-)^n + \cdots] \tag{17}$$

Such separation is possible as long as there are no polynuclear species, which would introduce high exponents for (M^{2+}). To resolve the question of metal speciation, it is then sufficient to evaluate the term in brackets which depends exclusively on the ligand concentrations.

Since the inorganic ligands are normally in large excess of the trace metals that they bind, the problem of evaluating the free ligand concentrations is independent of the nature and concentration of the metals. Only two considerations then enter in the calculation: (1) the effect of major ion interactions on the ligand speciation as discussed in Section 2, and (2) the acid-base chemistry of weak acid-weak base ligands as a function of pH. The use of apparent constants accounting for all ion-pairing effects allows the resolution of the acid-base speciation independently of the major ion interactions. For example, using the ionization fraction formalism for carbonate in seawater, one can write

$$(CO_3^{2-}) = \alpha_2 C_T$$

where α_2 is calculated on the basis of the appropriate apparent acidity constants. In general, Equation 17 can thus be rewritten

$$TOTM = (M^{2+})[1 + \beta_m(\alpha_A)^m(A_T)^m + \beta_n(\alpha_B)^n(B_T)^n + \cdots] \tag{18}$$

where the coefficient α's are uniquely determined for a given pH and a given degree of major ion association.

In most cases only a few of the terms inside the brackets are important, typically only one:

If $\qquad\qquad 1 \gg \beta_m(\alpha_A)^m(A_T)^m,\ \beta_n(\alpha_B)^n(B_T)^n, \quad$ etc.

then M^{2+} is the dominant species.

If $\qquad\qquad \beta_m(\alpha_A)^m(A_T)^m \gg 1,\ \beta_n(\alpha_B)^n(B_T)^n, \quad$ etc.

then $MA_m^{(m-2)-}$ is the dominant species, and so on.

Species that are not dominant but may be significant in the speciation of M^{2+} are those for which the expression in brackets is not too small compared to the largest one (say $>1\%$). Since L_T is an upper limit on (L), the coefficients α are always smaller than 1; thus ligands for which $\beta_n(L_T)^n$ is less than 1 can never form dominant complexes.

The methodology used to calculate the inorganic complexation of a trace metal is thus remarkably simple:

1   List all the species considered.

2 Calculate the free ligand concentrations on the basis of some ion inter-action model and of the acid-base speciation of the weak acids and bases, most simply by using apparent acidity constants that include the effects of major ion interactions.

3 Calculate the various terms in brackets, and retain only those that are significant.

4 Obtain the distribution of the trace metal among its major inorganic species by division.

Example 3.   Inorganic Speciation of Copper and Cadmium in Freshwater

Let us consider the freshwater system of Example 1 and the following data from Table 6.1

$CuOH^+$; $\log \beta_1 = 6.5$ $CuCO_3$; $\log \beta_1 = 6.7$ $CuSO_4$; $\log \beta_1 = 2.4$

$Cu(OH)_2$; $\log \beta_2 = 11.8$ $Cu(CO_3)_2^{2-}$; $\log \beta_2 = 10.2$ $CuCl^+$; $\log \beta_1 = 0.5$

$CdOH^+$; $\log \beta_1 = 3.9$ $Cd(SO_4)$; $\log \beta_1 = 2.3$ $CdCl^+$; $\log \beta_1 = 2.0$

$Cd(OH)_2$; $\log \beta_2 = 7.6$ $Cd(SO_4)_2^{2-}$; $\log \beta_2 = 3.2$ $CdCl_2$; $\log \beta_2 = 2.6$

$Cd(SO_4)_3^{4-}$; $\log \beta_3 = 2.7$ $CdCl_3^-$; $\log \beta_3 = 2.4$

$CdCl_4^{2-}$; $\log \beta_4 = 1.7$

From the calculations of Example 1 we already know the free ligand concentrations:

$$(OH^-) = 10^{-5.9}$$

$$(CO_3^{2-}) = 10^{-5.3}$$

$$(SO_4^{2-}) = 10^{-4.0}$$

$$(Cl^-) = 10^{-3.7}$$

A rapid examination of the thermodynamic data shows the largest complexes to be the following:

For copper

$$CuOH^+ : \beta_1(OH^-) = 10^{+0.6} \tag{19}$$

$$Cu(OH)_2 : \beta_2(OH^-)^2 = 10^{0.0} \tag{20}$$

$$CuCO_3 : \beta_1(CO_3^{2-}) = 10^{1.4} \tag{21}$$

$$Cu(CO_3)_2^{2-} : \beta_2(CO_3^{2-})^2 = 10^{-0.4} \tag{22}$$

For cadmium

$$CdOH^+ : \beta_1(OH^-) = 10^{-2.0} \tag{23}$$

$$CdSO_4 : \beta_1(SO_4^{2-}) = 10^{-1.7} = 2 \times 10^{-2} \tag{24}$$

$$CdCl^+ : \beta_1(Cl^-) = 10^{-1.7} = 2 \times 10^{-2} \tag{25}$$

Other complexes are negligible. The copper is thus mostly present as carbonate and hydroxide complexes and only about 3% as the free (hydrated) cupric ion:

$$TOTCu = (Cu^{2+})[1 + 10^{0.6} + 10^{0.0} + 10^{1.4} + 10^{-0.4}]$$

$$= (Cu^{2+}) \times 10^{1.5} = (Cu^{2+}) \times 31.5 = Cu_T \qquad (26)$$

therefore

$$\frac{(Cu^{2+})}{Cu_T} = \frac{1}{31.5} = 3\%$$

$$\frac{(CuOH^+)}{Cu_T} = \frac{10^{0.6}}{31.5} = 13\%$$

$$\frac{(Cu(OH)_2)}{Cu_T} = \frac{10^{0.0}}{31.5} = 3\%$$

$$\frac{(CuCO_3)}{Cu_T} = \frac{10^{1.4}}{31.5} = 80\%$$

$$\frac{(Cu(CO_3)_2^{2-})}{Cu_T} = \frac{10^{-0.4}}{31.5} = 1\%$$

On the other hand, only about 5% of the cadmium is complexed, and the major species is by far the free cadmium ion, Cd^{2+}:

$$TOTCd = (Cd^{2+})[1 + 10^{-2} + 2 \times 10^{-2} + 2 \times 10^{-2}]$$

$$= (Cd^{2+}) \times 1.05 = Cd_T \qquad (27)$$

therefore

$$\frac{(Cd^{2+})}{Cd_T} = \frac{1}{1.05} = 95\%$$

$$\frac{(CdOH^+)}{Cd_T} = \frac{0.01}{1.05} = 1\%$$

$$\frac{(CdSO_4)}{Cd_T} = \frac{0.02}{1.05} = 2\%$$

$$\frac{(CdCl^+)}{Cd_T} = \frac{0.02}{1.05} = 2\%$$

Example 4.   Inorganic Speciation of Copper and Cadmium in Seawater

Consider the seawater model of Example 2; the free ligand concentrations are

$$(OH^-) = 10^{-5.7}$$

$$(CO_3^{2-}) = 10^{-4.5}$$

$$(SO_4^{2-}) = 10^{-2.0}$$

$$(Cl^-) = 10^{-0.26}$$

Extrapolation of the previous copper and cadmium constants to $I = 0.5\ M$ according to the Davies equations yields

$CuOH^+; \log \beta_1 = 5.9 \qquad CuCO_3; \log \beta_1 = 5.5 \qquad CuSO_4; \log \beta_1 = 1.2$

$Cu(OH)_2; \log \beta_2 = 10.9 \quad Cu(CO_3)_2^{2-}; \log \beta_2 = 9.0 \quad CuCl^+; \log \beta_1 = -0.2$

$CdOH^+; \log \beta_1 = 3.3 \qquad CdSO_4; \log \beta_1 = 1.1 \qquad CdCl^+; \log \beta_1 = 1.4$

$Cd(OH)_2; \log \beta_2 = 6.7 \quad Cd(SO_4)_2^{2-}; \log \beta_2 = 2.0 \quad CdCl_2; \log \beta_2 = 1.7$

$Cd(SO_4)_3^{4-}; \log \beta_3 = 2.7 \quad CdCl_3^-; \log \beta_3 = 1.5$

$CdCl_4^{2-}; \log \beta_4 = 1.1$

For copper, the hydroxide and carbonate complexes are still the most important, and the speciation of copper is changed from that of Example 3 mostly because of ionic strength effects:

$$CuOH^+ : \beta_1(OH^-) = 10^{0.2} \tag{28}$$

$$Cu(OH)_2 : \beta_2(OH^-)^2 = 10^{-0.5} \tag{29}$$

$$CuCO_3 : \beta_1(CO_3^{2-}) = 10^{1.0} \tag{30}$$

$$Cu(CO_3)_2^{2-} : \beta_2(CO_3^{2-})^2 = 10^{0.0} \tag{31}$$

$$CuSO_4 : \beta_1(SO_4^{2-}) = 10^{-0.8}$$

$$CuCl^+ : \beta_1(Cl^-) \cong 10^{-0.5}$$

Copper is now about 7% in the free ionic form:

$$TOTCu = (Cu^{2+})[1 + 10^{0.2} + 10^{-0.5} + 10^{1.0} + 10^{0.0}$$
$$+ 10^{-0.8} + 10^{-0.5}] = Cu_T \tag{32}$$

$$\frac{(Cu^{2+})}{Cu_T} = \frac{1}{14.4} = 10^{-1.16} = 0.07$$

For cadmium, the chloride complexes are now obviously the most important:

$$CdCl^+ : \beta_1(Cl^-) = 10^{+1.14} \tag{33}$$

$$CdCl_2 : \beta_2(Cl^-)^2 = 10^{+1.18} \tag{34}$$

$$CdCl_3^- : \beta_3(Cl^-)^3 = 10^{+0.72} \tag{35}$$

$$CdCl_4^{2-} : \beta_4(Cl^-)^4 = 10^{+0.06} \tag{36}$$

Since the constants are not very precisely known, we can consider that Cd is present in seawater predominantly as chloro complexes and that the free cadmium ion is roughly 3% of the total metal:

$$TOTCd = (Cd^{2+})[1 + 10^{1.14} + 10^{1.18} + 10^{0.72} + 10^{0.06}] = Cd_T \tag{37}$$

therefore

$$\frac{(Cd^{2+})}{Cd_T} = 10^{-1.56} = 0.03$$

If we consider the results of Examples 3 and 4 we note that the speciation of metals that form important carbonate or hydroxide complexes (e.g., Cu) is a strong function of pH. For metals that form important chloride or sulfate complexes (e.g., Cd^{2+}, Hg^{2+}) in seawater, their speciation in estuaries as freshwater mixes with seawater provides an interesting and somewhat tedious case study of inorganic complexation, as illustrated in Figure 6.3.

Hydroxide complexes, which may be considered to be produced by the dissociation of the weakly acidic hydrated metal ions, are of course important for many metals in natural waters, as demonstrated in our study of ferric hydroxide precipitation (see Example 2 in Chapter 5). Many metal hydroxides form polymers (e.g., $Fe_n(OH)_n^{2n+}$) on the way to precipitation as hydrous oxide solids. These polynuclear complexes are poorly studied and often metastable. Note also that the presence of stoichiometric coefficients greater than one for the metal complicates markedly the type of analysis presented in the preceding examples. At the limit there is a continuum between large metal hydroxide polymers and small suspended colloids of hydrous oxides.

From examination of the thermodynamic data in Table 6.1 it appears that apart from the hydroxides there are relatively few inorganic complexes of trace

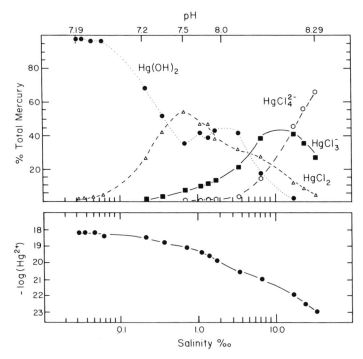

Figure 6.3 Inorganic speciation of mercury in an estuary as a function of salinity. This example has been calculated by considering the mixture of a freshwater $[(Cl^-) = 5 \times 10^{-4}\ M$, Alk $= 1.3 \times 10^{-4}\ M$, pH $= 7.2]$ with seawater (salinity $= 36\%_0$, pH $= 8.20$) and maintaining the total mercury concentration at $1\ nM$. Note the complicated changes in the inorganic mercuric species due to concomitant variations in pH (see top scale) and major ion (chiefly Cl^-) concentrations.

TABLE 6.6

Predominant Inorganic Species for Selected Trace Metals in Aquatic Systems

Ba
Ba^{2+}
$BaSO_4(s)$, $BaCO_3(s)$

Cr
$Cr(OH)_2^+$, $Cr(OH)_3$, $Cr(OH)_4^-$, $HCrO_4^-$, CrO_4^{2-}
$Cr(OH)_3(s)$

Al
$Al(OH)_3$, $Al(OH)_4^-$, AlF^{2+}, AlF_2^+
$Al(OH)_3(s)$, $Al_2O_3(s)$, $Al_2Si_2O_5(OH)_4(s)$, Al-silicates

Fe
Fe^{2+}, $FeCl^+$, $FeSO_4$, $Fe(OH)_2^+$, $Fe(OH)_4^-$
$FeS(s)$, $FeS_2(s)$, $FeCO_3(s)$, $Fe(OH)_3(s)$, $Fe_2O_3(s)$, $Fe_3O_4(s)$,
$FePO_4(s)$, $Fe_3(PO_4)_2(s)$, Fe-silicates

Mn
Mn^{2+}, $MnCl^+$
$MnS(s)$, $MnCO_3(s)$, $Mn(OH)_2(s)$, $MnO_2(s)$

Co
Co^{2+}, $CoCl^+$, $CoSO_4$
$CoS(s)$, $Co(OH)_2(s)$, $CoCO_3(s)$, $Co(OH)_3(s)$

Ni
Ni^{2+}, $NiCl^+$, $NiSO_4$
$NiS(s)$, $Ni(OH)_2(s)$

Cu
Cu^{2+}, $CuCO_3$, $CuOH^+$
$CuS(s)$, $CuFeS_2(s)$, $Cu_2CO_3(OH)_2(s)$, $Cu_3(CO_3)_2(OH)_2(s)$, $Cu(OH)_2(s)$,
$CuO(s)$

Zn
Zn^{2+}, $ZnCl^+$, $ZnSO_4$, $ZnOH^+$, $ZnCO_3$, ZnS
$ZnS(s)$, $ZnCO_3(s)$, $ZnSiO_3(s)$

Pb
Pb^{2+}, $PbCl^+$, $PbCl_2$, $PbCl_3^-$, $PbOH^+$, $PbCO_3$
$PbS(s)$, $PbCO_3(s)$, $Pb(OH)_2(s)$, PbO_2

Hg
Hg^{2+}, $HgCl^+$, $HgCl_2$, $HgCl_3^-$, $HgOHCl$, $Hg(OH)_2$, HgS_2^{2-}, $HgOHS^-$
$HgS(s)$, $Hg(liq)$, $Hg(OH)_2(s)$

Cd
Cd^{2+}, $CdCl^+$, $CdCl_2$, $CdCl_3^-$, $CdOH^+$, CdS, $CdHS^+$, $Cd(HS)_2$, $Cd(HS)_3^-$,
$Cd(HS)_4^{2-}$
$Cd(s)$, $CdCO_3(s)$, $Cd(OH)_2(s)$

Ag
Ag^+, $AgCl$, $AgCl_2^-$, AgS^-, $AgHS$, $AgHS_2^{2-}$, $Ag(HS)_2^-$
$Ag_2S(s)$, $AgCl(s)$, $AgBr(s)$

metals that are expected to be dominant in oxic waters. The major exceptions are the carbonate complex of copper in sufficiently alkaline systems and the chloride complexes of cadmium, silver, and mercury in the presence of high chlorinity. Generalizing to all species that account for a few percent of the total metal concentrations, Table 6.6 shows the inorganic aquatic species that are expected to be important for the trace metals of Table 6.1.

In anoxic waters the formation of sulfide, thiosulfate, and polysulfide complexes may be important in maintaining a fraction of some trace metals in solution despite the presence of very insoluble metal sulfides.[16,17]

4. ORGANIC COMPLEXATION

The concentration of dissolved organic matter in natural waters is typically in the range 10^{-1} to 10^{+1} mM C [dissolved organic carbon (DOC)]. In bogs the concentration may reach 50 mM C while in the ocean it varies from 40 μM C in deepwaters to six times this value in highly productive coastal surface waters. The chemical nature of this organic matter is poorly characterized. For example, in a typical attempt to fractionate organic compounds in seawater but 10% of the DOC could be identified, leaving some 90% to be broadly classified as humic and fulvic material (Table 6.7).[18] The term *humic acids* is normally used to indicate the fraction of DOC that precipitates at very low pH, while the term *fulvic acids* refers to acid soluble compounds, usually of lower molecular weight. Here we shall use the general terms humate, humic acid, or humic compounds to designate both fractions.

One could undoubtedly identify a myriad of other trace organic compounds from natural and pollution sources in all natural waters, but the compounds of Table 6.7 may in fact account for the bulk of the DOC in the samples. A majority of the organic matter is classified in the ill-defined class of humic acids because it possesses the inherent complexity and variability in structure characteristic of this class, probably not because we ignore the existence of some well-defined substance that accounts for a sizable fraction of the dissolved organic pool. Humic compounds represent typically some 40% to 99% of the DOC.

There is a major difference between the geochemical behavior of the humic organic fraction and the rest. Relatively small and reactive molecules such as sugars, amino acids, urea, or phenols provide readily available energy or nitrogen sources to aquatic microorganisms and are rapidly degraded. Their concentration is maintained in the water column by an equally rapid production rate from microorganisms. One should visualize this fraction of the total DOC as turning over very rapidly—on a time scale of minutes to days. The rapid turnover time of this material may be responsible in part for our ignorance of its chemical identity, and some researchers believe that compounds such as carbohydrates may represent in fact a sizable fraction of the DOC.[19,20] On the

TABLE 6.7

**Average Concentrations of Organic Compounds
in Baltic and North Sea Water**[a]

Components	Concentration (μg C liter^{-1})
Free amino acids	10
Combined amino acids	50 (to 100?)
Free sugars	20
Combined sugars	200
Fatty acids	10
Phenols	2
Sterols	0.2
Vitamins	0.006
Ketones	10
Aldehydes	5
Hydrocarbons	5
Urea	10
Uronic acids	18
Approximate identified total	340 μg C liter^{-1}
Approximate total	4000 μg C liter^{-1}

[a] Source: After Dawson (1976).[18]

other hand, the polymeric compounds that compose the humic fraction are refractory (a good definition for dissolved humic material is the refractory fraction of the DOC), and they are eliminated eventually, at least in part, by incorporation into the sediments rather than by degradation. Their residence time in the water column is on a time scale of weeks to years.

There is an ongoing scientific debate over the role that natural organic compounds play in complexing metal ions in aquatic systems.[21-23] On one side there are those who point to the imposing body of circumstantial experimental evidence that metal ions are largely, if not mostly, bound to organic ligands; e.g., nondialysability, apparent reduction of toxicity to assay organisms, and decrease of reduction kinetics at electrode surfaces. On the other side it is argued that very few important metal organic complexes have been directly demonstrated to exist, that stable colloids cannot easily be distinguished from soluble material, and that much of the experimental evidence cited above does not adequately discriminate between the various possible insoluble (precipitates and adsorbates) and macromolecular organic fractions. Furthermore, unless natural organic compounds possess unusual complexing specificity, protons and major cations such as Ca^{2+} and Mg^{2+} which are present in large excess of the organics should, in most instances, outcompete trace metals and occupy the available binding sites.

To examine quantitatively the question of organic complexation of trace metals in natural waters, the simplest approach is to make equilibrium calculations of model systems. For this purpose we can organize aquatic organic ligands into three categories:

1   Ligands of known composition that have been measured to form a sizable fraction of the DOC. Among these, amino acids have the most interesting complexing properties (Table 6.7).

2   Other ligands produced by the planktonic biota. Those may be present at only trace concentrations, but they may also exhibit much higher affinities for metals. The information concerning these ligands originates mostly from culture studies where high concentrations can be obtained, and this information often relates to the complexing characteristics of the ligands, not their exact chemical identity.

3   Humic substances isolated from a variety of aquatic systems and whose coordination properties have been characterized on concentrated samples by an assortment of methods.

Let us then examine each of these categories and obtain some quantitative estimates of organic complexation, recognizing that the question must ultimately be resolved through direct analysis and not model calculations.

4.1   Metal Complexation by Amino Acids

Of the organic compounds identified in natural waters (Table 6.7), amino acids are the major class with sizable complexation affinities for metals. The presence of both a carboxyl and an amino group gives all amino acids the ability to coordinate metals at two positions, and they are among the simplest chelating agents:

$$R-CH \begin{array}{c} NH_3^+ \\ \\ COO^- \end{array} + M^{2+} = R-CH \begin{array}{c} NH_2 \\ \\ COO \end{array} M^+ + H^+$$

In a few amino acids the radical R is highly reactive, and its specific chemistry controls the metal affinity of the ligand. However, in most common amino acids the radical is relatively unreactive and has little influence on the coordination characteristics of the amino or the carboxyl groups. As a result the acid-base and the metal complexation properties of many amino acids are very similar, leading to pK's that are within 0.5 units of each other (see Table 6.1).

The common trace metals that are most reactive with amino acids are iron (Fe·III), mercury, and copper, in accordance with our general discussion of complex formation. (In addition uncommon ions such as Zr^{4+}, Pd^{2+}, and Ga^{3+} are also quite reactive.) To examine the possible complexing role of amino acids in natural waters, let us then focus on copper which is much more soluble than iron and is naturally more abundant than mercury.

Example 5. Complexation of Copper by Glycine in a Freshwater System

Consider the model freshwater system of Example 1 to which we add $10^{-7}\ M$ copper. For what ligand concentration would the copper-glycine complex be the major aqueous copper species?

Major Inorganic Copper Complexes

$$Cu^{2+} + OH^- = CuOH^+; \qquad \log K = 6.5$$

$$Cu^{2+} + 2OH^- = Cu(OH)_2; \qquad \log K = 11.8$$

$$Cu^{2+} + CO_3^{2-} = CuCO_3; \qquad \log K = 6.7$$

$$Cu^{2+} + 2CO_3^{2-} = Cu(CO_3)_2^{2-}; \quad \log K = 10.2$$

$$Cu^{2+} + SO_4^{2-} = CuSO_4; \qquad \log K = 2.4$$

$$Cu^{2+} + Cl^- = CuCl^+; \qquad \log K = 0.4$$

Major Glycine Complexes

$$H^+ + Y^- = HY; \qquad \log K = 9.8$$

$$2H^+ + Y^- = H_2Y^+; \quad \log K = 12.1$$

$$Ca^{2+} + Y^- = CaY^+; \quad \log K = 1.4$$

$$Mg^{2+} + Y^- = MgY^+; \quad \log K = 2.7$$

Copper-Glycine Complexes

$$Cu^{2+} + Y^- = CuY^+; \quad \log K = 8.6$$

$$Cu^{2+} + 2Y^- = CuY_2; \quad \log K = 15.6$$

To solve such a problem, the simplifications that we made for studying inorganic complexation are not possible a priori: the organic ligand concentration need not be in large excess of the trace metal(s), and the speciation of the ligand may be markedly affected by complexation with the metal(s). Equilibrium problems involving several trace metals and ligands can be quite complicated and are a good field of application for iterative computer methods, although convenient hand methodologies can also be developed as we shall see.

To simplify matters in this example, let us hypothesize initially that the copper-glycine complexes will not be important for glycine speciation and verify this hypothesis a posteriori.

As previously, the mass law equations can be introduced in the copper mole balance to factor out the free cupric ion concentration:

$$TOTCu = (Cu^{2+})[1 + 10^{6.5}(OH^-) + 10^{11.8}(OH^-)^2 + 10^{6.7}(CO_3^{2-})$$
$$+ 10^{10.2}(CO_3^{2-})^2 + 10^{2.4}(SO_4^{2-}) + 10^{0.4}(Cl^-)$$
$$+ 10^{8.6}(Y^-) + 10^{15.6}(Y^-)^2] \qquad (38)$$

Introducing the free inorganic ligand concentrations calculated in Example 1,

$$(OH^-) = 10^{-5.9}$$
$$(CO_3^{2-}) = 10^{-5.3}$$
$$(SO_4^{2-}) = 10^{-4.0}$$
$$(Cl^-) = 10^{-3.7}$$

we find in accordance with Example 3 that sulfate and chloride complexes can be neglected and that the mole balance simplifies to

$$TOTCu = (Cu^{2+})[10^{1.5} + 10^{8.6}(Y^-) + 10^{15.6}(Y^-)^2] \qquad (39)$$

where $10^{1.5} = 31.5$ is the ratio of the total inorganic copper (including hydroxide and carbonate complexes) to the free cupric ion. The copper-glycine complexes will be dominant when

$$10^{8.6}(Y^-) + 10^{15.6}(Y^-)^2 > 10^{1.5}$$

therefore

$$(Y^-) > 10^{-7.3} \qquad (40)$$

The free ligand concentration, (Y^-), can be estimated as a function of the total ligand following the same methodology:

$$TOTY = Y_T = (Y^-)[1 + 10^{9.8}(H^+) + 10^{12.1}(H^+)^2 + 10^{2.7}(Mg^{2+})$$
$$+ 10^{1.4}(Ca^{2+}) + 10^{8.6}(Cu^{2+}) + 10^{15.6}(Y^-)(Cu^{2+})] \qquad (41)$$

Neglecting the copper species as hypothesized and introducing the known free metal concentrations, $(H^+) = 10^{-8.1}, (Ca^{2+}) = 10^{-3.45}$, and $(Mg^{2+}) = 10^{-3.82}$, yields

$$Y_T = 10^{1.7}(Y^-) \qquad (42)$$

(Note that the only important species is HY.) The total concentration of glycine necessary to complex most of the copper is then

$$Y_T > 10^{1.7} \times 10^{-7.3} = 10^{-5.6} = 2.5 \times 10^{-6} \ M \qquad (43)$$

This condition is also sufficient since it verifies our initial hypothesis that the ligand be in large excess of the metal $(Y_T \gg 10^{-7} \ M)$ and thus not be markedly affected by metal complexation.

Compared to natural concentrations of amino acids, the concentration of glycine calculated in Example 5 is large, and it appears that relatively weak ligands, such as amino acids, are not likely to be important metal complexing agents in most natural waters.

To generalize such a result over a reasonably wide range of pH, we can consider the reaction of copper with the major ligand species, HY, over the pH range 2.3 to 9.8:

$$HY + Cu^{2+} = CuY^+ + H^+; \quad K = 10^{-1.2} \tag{44}$$

$$2HY + Cu^{2+} = CuY_2 + 2H^+; \quad K = 10^{-4.0} \tag{45}$$

therefore

$$\frac{(CuY^+)}{(Cu^{2+})} = \frac{10^{-1.2}(HY)}{(H^+)} = \frac{10^{-1.2}Y_T}{(H^+)} \tag{46}$$

$$\frac{(CuY_2)}{(Cu^{2+})} = \frac{10^{-4.0}(HY)^2}{(H^+)^2} = \frac{10^{-4.0}Y_T^2}{(H^+)^2} \tag{47}$$

The "effective" binding constants

$$
\begin{aligned}
K_1^{eff} &= \frac{(CuY^+)}{(Cu^{2+})(Y_T)} \\
K_2^{eff} &= \frac{(CuY_2)}{(Cu^{2+})(Y_T)^2}
\end{aligned}
\tag{48}
$$

which include the acid-base speciation of the ligand are thus highly dependent on pH; in the pH range where HY is the dominant ligand

$$
\begin{aligned}
K_1^{eff} &= 10^{-1.2}(H^+)^{-1} \\
K_2^{eff} &= 10^{-4.0}(H^+)^{-2}
\end{aligned}
\tag{49}
$$

Since most trace organic ligands are present in less than micromolar concentrations in natural waters ($Y_T = 10^{-7} - 10^{-9}$ M), their effective copper binding constants need to be in excess of $10^{8.5}$ and $10^{15.5}$, respectively, in order to outcompete the carbonate complex. For glycine this does not occur below pH = 9.5, which is the maximum pH for which HY is the major ligand species and Equations 48 are applicable. As pH increases the carbonate and hydroxide complexes also become increasingly important, $(Cu \cdot inorganic)/(Cu^{2+}) \gg 10^{1.5}$, and a comparison of effective inorganic and organic constants shows that the organic complexes cannot be important at the low ligand concentrations found in natural waters.

This result is generalizable to all organic ligands with similar complexing characteristics. It is also generalizable to all common trace metals since those that form weaker inorganic complexes than copper also form much weaker organic complexes. For example, the effective binding constants of glycine for cadmium in the pH range 2.5 to 9.5 are given by

$$K_1^{\text{eff}} = \frac{(\text{CdY}^+)}{(\text{Cd}^{2+})(\text{Y}_T)} = 10^{-5.1}(\text{H}^+)^{-1} \tag{50}$$

$$K_2^{\text{eff}} = \frac{(\text{CdY}_2)}{(\text{Cd}^{2+})(\text{Y}_T)^2} = 10^{-11.2}(\text{H}^+)^{-2} \tag{51}$$

At a pH of 8, glycine would thus have to be in excess of $10^{-2.9}$ M to complex cadmium, even though the inorganic complexes are not important and the cadmium ion is practically free in freshwater.

For a metal such as iron the complexation constant with glycine is two orders of magnitude greater than that for copper, but the free metal ion concentration is reduced several orders of magnitude by formation of stable hydroxide complexes:

$$\text{Fe}_T = (\text{Fe}^{3+}) + (\text{FeOH}^{2+}) + (\text{Fe(OH)}_2^{2+}) + (\text{Fe(OH)}_4^-) + (\text{FeY}^{2+}) \tag{52}$$

$$\text{Fe}_T = (\text{Fe}^{3+})[1 + 10^{-2.2}(\text{H}^+)^{-1} + 10^{-5.7}(\text{H}^+)^{-2} + 10^{-21.6}(\text{H}^+)^{-4}$$
$$+ 10^{1.0}(\text{H}^+)^{-1}\text{Y}_T] \tag{53}$$

The Fe organic complex can thus never be dominant if $\text{Y}_T < 10^{-3.2}$ M. Note that Equation 53 only considers the soluble species and that iron is very insoluble at neutral pH.

In conclusion, over the range of pH encountered in natural waters trace concentrations of weak organic ligands simply cannot outcompete the inorganic ligands (e.g., Fe(OH)_x; CuCO_3; CdCl_x) for metal complexation.

4.2   Trace Metal Complexation by Strong Chelating Agents

Metal complexation by strong artificial chelating agents such as NTA or EDTA is probably the best studied topic in aquatic coordination chemistry, and the effects of these chelating agents in complex mixtures have been thoroughly analyzed. Over the past 20 years biologists have capitalized on this chemical knowledge and used artificial chelating agents rather than mysterious and vaguely cabalistic soil extracts in the recipes of a wide variety of growth media. Partly on the basis of the known beneficial effects of these strong chelating agents for the growth of aquatic microorganisms (which should synthesize what is good for them according to "teleo-biological" dogma), aquatic chemists have sailed the waters of the world searching for natural analogs of the man-made chelators. These modern argonauts have been largely frustrated on the diluting high seas, and it is principally in laboratory beakers and culture vessels that natural organic chelators have been found so far.

Practically all aquatic organisms produce complexing agents in their growth medium, and these complexing agents can be classified into two categories. The first category includes a variety of compounds whose affinity for metals ranges from that of simple amino acids up to that of strong artificial chelating agents such as EDTA. Table 6.8 shows some data on these chelating agents which are usually defined only by their coordinative properties with some metals, predominantly copper. The second category is that of iron siderophores, compounds of very high affinity for Fe(III) which are produced by some classes

TABLE 6.8

Copper Complexing Agents Released by Aquatic Organisms

log K_{Cu}^{app}	Organism	Method	pH	Reference
Phytoplankton				
pH + 0.5[a]	*Pandorina morum*	Cu^{2+} titration	3–9	24
	Gloeocystis gigas		3–9	24
	Chlamydomonas sp.		3–9	24
	Chlorella autotrophica		3–9	24
	Thalassiosira pseudonana		3–9	24
	Thalassiosira weissflogii		3–9	24
	Tribonema aequale		3–9	24
	Synura petersenii		3–9	24
8.1	*Navicula pelliculosa*	ion exchange	8.0	25
8.6	*Scenedesmus quadricauda*		8.0	25
Macroalgae				
10.15	*Ectocarpus siliculosus*	ion exchange	8.15	26
9.7	*Audouinella purpurea*		8.15	26
9.95	*Antithamnion spirographidis*		8.15	26
Cyanobacteria				
11[b]	*Anacystis nidulans*	Cu^{2+} titration	6.3	27
12[b]	*Microcystis aeruginosa*		6.3	27
>7.5	*Gloeocapsa alpicola*		6.3	24
	Anabaena cylindrica	Cu^{2+} titration		27
7.2	(Fe^{3+}-rich cultures)		6.25	27
10.2[c]	(Fe^{3+}-limited cultures)		6.25	27
	Anabaena flos-aquae			
8.4[c]	(Fe^{3+}-limited cultures)		6.25	27
	Synechococcus leopoliensus			
7.0	(Fe^{3+}-rich cultures)		6.25	27
7.7	*Anabaena cylindrica*	ion exchange	8.0	25
Bacteria				
4.20[d]	*Klebsiella aerogenes*	Cu^{2+} titration	ca. 5.7	28
3.40[e]	*Klebsiella aerogenes*	Cd^{2+} titration	ca. 5.7	28

(*continued*)

TABLE 6.8 (Continued)

log K_{Cu}^{app}	Organism	Method	pH	Reference
Fungi				
12.3, 6.0[b]	*Buergenerula spartinae*	Cu^{2+} titration	7.0	29
10.9, 7.1[b]			6.0	29
9.9, 6.2	*Leptosphaeria obiones*		7.0	29
11.7[b]	*Pleospora pelagica*		7.0	29
9.8, 6.5	*Asteromyces cruciatus*		7.0	29
11.6, 8.6, 6.6	*Curvularia sp.*		7.0	29
9.7, 7.1	*Dendryphiella salina*		7.0	29
9.3, 7.2	*Stagonospora sp.*		7.0	29
8.7, 7.6	*Varicosporina ramulosa*		7.0	29
12.0, 9.7, 7.9[b]	*Rhodosporidium sphaerocarpum*		7.0	29
Zooplankton				
8.6	*Daphnia magna*	Cu^{2+} titration	6.4	30

[a] Apparent constant written for H^+ exchange:

$$K = \frac{(CuL)(H^+)}{(Cu^{2+})(L)}$$

[b] Probably siderophores.
[c] Siderophores.
[d] Polysaccharides.
[e] Cadmium constant.

of microorganisms including bacteria, blue-green algae, and fungi, for transporting iron. The detailed structures and the exact properties of several of these siderophores have been characterized (see Figure 6.4) and one of them, desferri-ferrioxamin B, is included in Table 6.1. Conspicuous by its absence is any evidence for ligands of very high affinities for metals other than iron.

For the purpose of extrapolating from culture vessels to natural waters our information regarding the first category of ligands, we choose as a model compound the strong artificial complexing agent, EDTA, whose coordination properties are well-known and should provide a limiting case for the stronger complexing agents of that category. Again, according to the data of Table 6.1, copper is one of the metals expected to be most complexed by EDTA. For comparison purposes let us consider both copper and cadmium.

Example 6. Complexation of Copper and Cadmium by EDTA in Seawater

Since 10^{-6} M is the maximum concentration of natural organic ligand observed to be produced in dense laboratory cultures, 10^{-7} M EDTA is a reasonable upper limit to the possible ligand concentration in nature. To the recipe of our

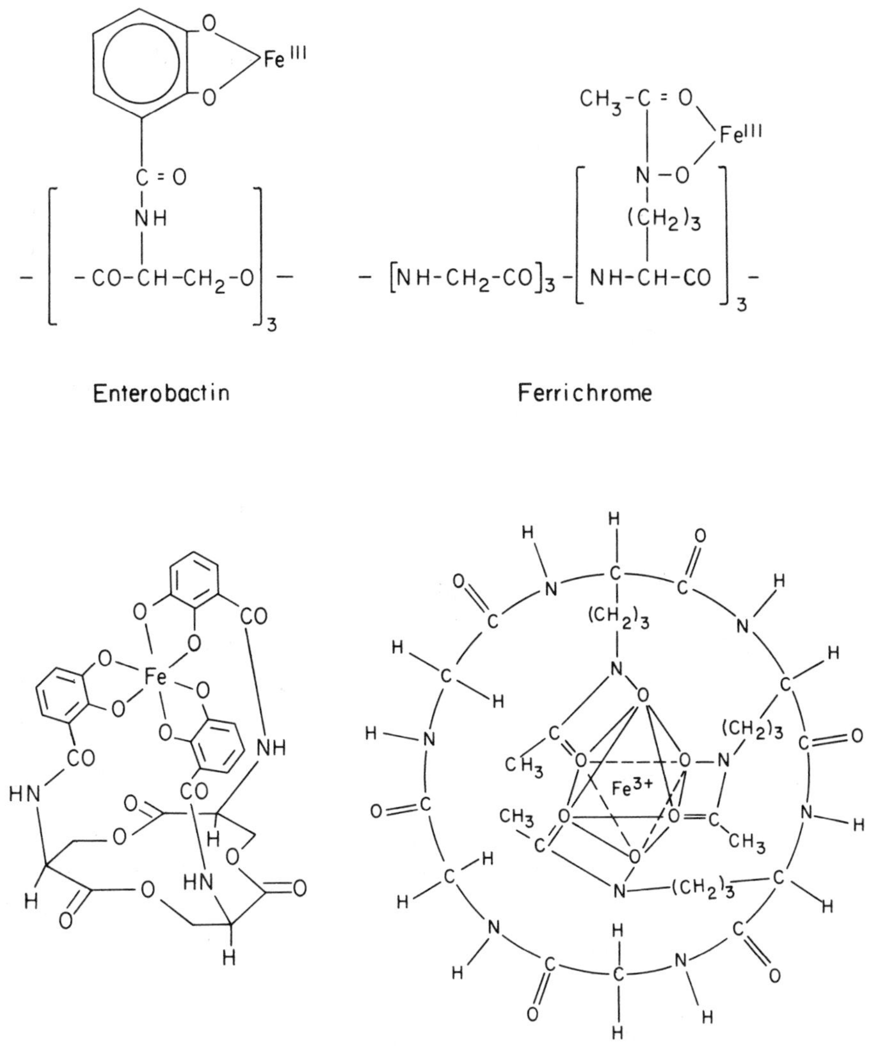

Enterobactin

Ferrichrome

Figure 6.4 Structures of two archetypical siderophores: the catechol enterobactin and the hydroxamate ferrichrome. Both of these compounds, and a large number of similar ones with equally strong affinity for Fe(III) ($K \cong 10^{20}$ to 10^{60}), are produced by microorganisms for transporting—and competing for—iron.

seawater model (Example 2), let us then add typical surface seawater concentrations of copper ($10^{-8.5}$ M), cadmium (10^{-9} M), and 10^{-7} M EDTA:

$$Y_T = 10^{-7} \ M$$

$$Cu_T = 10^{-8.5} \ M$$

$$Cd_T = 10^{-9} \ M$$

Recall that at pH = 8.1

$$(OH^-) = 10^{-5.7}$$

$$(CO_3^{2-}) = 10^{-4.5}$$

$$(SO_4^{2-}) = 10^{-2.0}$$

$$(Cl^-) = 10^{-0.26}$$

$$(Ca^{2+}) = 10^{-2.0}$$

$$(Mg^{2+}) = 10^{-1.3}$$

The relevant EDTA constants are obtained from Table 6.1 after appropriate ionic strength corrections:*

HY^{3-}; log $\beta_1' = 10.1$ CaY^{2-}; log $\beta_1 = 10.0$ CuY^{2-}; log $\beta_1 = 18.1$

H_2Y^{2-}; log $\beta_2' = 16.1$ MgY^{2-}; log $\beta_1 = 8.2$ CdY^{2-}; log $\beta_1 = 15.8$

To simplify the problem, let us recall the results of Example 4:

$$\sum \text{ inorganic Cu species} = 10^{1.16}(Cu^{2+})$$

$$\sum \text{ inorganic Cd species} = 10^{1.56}(Cd^{2+})$$

The mole balance equations for copper and cadmium thus simplify to

$$TOTCu = (Cu \cdot \text{inorganic}) + (CuY^{2-}) = (Cu^{2+})[10^{1.16} + 10^{18.1}(Y^{4-})]$$
$$= Cu_T = 10^{-8.5} \tag{54}$$

$$TOTCd = (Cd \cdot \text{inorganic}) + (CdY^{2-}) = (Cd^{2+})[10^{1.56} + 10^{15.8}(Y^{4-})]$$
$$= Cd_T = 10^{-9.0} \tag{55}$$

The free ligand concentration is obtained from the EDTA mole balance equation:

$$TOTY = (HY^{3-}) + (H_2Y^{2-}) + (CaY^{2-}) + (MgY^{2-}) + (CuY^{2-}) + (CdY^{2-})$$
$$= Y_T = 10^{-7}$$

The last two terms can obviously be neglected since Cu_T and $Cd_T \ll Y_T$:

$$TOTY = (Y^{4-})[10^{10.1}(H^+) + 10^{16.1}(H^+)^2 + 10^{10.0}(Ca^{2+})$$
$$+ 10^{8.2}(Mg^{2+})] = 10^{-7} \tag{56}$$

$$TOTY = (Y^{4-})[10^{2.0} + 10^{-0.1} + 10^{8.0} + 10^{6.9}] = 10^{-7} \tag{57}$$

Note that the first two terms are also negligible and that EDTA is predominantly in the calcium form. As a result

$$(Y^{4-}) = 10^{-8.0} \times 10^{-7} = 10^{-15.0} \tag{58}$$

---

* Ionic strength corrections according to the Davies equation have dubious validity for ions of charge 3 and 4. However, as seen in Example 6, the results of the calculations are controlled by the ratios K_{Cu}/K_{Ca} and K_{Cd}/K_{Ca} which are not affected by the ionic strength corrections.

Substituting into the copper and cadmium mole balances,

$$TOTCu = (Cu^{2+})[10^{1.16} + 10^{3.1}] = 10^{-8.5} \tag{59}$$

$$TOTCd = (Cd^{2+})[10^{1.56} + 10^{0.8}] = 10^{-9.0} \tag{60}$$

therefore

$$(Cu^{2+}) = 10^{-11.6}; \quad (Cu \cdot inorganic) = 10^{-10.4}; \quad (CuY^{2-}) = 10^{-8.5}$$

$$(Cd^{2+}) = 10^{-10.6}; \quad (Cd \cdot inorganic) = 10^{-9.1}; \quad (CdY^{2-}) = 10^{-9.9}$$

The copper is thus totally bound to EDTA while only some 12% of the cadmium is in the organic complex form. In this example there is of course no competition among the two trace metals for complexation with the organic ligand since EDTA is present in large excess.

Example 6 illustrates several important points concerning organic complexation of trace metals in natural waters. First and foremost, organic complexation is unlikely to be important for most trace metals in most natural waters. Despite the fact that EDTA is a very strong ligand and that $10^{-7} M$ is a high ligand concentration, the speciation of cadmium in seawater as calculated in Example 6 is barely affected by organic complexation. If the affinity and the concentration of the ligand were somewhat smaller, say, a factor of 10 each, even copper, one of the most reactive trace metals, would be present largely as inorganic species. For organic complexation to be important in the speciation of a trace metal, it is thus necessary either that the ligand exhibit a very high affinity for the metal (see next example) or that the ligand concentration and effective binding constants be higher than typically observed. Such situations are probably most often encountered in lakes supporting dense algal blooms or in freshwater systems of high humic content (see Section 4.3).

In addition to the effects of higher concentrations and lesser competition from inorganic ligands, a principal reason to expect organic complexation of trace metals to be more important in freshwater than in seawater is the dominant role of the major divalent cations, Ca^{2+} and Mg^{2+}, on the organic ligand speciation in seawater. As seen in Example 6, while copper is effectively 100% bound with EDTA, EDTA itself is 100% bound with calcium, which is itself 100% free. This is possible of course because of the wide differences in total concentrations: $Cu_T \ll Y_T \ll Ca_T$. The result is a large decrease in the apparent binding strength of EDTA for the metals:

$$(CuY^{2-}) = K_{Cu}(Cu^{2+})(Y^{4-}) \tag{61}$$

$$(CuY^{2-}) = \frac{K_{Cu}}{K_{Ca}(Ca^{2+})}(Cu^{2+})Y_T \tag{62}$$

therefore

$$K_{Cu}^{app} = \frac{K_{Cu}}{K_{Ca}(Ca^{2+})} = 10^{-8}K_{Cu} \tag{63}$$

In other words, the complexation of copper by EDTA is controlled by the equilibrium of the reaction

$$CaY^{2-} + Cu^{2+} = CuY^{2-} + Ca^{2+}; \quad K = \frac{K_{Cu}}{K_{Ca}} \qquad (64)$$

which is forced to the left by a large calcium concentration. Although sometimes the magnesium rather than the calcium complex is dominant, this result is applicable to most organic ligands in seawater, and it affects the complexation of all trace metals. In freshwaters, where the calcium and magnesium concentrations are one to three orders of magnitude smaller, the effective binding constants are increased accordingly. In some cases, depending on the particular ligand and the pH, the ligand speciation may in fact depend strictly on its acid-base chemistry. This is the case for weak ligands, as shown for glycine in Example 5. Even for EDTA the calcium and magnesium complexes become unimportant in soft acidic waters (e.g., where $pH = 4, (Ca^{2+}) = 10^{-4} M, I = 0 M$):

$$TOTY = (Y^{4-})[10^{11.1}(H^+) + 10^{17.9}(H^+)^2 + 10^{21.0}(H^+)^3 + 10^{12.4}(Ca^{2+})] \quad (65)$$

The ligand speciation is then dominated by the species H_2Y^{2-} and H_3Y^-.

Let us now consider the situation created by the presence of complexing agents with very high affinity for a particular metal, choosing the well-characterized trihydroxamate siderophore desferriferrioxamin B as our model compound.

Example 7.   Complexation of Iron (III) and Copper by a Trihydroxamate Siderophore in Freshwater

Consider a gradual addition of desferriferrioxamin B to the freshwater system of Example 1 containing $10^{-6} M$ Fe(III) and $10^{-7} M$ Cu. [Recall that at $pH = 8.1, (OH^-) = 10^{-5.9}, (CO_3^{2-}) = 10^{-5.3}, (SO_4^{2-}) = 10^{-4.0}, (Cl^-) = 10^{-3.2}, (Ca^{2+}) = 10^{-3.5}$, and $(Mg^{2+}) = 10^{-4.1}$.] The relevant organic complexation constants are given in Table 6.1:

$$HY^-; \log \beta_1 = 10.1 \quad CaY; \log \beta_1 = 3.5 \quad FeY^+; \log \beta_1 = 31.9$$

$$H_2Y; \log \beta_2 = 19.4 \quad MgY; \log \beta_1 = 5.2 \quad FeHY^{2+}; \log \beta = 32.6$$

$$H_3Y^+; \log \beta_3 = 27.8 \quad\quad\quad\quad\quad\quad\quad\quad CuY; \log \beta = 15.0$$

$$CuHY^+; \log \beta = 24.1$$

$$CuH_2Y^{2+}; \log \beta = 27.0$$

At a pH of 8.1, in the absence of chelating agents, iron is precipitated as $Fe(OH)_3(s)$ (see Example 2 in Chapter 5), and the free ferric ion concentration is given by

$$(Fe^{3+}) = 10^{3.2}(H^+)^3 = 10^{-21.1} \qquad (66)$$

The free cupric ion concentration in the absence of a chelator has been calculated as 3% ($10^{-1.5}$) of the total copper in Example 3:

$$(Cu^{2+}) = 10^{-1.5}Cu_T = 10^{-8.5}$$

The mole balance equation for desferriferrioxamin at low ligand concentration, low enough not to affect (Fe^{3+}) or (Cu^{2+}), can then be simplified:

$$
\begin{aligned}
TOTY = (Y^{2-})[1 &+ 10^{10.1}(H^+) + 10^{19.4}(H^+)^2 + 10^{27.8}(H^+)^3 \\
&+ 10^{3.5}(Ca^{2+}) + 10^{5.2}(Mg^{2+}) + 10^{31.9}(Fe^{3+}) \\
&+ 10^{32.6}(H^+)(Fe^{3+}) + 10^{15}(Cu^{2+}) + 10^{24.1}(H^+)(Cu^{2+}) \\
&+ 10^{27}(H^+)^2(Cu^{2+})] = Y_T
\end{aligned} \tag{67}
$$

All terms are negligible compared to the FeY^+ concentration:

$$TOTY = (FeY^+) = 10^{31.9}(Fe^{3+})(Y^{2-}) = 10^{10.8}(Y^{2-}) = Y_T \tag{68}$$

Given the enormous affinity of the organic ligand for the ferric ion, as long as $Y_T < Fe_T$, the speciation problem is simple: the ligand is entirely bound to iron; the free ferric ion is controlled by precipitation of $Fe(OH)_3(s)$; and the copper remains chiefly in the copper carbonate complex form $[K_{CO_3}(CO_3^{2-}) = 10^{1.4}]$.

As soon as the hydroxamate is in excess of Fe_T, however, it completely dissolves the iron, and the high affinity of the ligand for copper becomes important for copper speciation. Consider, for example, $Y_T = 10^{-5.8}$ M. In addition to FeY^+, the major species at pH = 8.1 are clearly H_2Y and H_3Y^+:

$$TOTY = (H_2Y) + (H_3Y^+) + (FeY^+) = Y_T = 10^{-5.8} \tag{69}$$

At this point a reasonable assumption to be verified subsequently is that all the iron is bound to the organic ligand:

$$(FeY^+) = Fe_T = 10^{-6} \tag{70}$$

therefore

$$(H_2Y) + (H_3Y^+) = Y_T - Fe_T = 10^{-6.23} \tag{71}$$

and

$$(Y^{2-})[10^{19.4}(H^+)^2 + 10^{27.8}(H^+)^3] = 10^{-6.23}$$

$$(Y^{2-})[10^{3.2} + 10^{3.5}] = 10^{-6.23} \tag{72}$$

$$(Y^{2-}) = 10^{-9.9}$$

We can verify that the iron is indeed dissolved:

$$(FeY^+) = 10^{31.9}(Y^{2-})(Fe^{3+}) = Fe_T = 10^{-6} \tag{73}$$

therefore

$$(Fe^{3+}) = 10^{-28.0} < 10^{-21.1}$$

The mole balance for copper is now written:

$$TOTCu = (Cu \cdot inorganic) + (CuY) + (CuHY^+) = Cu_T = 10^{-7} \qquad (74)$$

$$TOTCu = (Cu^{2+})[10^{1.5} + 10^{15.0}(Y^{2-}) + 10^{24.1}(H^+)(Y^{2-})] = 10^{-7}$$
$$TOTCu = (CuHY^+) = 10^{6.1}(Cu^{2+}) = 10^{-7} \qquad (75)$$

therefore

$$(Cu^{2+}) = 10^{-13.1}$$

The copper becomes immediately complexed by the hydroxamate, and its free concentration is lowered by a factor of approximately one million.

Since iron siderophores are thought to be ubiquitously produced by microorganisms, situations such as the one presented in the preceding example may be a more common occurrence than usually thought. It may well be that iron siderophores are responsible for complexing metals other than iron in some environments. Note, however, that this is possible only if the organic ligand is in excess of the iron, a situation that may be encountered exclusively in some lakes supporting dense blooms of blue-green algae. In seawater, as illustrated in Example 6, the major cations would affect (slightly) the complexing ability of the organic ligand toward copper since MgY is one of the major ligand species in seawater (once the iron is entirely complexed). In the most probable situation where iron siderophores are present at very low concentrations, lower than that of trace metals, their expected effect on metal speciation is a simple stoichiometric binding of Fe(III).

In Examples 5, 6, and 7 the problem of trace metal complexation has been kept simple by considering only one or two trace metals and one organic ligand at a time. The problem can become apparently quite complicated when one has to consider several metals and several organic ligands together. Determining the extent of competition among metals for the same ligand and among ligands for the same metal in a complex mixture can seem formidable, and computer programs have been used extensively for this purpose.

In fact examination of the results of many such computer calculations shows that the problems are often not as difficult as they may appear at first and that competition among metals for the same ligand seldom occurs. A few rules of thumb can be given:

1   Major cations (Ca^{2+}, Mg^{2+}), like the hydrogen ion (H^+), do not really "compete" with trace metals for strong ligands, they merely decrease the ligands apparent affinities for the metals (recall the $K_{Cu}/K_{Ca}(Ca^{2+})$ factor in Example 6).

2   Trace metals do not compete with each other for the same ligand as long as the ligand is in excess of the metals ($L_T > M_T$'s). This is the situation most often considered.

TABLE 6.9

Results of an Equilibrium Computation for Oxidizing Conditions ($p\varepsilon = 12$, $pH = 7$) in a System Containing Four Organic Ligands[a,b]

	Total Conc ↓	Free Conc ↓	CO_3	S	Cl	F	NH_3	PO_4	SiO_2	CIT	GLY	NTA	CYST	OH
Total Conc →			3.00	4.50	3.50	5.50	5.50	5.00	4.00	5.00	5.00	5.00	5.00	—
Free Conc →			6.41	4.60	3.50	5.52	7.60	11.84	12.91	17.68	8.10	10.06	11.07	7.00
Ca	3.00	*3.02*	*4.81*	5.32	—	—	—	—	—	5.90	—	5.28	—	—
Mg	3.50	*3.51*	*5.40*	5.71	—	7.43	10.82	7.55_s	—	7.09	9.92	6.77	—	8.52
Na	3.50	*3.50*	8.61	7.30	—	7.12	10.91	7.54	—	—	10.01	11.26	—	7.81
Fe[c]	5.00	16.60 / 15.40	—	16.90	17.80	16.18	21.60	15.44	13.41	*6.68*	14.00	*5.73*	*5.55*	*7.40_s*
Mn	5.50	10.00	11.41	12.30	12.30	—	16.80	12.94	—	12.38	15.50	11.26	16.37	*13.20_s*
Cu	6.00	10.05	9.75	12.35	12.85	14.16	11.75	12.58	—	*7.93*	9.27	*6.01*	—	11.05
Ba	7.00	*7.00*	—	—	—	—	—	—	—	26.98	14.10	10.96	—	13.00
Cd	6.00	*7.46*	10.76	9.76	8.36	11.87	12.46	15.89	—	9.34	10.56	*6.02*	—	9.56
Zn	7.00	*8.85*	—	11.16	11.05	12.97	14.16	12.29	—	13.34	11.01	*7.01*	8.92	11.45
Ni	6.50	9.14	—	11.45	12.24	13.46	13.94	—	—	10.42	10.84	*6.50*	11.39	10.94
Hg	9.00	31.95	—	34.16	24.75	35.87	29.63	—	—	—	27.63	26.50	*9.00*	57.21
Pb	7.00	10.20	—	12.21	11.90	—	—	—	—	11.31	12.60	*7.06*	*7.88*	10.40
Co[d]	7.50	*9.35* / 26.95	—	11.45	12.45	27.36	14.95	—	—	26.93	12.15	*7.51*	10.22	10.85
Ag	9.00	*9.22*	—	12.72	*9.41*	—	13.42	—	—	18.52	13.62	—	—	13.92
Al	5.00	13.49	—	15.30	—	11.67	—	—	*s*	*5.08*	—	7.94	—	9.29
H		*7.00*	*3.01*	9.30	—	9.32	5.50	6.08	4.01	7.69	5.00	6.66	5.37	—

[a] Source: Morel et al. (1973).[23]

[b] Note: This calculation was performed with a compilation of thermodynamic constants not entirely consistent with Table 6.1. Note in particular the inclusion of an extremely (unreasonably?) stable Al-citrate complex. CIT = citrate, GLY = glycine, NTA = nitrilotriacetate, CYST = cysteine. Numbers are negative logarithms of molar concentrations in solution. The presence of a solid is indicated by s. A blank signifies the absence of a computable species. Italic numbers indicate species or solids that amount to more than 1% of the total metal.

[c] The first number corresponds to Fe^{3+}, the second to Fe^{2+}.

[d] The first number corresponds to Co^{2+}, the second to Co^{3+}.

Table 6.9 presents the results of a computer calculation for a model freshwater system containing citrate, glycine, cysteine, and NTA, all in excess of all the trace metals but Fe. Except for mercury, which has a very high affinity for cysteine, and iron, which is partly present as the hydroxide solid, the reactive trace metals are found chiefly as NTA complexes. Half of NTA itself is present as the calcium complex. A series of hand calculations such as those presented in Examples 5, 6, and 7 would have yielded the same result ponderously but easily.

4.3   Complexation by Humic Compounds

As mentioned previously, a major fraction of the dissolved organic matter in natural waters is composed of refractory humic compounds, tridimensional polymers of variable composition. Although their overall structure may be complicated and ill-defined, humic compounds appear to contain reasonably simple and consistent functional groups for coordination: carboxyl, alcohol, and phenol.[31,32] Humic substances have sensibly different chemistry in the open

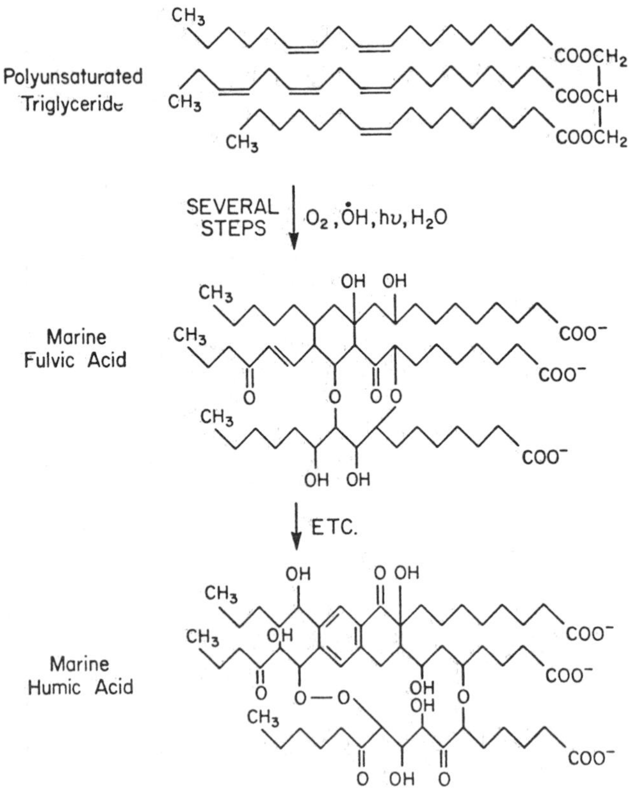

Figure 6.5 A possible pathway for the formation of marine humic acids from a triglyceride. From Harvey et al., 1983.[34]

ocean and in fresh or coastal waters. In the open ocean where organic matter
is almost entirely autochthonous (formed *in situ*), humic compounds are formed
by condensation, polymerization, and partial oxidation of smaller molecules
such as triglycerides, sugars, or amino acids, and they exhibit little aromatic
character.[33] This is illustrated in Figure 6.5, which shows a possible pathway of
formation of marine humates from polyunsaturated triglycerides, leading to
crosslinked aliphatic acids.[34] Near land, on the other hand, the DOC is largely
allochthonous (imported from foreign sources) and derived from the decaying
material of higher plants, particularly lignins (Figure 6.6). The resulting terrigenous humic compounds retain a relatively high degree of aromaticity from

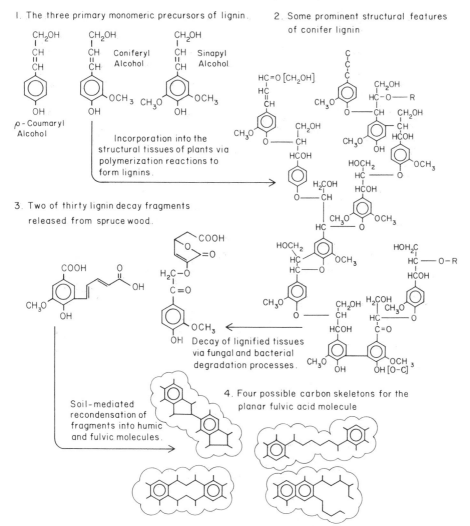

Figure 6.6 A possible schematic pathway for the formation of terrigenous humic acids from
lignins. 1, 2, and 3 from Crawford, 1981;[37] 4 from Schnitzer and Khan, 1972.[32]

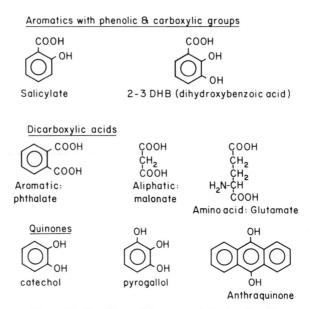

Figure 6.7 Possible model compounds for humic acids.

their precursors.[35] Thus, while carboxylic and alcoholic functional groups are characteristic of all humates, phenolic groups are found predominantly in the humic compounds near land.

Because of the limited available information on oceanic humates and the greater demonstrated importance of terrigenous compounds in trace element speciation, this discussion centers on metal complexation by humic matter in fresh and coastal waters. Chelation by carboxyl and phenolic groups in ortho position on a benzene ring is generally considered to be the major mode of metal complexation by such humates, thus pointing at salicylic acid as a convenient model compound for the coordinative properties of humic acids. Other model compounds include dicarboxylic acids or simple quinones, as illustrated in Figure 6.7. The latter might be better model compounds to represent the photochemical properties of humic substances (see Chapter 7).

While they may be convenient for didactic or illustrative purposes, model compounds do not provide an accurate image of the complications involved in studying the coordinative properties of humic substances. Reduction of data from typical acid-base titrations or metal-coordination experiments with humates usually do not yield a simple set of discrete constants; the titration curves are "smeared," and the resulting constants vary continuously with the experimental conditions. There are three principal causes for this confusing state of affairs:

1 Due to chemical and steric differences in their neighboring groups, co-ordination sites on humates may well have a continuous range of affinities for metals including H^+.

2 Conformational changes resulting from electrostatic interactions among the various functional groups on a single molecule may make the co-ordination properties of the molecule a strong function of the extent of cation binding and of the ionic strength of the solution.

3 Electrostatic attraction and repulsion from neighboring groups may affect markedly the metal affinity of a given group, thus making the free energy of complexation, and hence the complexation constant, a direct function of the extent of cation complexation of the ligand, particularly its protonation.

Complications similar to (1) and (3) are encountered when studying solute adsorption on suspended solids, and we shall examine them in more detail in Chapter 8. Here we take a pragmatic approach and consider that the co-ordination properties of humates can be described by conditional constants valid for a limited range of solution composition. Before embarking on a study of metal complexation by humic compounds, let us then review briefly the different types of conditional constants and provide a systematic terminology for such constants.

Conditional Stability Constants

It is often convenient to describe the thermodynamics of equilibrium reactions with conditional stability constants that are applied to some analytical fraction of the reactants and products rather than their activities and are valid under a given set of conditions. Consider the complexation reaction

$$M^{2+} + L^- = ML^+ \tag{76}$$

The thermodynamic constant for this reaction is written

$$K = \frac{\{ML^+\}}{\{M^{2+}\}\{L^-\}} \quad \text{(thermodynamic constant)} \tag{77}$$

In Chapter 2 we defined the corresponding concentration constant that we are using throughout the text:

$$K_c = \frac{(ML^+)}{(M^{2+})(L^-)} \quad \text{(concentration constant)} \tag{78}$$

This is a conditional constant valid for a particular ionic strength, and it includes implicitly all nonspecific long-range (mostly electrostatic) interactions among ions, namely, activity coefficients.

Going one step further, in Section 2 of this chapter we defined apparent constants for carbonate species in seawater, including in the constants all the specific and unspecific interactions among the major ions:

$$K^{app} = \frac{(\Sigma ML^+)}{(M^{2+})(\Sigma L^-)} \quad \text{(apparent constant)} \tag{79}$$

The symbol Σ indicates that all complexes formed with the major metals, Na^+, K^+, Ca^{2+}, and Mg^{2+}, are included in the formulation. The domain of validity of such apparent constants is of course restricted to solutions with similar major ion composition. For dilute solutions such as freshwaters, $K_c = K^{app}$.

In our study of organic complexation of metals (see Section 4.1) we have introduced yet another type of conditional constant, an effective constant that includes the acid-base speciation of the ligand $[(\Sigma H_x L) = (L^-) + (HL) + (H_2 L^+) + \cdots]$:

$$K^{eff} = \frac{(ML)}{(M^{2+})(\Sigma H_x L)} \quad \text{(effective constant)} \tag{80}$$

Such a constant is of course highly pH dependent and is obtained simply by introducing the ligand-ionization fractions in the mass law expression. It can be generalized slightly by considering also all the various acid-base species of the complex $[(\Sigma ML) = (ML^+) + (MHL^{2+}) + (MOHL) + \cdots]$:

$$K^{eff} = \frac{(\Sigma ML)}{(M^{2+})(\Sigma H_x L)} \tag{81}$$

Over a limited pH range, the (H^+) dependency of such a constant can often be made explicit:

$$K^{eff} = K_x^{eff}(H^+)^x \tag{82}$$

where K_x^{eff} is constant over some pH range; see Section 4.1. In our study of inorganic speciation, we have used effective constants as the product $\beta \alpha^m$ (see Section 3).

We can now combine the notion of apparent constant with that of effective constant to obtain an overall conditional constant which includes the interactions of the ligand with all major cations, H^+, Ca^{2+}, and Mg^{2+}:

$$K^{cond} = \frac{(\Sigma ML)}{(M^{2+}) \times L_T} \tag{83}$$

In cases where the calcium or magnesium complexes of the ligand are dominant $K^{cond} = K^{app}$; in cases where the acid-base species dominate the ligand speciation, $K^{cond} = K^{eff}$.

Note that, in general, the inorganic metal speciation is *not* included in the expressions of conditional constants and the free metal ion concentration, (M^{2+}), appears in all the foregoing mass law expressions. In situations where the complex is a sizable fraction of the ligand, only the unbound ligand should appear in the denominator of the mass law expression:

$$K^{cond} = \frac{(\Sigma ML)}{(M^{2+})(L_T - \Sigma ML)} \tag{84}$$

Conditional constants are the most directly attainable quantities in complexation experiments where the measured concentrations are simply the fractions of bound and free metal:

$$K^{cond} = \frac{(M \cdot bound)}{(M \cdot free)(L_T - M \cdot bound)} \qquad (85)$$

While such constants may have little thermodynamic meaning and only a limited range of applicability, they are very convenient to use. In addition, for humates whose thermodynamic constants are difficult to define, they provide a direct link between experimental data and calculations, without an intervening and often misunderstood extrapolation procedure. When using conditional constants, one should always remember the particular conditions of applicability and be wary that different authors may include implicitly different factors in their constants (e.g., acid-base chemistry, major ion complexation). As noted earlier, consistency in thermodynamic data sets is as important as the absolute values of the individual constants. Often, to increase the range of applicability of conditional constants for humates, a pH dependency is given explicitly:

$$K^{cond} = K_x^{cond}(H^+)^x = \frac{(H^+)^x \times (ML)}{(M^{2+})(L_T - ML)} \qquad (86)$$

In this case the exponent x reflects not only the acid-base chemistry of the ligand and the complex over a certain pH range but also the pH dependency of the electrostatic effects on the metal affinity for the ligand; that is, the value of x can be interpreted as the average number of protons released per bound metal ion, or it can be a more generalized fitting parameter.

Acid-Base Chemistry and Major Cation Complexation of Humic Compounds

By using conditional stability constants, one may obviate the need to calculate the degree of ligand protonation and binding by the major cations to study trace metal complexation by humates in a particular system. Nonetheless, it is instructive, as a point of comparison, to examine briefly what the speciation of humic acids in water may be, leaving aside the issue of adsorption on particles.

A typical acid-base titration of a humic acid sample is shown in Figure 6.8. A comparison with a titration curve for salicylic acid illustrates the difficulty in extracting thermodynamic constants from such data. What is more easily obtained is the total number of acid groups titratable over a given pH range, yielding typically 10 to 20 meq $(gC)^{-1}$. Terrigenous humic compounds are usually close to 50% carbon by weight. Using calorimetric techniques, it has been estimated that 60 to 90% of these acid groups are carboxyl groups and the rest phenolic groups.[38] The "smeared" aspect of the titration curve of Figure 6.8 is explained by a wide range of conditional acidity constants, the carboxyl groups (pK_a's 2.5 to 7) almost overlapping the phenolic groups (pK_a's 8 to 11).

As may be expected by analogy with well-characterized chelating agents, the calcium or magnesium, or even the sodium or potassium humate complexes may control the chemistry of humic compounds in freshwaters. The

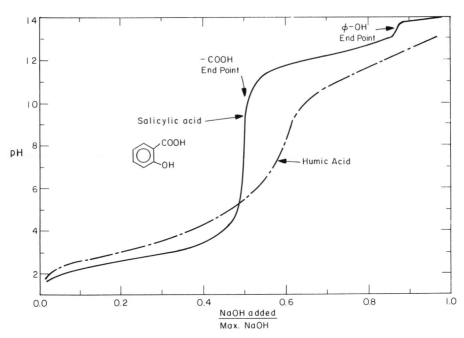

Figure 6.8 Comparison of alkalimetric titrations of salicylic and humic acids. The humic acid yields a smeared titration curve corresponding to a continuously varying apparent acidity constant.

partitioning of these metals between the water and the humic compounds is often viewed as an ion-exchange process (see Chapter 8) where the positive ions serve as counter ions for fixed negative charges distributed in a tridimensional matrix which defines practically a separate phase. In this case, the apparent binding constants—then called selectivity coefficients—are found to vary with the degree of saturation of the ion exchanger. This is illustrated in Figure 6.9 which shows the variations in apparent binding constants for sodium and calcium over a range of metal saturation of the humates.[39] The constants are given for the proton exchange reactions, considering one proton to be exchanged for Na^+ and two for Ca^{2+}. Taking, for example, the point of 50% saturation of the ligand by the metals at pH = 7, the overall apparent constants for sodium and calcium are found to be approximately 10^3 and 10^7, respectively ($I = 0.3\ M$). These constants are certainly high enough to suggest that, for this particular sample of humic material, extracted from Chesapeake Bay sediments, major ions should control largely the speciation of the humates. In general, one might expect that this situation would be more prevalent for the larger, sometimes condensed, fraction of the humic matter than for fulvic acids. In any case the role of the major ions in the chemistry of humic material cannot be ignored, as it probably affects drastically the apparent trace metal affinity of the ligands in waters of varying ionic content such as estuaries.

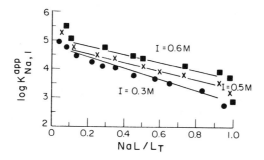

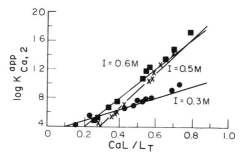

Figure 6.9 Apparent sodium and calcium humate stability constants as a function of the fraction of metal bound humate. These constants, which are defined as $K = (H^+)(NaL)/(Na^+)(L_T - NaL)$ and $K = (H^+)^2(CaL)/(Ca^{2+})(L_T - CaL)$ and are akin to selectivity coefficients for ion exchange (see Chapter 8), are seen to vary markedly with the degree of metal binding. From Frizado, 1979.[39]

Trace Metal Complexation by Humates

There are large differences among the literature data on humic acid affinities for trace metals reported by various authors using different techniques and material extracted by several methods from samples of diverse origins: seawater, freshwaters, bogs, soils, sediments, sewage, and so on. Despite this variability, which is probably ascribable as much to the techniques as to the humic compounds themselves, it appears that the affinity of humates for reactive metals such as copper is moderately high, as is expected from the salicylic acid model of humates. Both the data variability and the magnitude of the metal affinities are illustrated in Table 6.10 which shows a selection of data from the literature.

Because they presumably possess functional groups in several different configurations, one would expect that humic acids would complex metals at several sites with different affinities. Complexation experiments which are most often conducted at high metal concentrations may reveal only the characteristics of the more abundant and weaker complexing sites. This is because upon titration with a metal the ligand sites with higher affinities are occupied first. Unless complexation experiments are run at sufficiently low metal to ligand ratios, the strongest binding sites cannot be identified. The possible existence of high affinity humate sites at low concentrations is a major issue in assessing the extent of trace metal complexation by humic compounds in natural waters where the metal concentrations themselves are very low. For this reason we choose to study an example based on the results of Sunda and Hanson[5] who report the existence of such sites (Table 6.10). It is interesting to note that these authors used an initial metal concentration (10^{-8} M) that was an order of magnitude lower than most other workers. This is likely to be responsible for their finding of strong binding sites.

TABLE 6.10

A Synopsis of the More Reliable Stability Constants for Metal-Humate Complexes as Compiled from the Literature

Metal	Valid pH or range[a]	Source	Log K_i or Log β_i[b]				Reference
Cu^{2+}	$3 \rightarrow 7$	Lake water	β_1' 4.8	β_2' 9.5			
	"	Marsh water	5.3	9.9			40
	"	Soil water	4.9	9.5			
	4.0	Lake water	K_1 5.48	K_2 4.00			
	6.0	"	6.11	3.85			
							41
	4.0	Soil extract	5.60	3.95			
	6.0	"	6.30	3.78			
	6.8	"Aldrich" humic	K_1 6.20	K_2 5.08			42
Cu^{2+}	8.0	Seawater loch sediment (a)	K_0 11.37				
	"	" (b)	K_0 9.91				
							43
	8.0	Lake water	K_0 9.35				
	8.0	Peat: HA	K_0 8.29				
	"	Peat: FA	K_0 7.85				
	5.95	River water	K_1 9.0	K_2 6.2	K_3 4.7		
	7.00	"	9.75	7.85	5.65		5
	8.00	"	10.85	8.85	6.90		
	7.4	Lake Ontario	K_0 8.6				
	8.4	"	K_0 9.5				
							44
	8.3	Lake Huron	K_0 9.2				
	7.6	Fulvic extract	K_0 7.8				

TABLE 6.10 (Continued)

Metal	Valid pH or range[a]	Source	Log K_i or Log β_i[b]	Reference
Zn^{2+}	6.8	"Aldrich" humic	K_0 5.00	42
	8.0	Lake water	K_0 5.14	45
Ni^{2+}	8.0	Lake water	K_0 5.15	45
	8.0	Peat extract	K_1 5.64 K_2 4.32	
Pb^{2+}	6.8	"Aldrich" humic	K_1 6.53 K_2 5.30	42
Cd^{2+}	6.8	"Aldrich" humic	K_0 5.04	

[a] The "valid" pH, when given as a single value, means the pH at which the determination was made. When a range is given (e.g., Buffle et al., 1980),[40] it is implied that theoretical and empirical results suggest that β values are constant over this range.

[b] K_0 is the overall conditional constant, assuming an aggregate site-type:

$$K_0 = \frac{(ML)}{(M^{2+})(L_T - ML)}$$

K_i is the conditional constant for binding sites of type i:

$$K_i = \frac{(ML)}{(M^{2+})(L_{i_T} - ML_i)}$$

where L_i is the concentration of sites of type i. β_i is the cumulative stability constant for the addition of the ith ligand (assuming all ligands are of equal strength) to the metal ion, which implies a $M:L_n$ complex:

$$\beta_i' = \frac{(ML_i)}{(M^{2+})(LH_x)^i} = \frac{\beta_i^*}{(H^+)^x}$$

where

$$\beta_i^* = \frac{(ML_i)(H^+)^x}{(M^{2+})(LH_x)^i}$$

and x is the average number of protons released per metal ion complexed ($x = 0.6 - 0.8$ is used in Buffle et al., 1980).[40]

Example 8.   Copper Speciation in River Water of High Humic Content

Consider a river containing 10 mg liter^{-1} organic carbon, all of which is humic material. Following Sunda and Hanson, we consider that copper complexation by this material can be represented at pH-8 by three ligands:

X: 8×10^{-5} sites gC^{-1}; $X_T = 10^{-6.1}$;   log $K_X^{cond} = 10.9$

Y: 7.5×10^{-4} sites gC^{-1}; $Y_T = 10^{-5.1}$;   log $K_Y^{cond} = 8.8$

Z: 8×10^{-3} sites gC^{-1}; $Z_T = 10^{-4.1}$;   log $K_Z^{cond} = 6.9$

Let us consider the distribution of copper among the various species as a function of total copper. Assuming the same major ion composition as in our previous freshwater examples, we can write the mole balance equation for copper:

$$TOTCu = (Cu \cdot inorganic) + (CuX) + (CuY) + (CuZ) = Cu_T \qquad (87)$$

$$TOTCu = (Cu^{2+})[10^{1.5} + K_X(X) + K_Y(Y) + K_Z(Z)] \qquad (88)$$

$$TOTCu = (Cu^{2+})[10^{1.5} + 10^{10.9}(X_T - CuX) + 10^{8.8}(Y_T - CuY)$$
$$+ 10^{6.9}(Z_T - CuZ)] = Cu_T \qquad (89)$$

At low copper concentrations $(Cu_T < 10^{-6.1})$ the copper-ligand complexes are negligible for the ligand speciation $(CuX \ll X_T$, etc.):

$$TOTCu = (Cu^{2+})[10^{1.5} + 10^{4.8} + 10^{3.7} + 10^{2.8}] = Cu_T \qquad (90)$$

The strongest of the three ligands then dominates the copper speciation, and the free cupric ion is decreased more than three orders of magnitude as compared

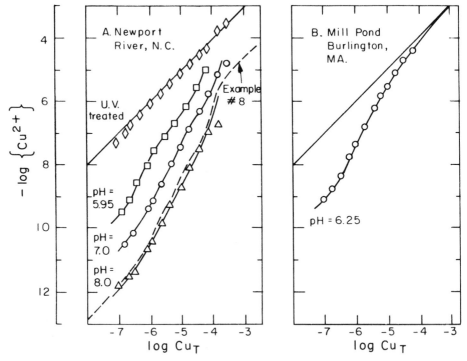

Figure 6.10 Copper titrations of humic acid. These titration curves were obtained by adding copper to humic acid samples and measuring the cupric ion activity with a Cu^{2+} sensitive electrode. The dashed line in panel A shows the results of the calculation of Example 8. Panel A from Sunda and Hanson, 1979;[5] panel B from McKnight, 1979.[46]

to the purely inorganic system. As the total copper concentration is increased in the system, the organic ligands are gradually titrated. To obtain the complexometric titration curve (Cu^{2+} versus Cu_T), we express the free ligand concentrations as functions of (Cu^{2+}) from the ligand mole balance equations:

$$TOTX = (X) + (CuX) = X_T \tag{91}$$

$$TOTX = (X)[1 + K_X(Cu^{2+})] = X_T \tag{92}$$

therefore

$$(X) = \frac{X_T}{1 + K_X(Cu^{2+})} = \frac{10^{-6.1}}{1 + 10^{10.9}(Cu^{2+})} \tag{93}$$

Introducing this and similar expressions for (Y) and (Z) into the copper mole balance yields the necessary formula:

$$TOTCu = (Cu^{2+})\left[10^{1.5} + \frac{10^{4.8}}{1 + 10^{10.9}(Cu^{2+})} + \frac{10^{3.7}}{1 + 10^{8.8}(Cu^{2+})} \right.$$

$$\left. + \frac{10^{2.8}}{1 + 10^{6.9}(Cu^{2+})} \right] = Cu_T \tag{94}$$

The corresponding plot of the complexometric titration curve is shown in Figure 6.10 in the left-hand graph along with the original data of Sunda and Hanson. The small differences are due to different chemical conditions in our model freshwater and in the Newport River from which the humate sample was obtained. To put this example into perspective, another copper titration curve obtained with the same experimental technique for another water sample is shown in Figure 6.10 in the graph to the right. Note that the cupric ion activity is not reduced nearly as much in this second case.

Example 8 provides an extreme case of metal complexation by humates since the humic acid content of the system is large, the pH is high (the apparent complexation constants would decrease by more than one order of magnitude for each pH unit in this range: see Table 6.10)., and the result of Sunda and Hanson demonstrate more copper complexation than any other data set (compare Figures 6.10 a and b). Nonetheless, it is clear that humic compounds may control the speciation of some trace metals in waters of high humic content, often conspicuous by their color. According to Example 8, some four orders of magnitude decrease in the product $K_X X_T$ is necessary for the metal-humate complex to be unimportant. If the data of Table 6.10 are representative of the relative metal affinities of all complexing sites on humates, metals such as Cd^{2+}, Ni^{2+}, or Zn^{2+} should have humate affinities roughly 10 times less than that of copper. Depending on the pH and the humic content of the water, the complexation of such metals by humic material may thus vary from 0% (pH < 7, DOC < 1 mg liter^{-1}) to 100% in situations such as that of the example.

5.   CHEMOSTASIS

Through solid and complex formation, metals and ligands in natural waters constitute a sort of interacting chemical network. In principle, the speciation and the free concentration of any constituent of the system is dependent to some degree on the total concentration of any other constituent through a series of more or less direct interactions. For example, increasing the total concentration of a metal $(M_1)_T$ may decrease a free ligand concentration (L_1) through complex formation; in turn the decreasing (L_1) may lead to an increase in the free concentration of a second metal (M_2) to satisfy a solubility product, and so on. How sensitive a particular component is to variations in another one, how interdependent and homeostatic the system is, are questions particularly relevant to natural waters subjected to increasing anthropogenic inputs.

Clearly we are dealing here with a generalization of the concept of buffer capacity, and one can define interaction intensity parameters according to

$$\delta_{X,Y} = \frac{\partial pX}{\partial pY_T} \quad \text{(interaction intensity)} \tag{95}$$

where $p = -\log$. This parameter gives a measure of the effect of an incremental addition of the component Y on the free concentration of X through all possible interactions from the most direct to the most circuitous. General methods for calculating interaction intensities can be included in computer algorithms for equilibrium calculations, but simple approximate formulae may also be obtained in the manner described for the minor species theorem of buffer capacity. Without going into too much theory, a few rather intuitive results can be mentioned:

1.   The free concentration of a given component cannot be affected by the total concentration of another component present at much lower concentration. For example, the free concentration of calcium in seawater cannot be affected by the trace metal content of the water: $\delta_{Ca,M} \cong 0$. This reasoning allowed us in this chapter to study the issue of major ion interactions with no regard for the trace element composition of the water.

2.   Almost all sizable interactions $(|\delta_{X,Y}| > 0.1)$ among chemical components are very simple and direct; for example, a metal affects a ligand by forming a complex or a solid with this ligand. The inversely proportional dependence of the free EDTA concentration upon the total calcium concentration in Example 6 is typical of such interactions: $\delta_{EDTA,Ca} = -1$; also $\delta_{Cu,EDTA} = -1$.

3.   Very strong interactions $(|\delta_{X,Y}| > 1)$ are only observed when a "titration" phenomenon is taking place, for example, when a metal and a ligand forming a strong complex are present in similar concentrations. The steep region of the titration curve of Figure 6.10 exemplifies such strong interactions: a small increase in total copper results in a relatively large increase in the free cupric ion concentration: $\delta_{Cu,Cu} > 1$.

TABLE 6.11

Interaction Intensities for a Selected Set of Metal Ions with Respect to a Chosen Set of Metals and Ligands[23,a]

	$TOTCa$	$TOTMg$	$TOTCu$	$TOTCd$	$TOTZn$	$TOTHg$	$TOTPb$
Ca	1.0	5.4×10^{-3}	7.0×10^{-4}	6.8×10^{-4}	6.9×10^{-5}	1.2×10^{-8}	6.1×10^{-5}
Mg	1.7×10^{-3}	1.0	7.2×10^{-5}	6.9×10^{-5}	7.0×10^{-6}	1.2×10^{-10}	6.3×10^{-6}
Cu	0.7	2.3×10^{-2}	1.1	0.13	1.3×10^{-2}	2.1×10^{-7}	1.1×10^{-2}
Cd	0.68	2.2×10^{-2}	0.13	1.1	1.2×10^{-2}	2.1×10^{-7}	1.1×10^{-2}
Hg	1.1×10^{-3}	3.7×10^{-5}	2.1×10^{-3}	2.1×10^{-4}	1.7×10^{-4}	1.0	1.7×10^{-3}
Pb	0.61	2.0×10^{-2}	0.11	0.11	1.1×10^{-2}	1.7×10^{-5}	1.0
Ag	3.3×10^{-5}	1.2×10^{-5}	7.2×10^{-7}	6.3×10^{-6}	8.2×10^{-8}	1.2×10^{-12}	6.5×10^{-8}
Al	1.1×10^{-2}	7.0×10^{-4}	1.4×10^{-4}	3.9×10^{-5}	3.5×10^{-6}	6.0×10^{-11}	3.2×10^{-6}

	$TOTAl$	$TOTCO_3$	$TOTS$	$TOTCl$	$TOTNTA$	$TOTCYST$
Ca	1.1×10^{-4}	-1.6×10^{-2}	-4.8×10^{-3}	-3.0×10^{-6}	-7.1×10^{-3}	-5.8×10^{-6}
Mg	2.2×10^{-5}	-1.3×10^{-2}	-6.1×10^{-3}	-3.1×10^{-7}	-7.2×10^{-4}	-5.9×10^{-7}
Cu	1.4×10^{-3}	-1.1×10^{-2}	-3.5×10^{-3}	-5.6×10^{-4}	-1.3	-1.1×10^{-3}
Cd	3.9×10^{-4}	-1.1×10^{-2}	-3.6×10^{-3}	-5.0×10^{-3}	-1.3	-1.0×10^{-3}
Hg	6.0×10^{-7}	-1.8×10^{-5}	-5.8×10^{-6}	-9.5×10^{-7}	-1.3	-1.3
Pb	3.2×10^{-4}	-9.7×10^{-3}	-3.1×10^{-3}	-5.0×10^{-4}	-1.1	$-8.6 - 10^{-2}$
Ag	5.7×10^{-9}	-6.6×10^{-7}	-1.9×10^{-4}	-0.4	-7.3×10^{-6}	-6.2×10^{-9}
Al	0.1	-1.8×10^{-4}	-5.7×10^{-5}	-1.7×10^{-7}	$-3.6 - 10^{-4}$	-3.0×10^{-7}

Source: After Morel et al. 1973.[23]

[a] Note: The interaction intensities here correspond to the equilibrium calculation of Table 6.9. The interaction intensities are partial derivatives of $(-\log)$ the free metal ion concentrations (listed vertically) with respect to $(-\log)$ the total concentrations (listed horizontally).

4. Indirect interactions can only be weaker (lower δ's) than the direct interactions that mediate them. As a result strong "high-order" interactions are only observed where strong "first-order" interactions are involved. This is demonstrated in Example 6 where the total calcium affects the free cupric ion concentration proportionally: $\delta_{EDTA,Ca} = -1$ and $\delta_{Cu,EDTA} = -1$ yields $\delta_{Cu,Ca} = +1$. However, cadmium has no effect on copper: $\delta_{EDTA,Cd} \cong 0$ leads to $\delta_{Cu,Cd} \cong 0$.

Because important inorganic ligands in natural waters are typically in large excess of the metals, the interactions mediated by inorganic complexes or solids are simple:

$$\delta_{L,M} \cong 0$$

$$\delta_{M,L} \cong 1 \quad \text{if ML is the major metal species}$$

$$\delta_{M,L} \cong 0 \quad \text{otherwise}$$

Also no sizable interactions among metals or ligands are mediated through inorganic complexes and solids:

$$\delta_{M,M} = \delta_{L,L} = 1$$

$$\delta_{M_1,M_2} \quad \text{and} \quad \delta_{L_1,L_2} \cong 0$$

The presence of organic ligands, which form many more strong complexes than inorganic ligands and which may be present at total concentrations comparable to that of the metals, results in an increase of interdependency among components in chemical systems. Consider in Table 6.11, for example, some of the interaction intensities calculated for the model of Table 6.9 (The pH is supposed to be fixed, and as a result H^+ mediates no interactions.) The free concentrations of the trace metals are more than proportionally dependent on the total NTA concentration ($\delta_{M,NTA} > 1$). This is due to the fact that NTA is almost titrated by the trace metals, about half of the ligand being bound to Fe, Cu, and Cd. Furthermore the free concentration of a metal such as cadmium is sensitive to the total concentration of other metals that bind a sizable fraction of the NTA: $\delta_{Cd,Ca} = 0.68$; $\delta_{Cd,Cu} = 0.13$. These sizable interaction intensities effectively represent the degree to which cadmium is displaced from the NTA complex by calcium or copper. The metals are competing with each other for the strong ligand NTA whose presence in the system mediates more pathways of strong interactions; as a result more components are substantially interdependent. Chemical systems are made less "chemostatic" by the presence of complexing organic compounds.

6. COMPLEXATION KINETICS

The preceding discussion of chemostatic processes in natural waters is based strictly on equilibrium arguments. All reactions that are considered to take place in the system are also considered to have reached equilibrium. In many cases complexation of metals by ligands is indeed rapid relative to other processes of interest, and equilibrium models provide then a good representation of complexation phenomena in natural waters. However, we are sometimes interested in relatively rapid processes and at least some understanding of complexation-dissociation kinetics is necessary (see Section 7 for an example). Note also that the effective rate of concentration change due to complexation is the product of two or three terms $[$e.g., $k_f(M)(L)]$ and that very low metal and ligand concentrations $(10^{-8}–10^{-15}\ M)$ may easily compensate for large rate constants and lead effectively to a slow reaction process.

As pointed out earlier in this chapter, stable metallic species in solution have typically all their coordination sites satisfied, either bound to water molecules (primary hydration shell) or attached to some other inorganic or organic ligands. In order for any of the bound ligands to be replaced by another—such as water for water, water for a complexing agent, or a complexing agent for water—there is very probably an intermediate state where the coordination number of the metal is either momentarily decreased or increased. For example, a metal with a coordination number of six may become bound to only five water molecules,

$$M(H_2O)_6^{2+} \rightarrow M(H_2O)_5^{2+} \xrightarrow{+L^-} M(H_2O)_5L^+ \tag{96}$$

or a fully hydrated metal may become bound also to some other ligand,

$$M(H_2O)_6^{2+} \xrightarrow{+L^-} M(H_2O)_6L^+ \rightarrow M(H_2O)_5L^+ + H_2O \tag{97}$$

In some cases the ligands may be exchanged simultaneously:

$$M(H_2O)_6^{2+} \xrightarrow{+L^-} M(H_2O)_5L^+ + H_2O \tag{98}$$

Much of the literature on complexation-dissociation kinetics focuses on the question of which of these two or three mechanisms is the pertinent one and on the nature of the intermediate species.[47]

For most metals of interest in aquatic chemistry, a simplified description of the exchange of hydration water for a ligand involves the formation of an outer sphere complex (an ion pair) between the hydrated metal and the ligand, followed by expulsion of a water molecule:[48]

$$M(H_2O)_6^{2+} + L^- \underset{k_{-L}}{\overset{k_L}{\rightleftarrows}} M(H_2O)_6 \cdot L^+ \tag{99}$$

$$M(H_2O)_6 \cdot L^+ \underset{k_w}{\overset{k_{-w}}{\rightleftarrows}} M(H_2O)_5L^+ + H_2O \tag{100}$$

The first reaction is typically fast ($k_L \cong 10^9$ to $10^{10}\ M^{-1}\ sec^{-1}$; $k_{-L} \cong 10^7$ to $10^{10}\ sec^{-1}$) so that pseudoequilibrium is usually achieved, and the rate constants for the overall reaction

$$M(H_2O)_6^{2+} + L^- \underset{k_b}{\overset{k_f}{\rightleftarrows}} M(H_2O)_5L^+ + H_2O \tag{101}$$

are given by

$$k_f = k_{-w}K \tag{102}$$

$$k_b = k_w \tag{103}$$

where $K = k_L/k_{-L}$.

The formation of the intermediate ion pair is due principally to electrostatic interactions, and the stability constant K depends primarily on the charge numbers of the ions, not on their chemical nature. It is considered that the loss of the water molecule from the ion pair corresponds to a momentary decrease in the coordination number of the metal ion. Thus the corresponding rate constant depends primarily on the nature of the metal ion, not on the nature of the ligand, OH^- being a common exception, and it is roughly equal to the second-order constant for water exchange in the hydration sphere of the metal ion. As a result complexation rate constants for a given metal with many different ligands are often approximately equal to each other (depending on the charge of the ligand) and can be predicted from water-exchange rates (see Table 6.12). Conversely, dissociation constants are then inversely proportional to stability constants and can be predicted by using linear free energy relationships[49] (see Figure 2.4).

This general result which shows that the loss of a water molecule from the primary hydration sphere of a metal ion is the rate-determining step in com-

TABLE 6.12

**Characteristic Second-Order Rate Constant
for Water Exchange in the Primary
Solvation Shell of Metal Ions**[a,b]

	$k_w (M^{-1} sec^{-1})$
Cr^{3+}	3.6×10^{-5}
Mn^{2+}	3.4×10^6
Fe^{3+}	3.3×10^2
Fe^{2+}	3.5×10^5
Co^{3+}	$\leq 10^2$
Co^{2+}	1.2×10^5
Ni^{2+}	2.9×10^3
Cu^{2+}	$\geq 9 \times 10^8$
Zn^{2+}	5×10^6
Cd^{2+}	9×10^7
Hg^{2+}	4×10^8

[a] Source: Adapted from Sutin 1966[48], with permission from Annual Reviews Inc.

[b] Note: The second-order rate constant for water exchange is equal to 6/55 times the first-order rate constant (sec^{-1}) for replacement of a particular water molecule.

plexation reactions is particularly convenient for estimating reaction rates, even in the absence of specific information. For example, one expects the second-order rate constants for the formation of most complexes of copper to be of the order of $10^9 \ M^{-1} \ \text{sec}^{-1}$, about a million times larger than those for nickel complexes (see Table 6.12). Somewhat more precise estimates of these constants are obtained by multiplication with the appropriate ion pair stability constants (from ca. 10^{+2} for a divalent anionic ligand to ca. 10^{-1} for a neutral ligand binding to a divalent metal).

In the case of chelate formation more than one water molecule must be replaced by a reacting ligand group, and the complexation reaction can be represented by a stepwise water-ligand exchange;

$$M(H_2O)_6^{2+} + L-L^{2-} \rightleftarrows (H_2O)_5M-L-L + H_2O \tag{104}$$

$$(H_2O)_5M-L-L \rightleftarrows (H_2O)_4M\begin{matrix} L \\ \diagup \\ \diagdown \\ L \end{matrix} + H_2O \tag{105}$$

Often the presence of the first coordinated ligand facilitates the replacement of the remaining water molecules, and the second reaction is relatively fast. The situation is then similar to that described for a unidentate ligand, and the exchange of the first coordinated water controls the overall complexation rate. Note, however, that this is not always the case and rate-determining steps in the formation of chelates involving several kinds of ligand groups can correspond to complicated intermediate species.

As schematized so far the subject of complexation kinetics appears relatively simple and coherent, and it seems that one could attempt a systematic application of fundamental kinetic data to natural waters. However, the bulk of chemical kinetic studies are performed under chemical conditions that are chosen for simplifying data interpretation, not for mimicking natural waters. In particular, very low pH's are often utilized to avoid the complications posed by hydrolysis of metal ions. It is inappropriate, or at least difficult, to extrapolate to multiligand systems at high pH's data that are strictly applicable to hydrated metal ions at low pH's. Many complexation-dissociation reactions are very much accelerated at high or low pH. This may be due to increased reactivity of stable hydrolyzed or protonated species. For example, $Fe(H_2O)_5OH^{2+}$ is known to react about 100 times faster than $Fe(H_2O)_6^{3+}$. The kinetic effect of low or high pH may also be due to the increased stability of some intermediary species by hydrolysis or protonation; it is then referred to as acid or base catalysis.

The presence of many reactive ligands, other than OH^-, in natural waters can also affect complexation kinetics. The replacement of H_2O by any ligand in the coordination sphere of a metal usually "labilizes" the other water molecules and increases the water exchange rate. In many cases the exchange of one ligand for another ($\neq H_2O$) may be achieved without intermediate formation of the stable hydrated complex: intermediary ternary complexes (ML_1L_2) may be formed, or the unstable metallic species with low coordination number $[$e.g., $M(H_2O)_5^{2+}]$ may react directly with the second ligand. During the exchange of metals in chelates, dinuclear species may be formed, resulting in much higher reactions rates than would be predicted from a dissociation-recomplexation mechanism; for example,

$$NiEDTA^{2-} \rightleftharpoons Ni\underset{\begin{subarray}{l}\diagdown\\OOC—CH_2\end{subarray}}{\overset{\begin{subarray}{l}OOC—CH_2\\\diagup\end{subarray}}{\underset{}{}}}N—CH_2—CH_2—N\begin{subarray}{l}CH_2—COO^-\\|\\CH_2—COO^-\end{subarray} \tag{106}$$

$$+Cu^{2+} \rightleftharpoons Ni\underset{\begin{subarray}{l}\diagdown\\OOC—CH_2\end{subarray}}{\overset{\begin{subarray}{l}OOC—CH_2\\\diagup\end{subarray}}{\underset{}{}}}N—CH_2—CH_2—N\overset{\begin{subarray}{l}CH_2—COO\\|\quad\diagdown\end{subarray}}{\underset{\begin{subarray}{l}CH_2—COO\\\diagup\end{subarray}}{}}Cu$$

$$\rightarrow CuEDTA^{2-} + Ni^{2+} \tag{107}$$

Such a process can be viewed as a Cu-catalyzed dissociation of NiEDTA. It is probably important in promoting reasonably fast complexation of trace metals by strong complexing agents in systems such as seawater where the ligands may be bound to calcium or magnesium. For example, upon addition of copper to a micromolar seawater solution of EDTA, the CuEDTA complex achieves equilibrium within one or two hours, much faster than the uncatalyzed dissociation of CaEDTA could proceed.

Overall we may note on the subject of complexation-dissociation kinetics that (1) fundamental information is available from the chemical literature so that at least limit cases can be calculated; (2) the multitude of chemical species in natural waters may provide many alternate reaction pathways and accelerate the reactions beyond what is expected from studies in simple systems (the converse may also be true, and reaction pathways may sometimes be effectively blocked); and (3) the very low concentrations of many reactants in natural waters may result in slow complexation rates despite reasonably high second-order rate constants, and the dissociation of strong chelates is typically slow.

□ □ □

7. TRACE METALS AND MICROORGANISMS

As mentioned early in this chapter, complexation of trace metals in natural waters modulates their effects on the aquatic biota. The growth of all organisms, including man, is dependent on the acquisition of the proper quantities of many trace elements. Some trace metals such as iron, zinc, manganese, and copper are required for growth, but too high a concentration of these metals or others produce toxic effects. While higher organisms maintain a finely poised *internal milieu* for the proper operation of the various physiological tasks of the individual cells, aquatic microorganisms, particularly unicellular organisms such as phytoplankton and bacteria, must function in an *external milieu* whose chemistry is governed chiefly by geochemical processes with little regard for their needs or sensitivities. The growth of aquatic organisms is linked directly to the chemistry of trace elements in natural waters. Variations in trace element composition beyond the normal physiological range of indigenous organisms must thus result in some physiological adaptation, selection of genetic mutants, or ecological succession of species.

The major progress in our understanding of trace metal interactions with aquatic microorganisms over the past decade has come from a better appreciation of the importance of the chemical speciation of the metals in the medium and the ability to separate purely physiological effects from chemical ones. For example, it is now understood that the decrease in metal toxicity observed in the presence of chelating agents is simply the result of the chelation of the metals in the medium, not a physiological effect of the chelating agents. In fact the most important result to emerge—one that is now a tenet of the field—is the universal importance of the free metal ion activities in determining the uptake, nutrition, and toxicity of all cationic trace metals for bacteria, phytoplankton, fungi, zooplankton, and even to some degree for fish.[50-54]

To demonstrate that the free ion activity of a metal is the parameter that determines its biological effects, it is necessary to measure the same physiological responses for the same free metal ion activities obtained by several combinations of metal and ligand concentrations. For example, Figure 6.11 shows how different combinations of chelating agents and total metals in algal culture media result in a unique physiological response to $\{Cu^{2+}\}$ (negative) or $\{Zn^{2+}\}$ (positive). The first step in designing experiments to study the biological effects of trace metals is thus to calculate the trace metal speciation in the medium and tailor it to the particular purposes of the experiments.

In so doing, one must be aware of the interactive nature of chemical speciation in multimetal and multiligand systems and take into account not only first-order interactions [e.g., effects of M_T and L_T on $\{M^{2+}\}$] but also higher-order ones [e.g., effects of $(M_1)_T$ on $\{M_2^{2+}\}$]. Such indirect effects mediated by purely chemical processes are easily misinterpreted as having physiological significance. For example, increased uptake rates of iron due to increasing cadmium concentrations have been observed in diatom cultures.[57] This was due to a displacement of iron from EDTA by cadmium and could be quantitatively accounted for by the increased ferric ion activity due to the cadmium additions. Cadmium does not produce a beneficial *physiological* effect on iron uptake rates in diatoms, but indirect chemical effects made it appear that way.

The central importance of free metal ion activities in controlling the biological effects of trace metals is often misinterpreted to mean that the free hydrated metal ions are the chemical species actually taken up by aquatic organisms. Questions are then asked about the possible uptake or biological effects of other species such as hydroxides, carbonates, or other metal complexes that may be difficult to distinguish experimentally from the free metal ions. Such questions arise from a fundamental misunderstanding of the processes at play. The importance of free metal ion activities in determining physiological effects does not reflect the unique role of the metal coordinated to water molecules rather than to other ligands. It reflects the fact that the chemical

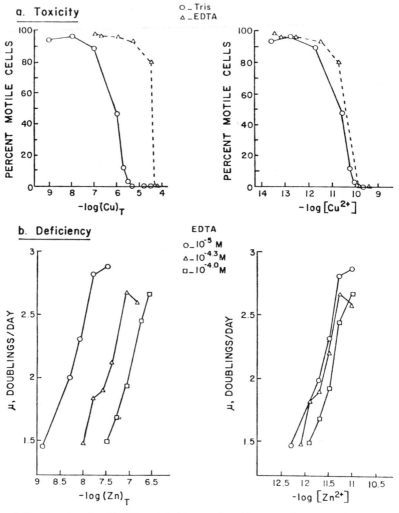

Figure 6.11 Experiments with different chelators and a wide range of trace metal concentrations demonstrate that trace metal toxicity and deficiency are determined by metal ion activities and not total concentrations. (*a*) Motility data of the dinoflagellate *Gonyaulax tamarensis* as a function of total copper $(Cu)_T$ and cupric ion activity $[Cu^{2+}]$ for 2 chelators, Tris and EDTA. After Anderson and Morel, 1978.[55] (*b*) Growth rate of the diatom *Thalassiosira weissflogii* as a function of total zinc $(Zn)_T$ and zinc ion activity $[Zn^{2+}]$ for 3 concentrations of EDTA. After Anderson *et al.*, 1978[56] with permission from Macmillan Journal Limited.

reactivity of the metal is measured by the free metal ion activity and that the physiological effects of a metal are mediated by chemical reactions between the metal and the various cellular ligands.

To make this point clearer, consider a cellular site X that is the critical site of action of a metal M. For example, X may be a transport protein for M, or it may be a reactive group on an enzyme that is deactivated by M. However the free metal ion activity $\{M\}$ may be maintained, the extent of the reaction with X will be directly proportional to that activity:

$$(MX) = K_M(X)\{M\} \tag{108}$$

Hence the physiological effect—such as transport rate or extent of enzyme inhibition—will be directly dependent upon $\{M\}$ regardless of what particular species, hydrated free metal or other complexes, actually react with X. In other words, the free metal ion activity provides a measure of the extent of the reactions with cellular sites independently of the mechanisms of the reactions. Note that the MX complex may represent only a fraction—perhaps even a very small fraction—of the total cellular metal. Many cellular sites present at various concentrations and possessing various affinities may bind the metal of interest. Yet $\{M\}$ is, in any case, the critical parameter determining the biological effect because the extent of metal complexation by all these cellular sites, including those that promote a physiological response and that we denote X, are determined by the free metal ion activity. In this framework the general validity of the paradigm that the free metal ion activity is the parameter that determines the physiological effect of the metal is easily understood: the reactivity of a metal is determined by its free ion activity independently of the particular metal considered, of its role as a nutrient or as a toxicant, of the nature of the inorganic or organic complexing ligands, or of the taxonomic classification of the affected organisms. Very simply, trace metal interactions with aquatic microorganisms can be understood by considering the organisms as assemblages of reactive ligands whose degrees of complexation with particular metals are reflected in measurable physiological effects.

A number of important questions are implicitly raised by this simplified view of trace metal-organism interactions: Are metal ions similarly buffered in experimental systems and natural waters? Is (pseudo)equilibrium achieved or approached in most cases? Are metals competing with each other for the same ligand sites?

In natural waters the concentration of organisms is usually so low that despite the prevalence of high metal concentration ratios, (M · organisms)/(M · water), only a small fraction of any trace metal is normally found in the biota. In such cases the uptake of a metal by organisms does not affect appreciably the speciation of that metal, and the free metal ion activity is effectively buffered by the large pool of metal in the medium relative to that in the biota. This situation is not necessarily encountered in the midst of dense blooms of algae or in laboratory beakers that may contain 10^3 to 10^5 times the normal cell concentrations. In these cases the speciation of the metal can be greatly affected by biological uptake or by release of complexing exudates, and one should consider the final, not the initial, free metal ion activity as the parameter governing the physiological effects of a metal. Since the free metal ion activity is a function of the concentration of organisms in the medium, the physiological effects are time dependent if the organisms are proliferating or producing complexing agents. To avoid this complication, laboratory investigators find it convenient to buffer the metal by using complexing agents—typically EDTA, NTA, or citrate—that allow an increase in the total metal concentration for the same metal ion activity (see Section 5). In cultures of marine organisms such complexing agents have the additional advantage to provide low free metal ion activities in the medium without the need to maintain in culture vessels, at the cost of extraordinary care, the low total metal concentrations that are typical of marine waters. It is a major experimental paradox that high metal and organic ligand concentrations are necessary in order to mimic in the laboratory the very low concentrations of trace metals and organic ligands in marine waters.

In our discussion of metal interactions with cellular ligands, we have implicitly assumed that equilibrium was reached. Since organisms are continuously growing and metals are continuously taken up, the situation should more properly be described as a pseudoequilibrium where the various metal species, in the medium and bound to the cellular sites, are at equilibrium with each other despite the continuous depletion of the metal from the medium and its accumulation into the organisms. As discussed in Chapter 2, the achievement of such a condition is dependent on the relative kinetics of the various processes at play and on the time scale of the observations. The critical process whose kinetics need to be rapid to allow pseudoequilibrium to be reached is the reaction between the metal species in solution and the cellular sites. In phytoplankton, when it has been studied, that process has been observed to be fast indeed, as compared, for example, to the characteristic time scale of the experiments. These observations have invited speculation that the critical cellular sites are on the surface of the cell or that equilibration of trace metals between the medium and internal pools is rapid. Paradoxically, it is in the medium itself that chemical reactions have been sometimes observed to be relatively slow, leading to transient responses or to a nonequilibrium steady-state free metal ion activity governing the physiological response. For example, upon the addition of copper to a seawater medium buffered with EDTA, the free cupric ion activity does not reach equilibrium for several hours. This is due to the sluggish kinetics of the reaction:

$$CaY^{2-} + Cu^{2+} \rightarrow CuY^{2-} + Ca^{2+} \tag{109}$$

(Recall from Example 6 that EDTA is entirely bound to calcium in seawater.) As a result, if unchelated copper is added to culture vessels[55] (see Figure 6.12), short-term physiological responses are observed in laboratory experiments. Often these short-term effects are reversible, and the physiological responses follow the progress of the chemical reactions.

The same slow metal exchange reaction is the cause of another disequilibrium

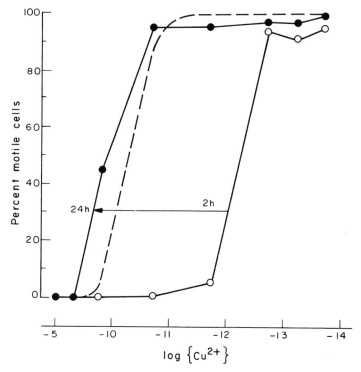

Figure 6.12 Motility of the dinoflagellate *G. tamarensis* as a function of the calculated equilibrium cupric ion activity in the medium. Upon addition of copper to an EDTA-containing medium, it takes several hours for the cupric ion activity to reach its equilibrium value. This is reflected in the physiological response of the organism which initially *appears* more sensitive than in preequilibrated medium (dashed line; see Figure 6.11). Adapted from Anderson and Morel, 1978.[55]

condition observed in iron-uptake experiments.[58] In the presence of either light or a reducing agent such as ascorbic acid, Fe(III)EDTA is known to reduce, leading to the production of Fe^{2+} in the medium. Because of the slow recomplexation of Fe^{2+} by EDTA in seawater

$$Ca Y^{2-} + Fe^{2+} \rightarrow Fe Y^{2-} + Ca^{2+} \quad \text{(slow)} \qquad (110)$$

and of the relatively slow reoxidation rate of Fe(II); an appreciable steady-state ferrous ion activity is maintained in the medium leading to a large enhancement of Fe uptake due to light or reducing agents (Figure 6.13).

If one considers the slow dissociation of metal chelates that promote disequilibrium conditions in these two examples, it is surprising that the reactions of the metals of interest (e.g., Cu, Fe) with the relevant cellular sites are as rapid as they seem to be. A possible explanation is that the metals are transferred from the chelator Y to the cellular site X via formation of a ternary complex XMY:

$$X + MY \rightarrow XMY \rightarrow XM + Y \quad \text{(fast)} \qquad (111)$$

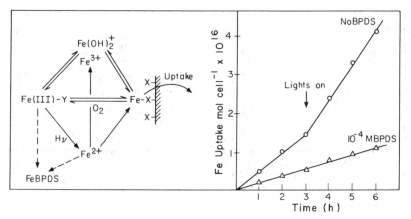

Figure 6.13 Fe uptake by the diatom *Thalassiosira weissflogii*; importance of the formation of unstable Fe(II) in oxic medium by photoreduction of the Fe(III) EDTA complex. Light enhances the rate of iron uptake by producing Fe(II) which is bound in part to the transport site X before being oxidized. This effect is not seen in the presence of the strong Fe(II) complexing agent BPDS (bathophenantroline disulfonate) which scavenges the Fe(II) produced. [BPDS also partially inhibits Fe(III) uptake by some unknown mechanism, possibly complexation of Fe(III).] Adapted from Anderson and Morel, 1982.[58]

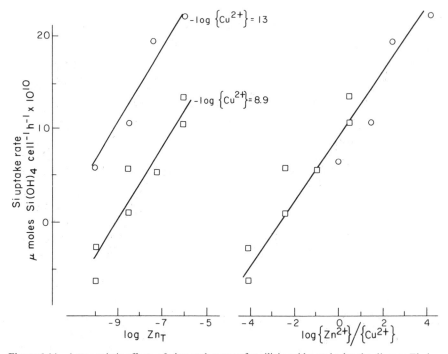

Figure 6.14 Antagonistic effects of zinc and copper for silicic acid uptake by the diatom *Thalassiosira pseudonana*. The silicic acid uptake rate, which is somehow mediated by zinc, is dependent on the ratio of the zinc and cupric ion activities in accordance with a simple model of competitive inhibition by copper. Adapted from Rueter and Morel 1981.[59]

Generalizing boldly from this observation, it appears that for the type of reactions encountered in these systems, metal exchange reactions are often slow while ligand-exchange reactions are typically fast.

So far we have considered that only one metal reacted with the cellular site X. From what we know of even the most specific chelating agents, it seems appropriate to consider that several metals, including the hydrogen ion and the major divalent cations, may compete with M for complexation with X:

$$HX = H^+ + X$$

$$CaX = Ca^{2+} + X$$

$$M_1X = M_1 + X$$

$$M_2X = M_2 + X$$

On this basis, synergistic and antagonistic interactions among trace metals and modulation of metal effects by pH or major ion composition can be readily understood. For example, if two metals M_1 and M_2 inactivate the same enzyme system, the physiological effect should be governed by the expression

$$\% \text{ inhibition} = \frac{(M_1X) + (M_2X)}{X_T} = (K_1\{M_1\} + K_2\{M_2\}) \frac{(X)}{X_T} \qquad (112)$$

Conversely, if the presence of one metal M_1 is necessary for the proper functioning of the cellular site X, the presence of a competing metal M_2 should result in an antagonism governed by the expression:

$$\% \text{ inhibition} = \frac{(X) + (M_2X)}{X_T} = \frac{(X) + (M_2X)}{(X) + (M_1X) + (M_2X)} \qquad (113)$$

$$= \frac{1 + K_2\{M_2\}}{1 + K_1\{M_1\} + K_2\{M_2\}} \qquad (114)$$

If the cellular sites are entirely bound to the metals $[(X) \ll (M_1X) + (M_2X)]$, the inhibition is only a function of the ratio of the metal ion activities:

$$\% \text{ inhibition} = \frac{K_2/K_1\{M_2\}/\{M_1\}}{1 + K_2/K_1\{M_2\}/\{M_1\}} \qquad (115)$$

Such competitive interactions among metals have been observed in algae with Zn^{2+}/Cu^{2+} (Figure 6.14), Mn^{2+}/Cu^{2+}, and Fe^{3+}/Cd^{2+}. In these studies zinc,[59] manganese,[60] and iron[57] serve as essential nutrients, and copper and cadmium as toxicants. The addition of a chelating agent to a growth medium can thus have positive or negative effects on the organisms, depending on the relative affinities of the chelating agent and the cellular site for the two metals. Similarly, observed differences in metal toxicity to euryhaline phytoplankton in freshwater and seawater may well be controlled by such ratios as $\{H^+\}/\{M\}$ or $\{Ca^{2+}\}/\{M\}$.

In general, because of the inherent difficulty in synthesizing transport molecules of very high specificity (energetic cost to the organisms), one would expect the principal mode of action of toxic metals at low concentrations (activities) to be a competition with essential metals. Toxic metals act effectively as co-ordinative analogs of essential ones. In nature aquatic microorganisms may be continuously functioning at the thin edge between limitation by essential elements and inhibition by toxic ones.

□ □ □

REFERENCES

1 J. Burgess, *Metal Ions in Solution*, Ellis Horwood-Wiley, New York, 1978.
2 A. E. Martell and R. M. Smith, *Critical Stability Constants*, vol. 1, *Amino Acids*, Plenum, New York, 1974; R. M. Smith and A. E. Martell, vol. 2, *Amines*, Plenum, New York, 1975; A. E. Martell and R. M. Smith, vol. 3, *Other Organic Ligands*, Plenum, New York, 1977; R. M. Smith and A. E. Martell, vol. 4, *Inorganic Ligands*, Plenum, New York, 1976.
3 M. Whitfield, *Limnol. Oceanogr.*, **19**, 235–248 (1974).
4 C. F. Baes, Jr., and R. E. Mesmer, *The Hydrolysis of Cations*, Wiley, New York, 1976.
5 W. G. Sunda and P. J. Hanson, in *Chemical Modeling in Aqueous Systems*, E. A. Jenne, Ed., ACS Symposium Series 93, American Chemical Society, Washington, D.C., 1979.
6 J. Sainte Marie, A. E. Torma, and A. O. Gübeli, *Can. J. Chem.*, **42**, 662 (1964).
7 J. J. Morgan, in *Principles and Applications in Water Chemistry*, S. D. Faust and J. V. Hunter, Eds., Wiley, New York, 1967.
8 W. Stumm and J. J. Morgan, *Aquatic Chemistry*, 2nd ed., Wiley, New York, 1981, p. 332.
9 H. Irving and R. J. P. Williams, *J. Chem. Soc.*, p. 3192 (1953).
10 F. A. Cotton and G. Wilkinson, *Basic Inorganic Chemistry*, Wiley, New York, 1976.
11 H. D. Holland, *The Chemistry of the Atmosphere and Oceans*, Wiley, New York, 1978.
12 R. M. Garrels and M. E. Thompson, *Am. J. Sci.* **260**, 57 (1962).
13 K. S. Johnson and R. M. Pytkowicz, *Mar. Chem.*, **8**, 87–93 (1979).
14 R. A. Horne, *Marine Chemistry*, Wiley, New York, 1969.
15 W. Stumm and J. J. Morgan, *Aquatic Chemistry*, 2nd ed., Wiley, New York, 1981, p. 364.
16 J. Boulègue, J. P. Ciabrini, C. Fouillac, G. Michard, and G. Ouzounian, *Chem. Geol.*, **25**, 19–29 (1979).
17 J. Boulègue and G. Michard, in *Chemical Modeling in Aqueous Systems*, E. A. Jenne, Ed., ACS Symposium 93, American Chemical Society, Washington, D.C., 1979.
18 R. Dawson, *Symposium on Concepts in Marine Organic Chemistry*, unpublished manuscript, Edinburgh (1976).
19 C. M. Burney, K. M. Johnson, D. M. Lavoie, and J. McN. Sieburth, *Deep Sea Res.*, **26**, 1267–1290 (1979).
20 G. Billen, C. Joiris, J. Wijnant, and G. Gillain, *Estuar. Coastal Mar. Sci.*, **11**, 279–294 (1980).
21 W. Stumm and P. A. Brauner, in *Chemical Oceanography*, J. P. Riley and G. Skirrow, Eds., 2nd ed., Academic, London, 1975.
22 R. F. C. Mantoura, in *Marine Organic Chemistry*, E. K. Duursma and R. Dawson, Eds., Elsevier, Amsterdam, 1981.
23 F. M. M. Morel, R. E. McDuff, and J. J. Morgan, in *Trace Metals and Metal-Organic Interactions in Natural Waters*, P. C. Singer, Ed., Ann Arbor Science, Ann Arbor, 1973.
24 D. M. McKnight and F. M. M. Morel, *Limnol. Oceanogr.*, **24**, 823–837 (1979).
25 C. M. G. Van den Berg, P. Wong, and Y. Chau, *J. Fish. Res. Board Can.*, **36**, 901–905 (1979).
26 S. N. Sueur, C. M. G. Van den Berg, and J. P. Riley, *Limnol. Oceanogr.*, **27**, 536 (1982).
27 D. M. McKnight and F. M. M. Morel, *Limnol. Oceanogr.*, **25**, 62–71 (1980).
28 G. Britton and V. Freihofer, *Microbial Ecol.*, **4**, 119–125 (1978).
29 W. G. Sunda and R. V. Gessner, unpublished data.
30 W. Fish and F. M. M. Morel, *Can. J. Fish. Aquatic Sci.* in press.

31 M. Schnitzer, in *Compounds in Aquatic Environments*, S. D. Faust and J. V. Hunter, Eds., Marcel Dekker, New York, 1971.

32 M. Schnitzer and S. U. Khan, *Humic Substances in the Environment*, Marcel Dekker, New York, 1972.

33 D. H. Stuermer and G. R. Harvey, *Mar. Chem.*, **6**, 55–70 (1978).

34 G. R. Harvey, D. A. Boran, L. S. Chesal, and J. M. Tokar, *Mar. Chem.*, in press.

35 M. A. Wilson, P. F. Barron, and A. H. Gillam, *Geochim. Cosmochim. Acta*, **45**, 1743–1750 (1981).

36 E. Adler, *Wood Sci. Technol.*, **11**, 169–218 (1977).

37 R. L. Crawford, *Lignin Biodegradation and Transformation*, Wiley, New York, 1981.

38 E. M. Perdue, *Geochim. Cosmochim. Acta*, **42**, 1351–1358 (1978).

39 J. P. Frizado, in *Chemical Modeling in Aqueous Systems*, E. A. Jenne, Ed., ACS Symposium Series 93, American Chemical Society, Washington, D.C., 1979.

40 J. Buffle, P. Deladoey, F. L. Greter, and W. Haerdi, *Anal. Chim. Acta*, **116**, 255–274 (1980).

41 W. T. Bresnahan, G. L. Grant, and J. H. Weber, *Anal. Chem.*, **50**, 1675–1679 (1978).

42 R. D. Guy and C. L. Chakrabarti, *Can. J. Chem.*, **54**, 2600–2611 (1976).

43 R. F. C. Mantoura, A. Dickson, and J. P. Riley, *Est. Coastal Mar. Sci.*, **6**, 387 (1978).

44 C. M. G. Van den Berg and J. R. Kramer, *Anal. Chim. Atac.*, **106**, 113–120 (1979).

45 R. F. C. Mantoura and J. P. Riley, *Anal. Chemica Acta*, **78**, 193–200 (1975).

46 D. M. McKnight, "Interactions between Freshwater Plankton and Copper Speciation," Ph.D. Dissertation, Department of Civil Engineering, Massachusetts Institute of Technology, Cambridge, 1979.

47 H. B. Gray and C. H. Langsford, *Chem. Eng. News*, **46** (15), 68–75 (1968).

48 N. Sutin, *Ann. Rev. Phys. Chem.*, **17**, 119–172 (1966).

49 M. R. Hoffman, *Env. Sci. Technol.*, **15**, 345–353 (1981).

50 W. G. Sunda and P. A. Gillespie, *J. Mar. Res.*, **37**, 761 (1979).

51 W. G. Sunda and R. R. L. Guillard, *J. Mar. Res.*, **34**, 511 (1976).

52 R. W. Andrew, K. E. Biesinger, and G. E. Glass, *Water Res.*, **11**, 309 (1977).

53 W. G. Sunda, D. W. Engel, and R. M. Thuotte, *Environm. Sci. Technol.*, **12**, 409 (1978).

54 C. Chakoumakos, R. C. Russo, and R. V. Thurston, *Environm. Sci. Technol.*, **13**, 213 (1979).

55 D. A. Anderson and F. M. M. Morel, *Limnol. Oceanogr.*, **23**, 283–295 (1978).

56 M. A. Anderson and F. M. M. Morel, and R. R. L. Guillard, *Nature*, **276**, 70–71 (1978).

57 G. Harrison and F. M. M. Morel, *J. Phycology*, in press.

58 M. A. Anderson and F. M. M. Morel, *Limnol. Oceanogr.*, **27**, 789–813 (1982).

59 J. G. Rueter and F. M. M. Morel, *Limnol. Oceanogr.*, **26**, 67–73 (1981).

60 W. G. Sunda, R. T. Barber, and S. A. Huntsman, *Mar. Res.*, **39**, 567–586 (1981).

PROBLEMS

6.1 a. Using the same major ion composition as in Example 1, calculate the inorganic speciation of nickel and mercury in freshwater at pH = 8.

b. Maintaining C_T constant, calculate the effect of pH (from 5 to 12) on the inorganic speciation of nickel and mercury.

c. At what pH's would the carbonate or hydroxide solids precipitate given $(Ni)_T = (Hg)_T = 10^{-7} M$?

6.2 Consider the usual approximate stoichiometric reaction of photosynthesis: $CO_2 + H_2O \rightarrow$ "CH_2O" $+ O_2$. Note that this reaction changes both C_T and pH. Assuming the system to be closed to the atmosphere, discuss quantitatively the effect of an algal bloom on the formation of carbonate and bicarbonate trace metal complexes [e.g., $CuCO_3(aq)$ and $CuHCO_3^+$].

6.3 What is the effect of major ion associations in seawater on (a) pH (in equilibrium with atmosphere), and (b) carbonate and bicarbonate trace metal complexes.

6.4 a. Redo Example 8 of Chapter 5 adding $Ca_T = 10^{-3}$ M and $(NTA)_T = 10^{-4}$ M to the recipe of the system and letting $CaCO_3(s)$ supersaturate.

b. How much NTA would have to be added at any pH to avoid precipitation of the iron $[Y_T = Y_T (pH)]$?

c. Redo parts a and b for $CaCO_3(s)$ precipitation.

6.5 Consider a seawater culture medium containing $(EDTA)_T = 5 \times 10^{-5}$ M, $Fe(III)_T = 10^{-8}$ M and $Cd_T = 10^{-10}$ M. ($I = 0.5$ M, $Ca_T = 10^{-2}$ M, and pH = 8.2.)

a. Calculate the speciation and the free concentrations of EDTA, Fe, and Cd.

b. 4×10^{-5} M $CdCl_2$ is added to the system. Calculate the new speciation and new free concentrations. What would the result have been if 4×10^{-5} M CdEDTA had been added to the system?

c. The algae in the cultures (10^6 cells liter^{-1}) have on their surface an iron-transport molecule characterized by $X_T = 10^{-16}$ mol cell^{-1}; $K_{Fe} = 10^{19}$; $K_{Cd} = 10^8$. How much Fe and Cd are bound to the cells before and after $CdCl_2$ addition? What is the net effect of the Cd addition on the Fe transport rate?

6.6 A lake with the simple composition

$$Na_T = Ca_T = Mg_T = K_T = Cl_T = (SO_4)_T = (CO_3)_T = 10^{-3} \ M$$

contains an organic complexing agent Y, characterized by

$$HY = H^+ + Y^-; \qquad pK_a = 6.0$$

$$CuY^+ = Cu^{2+} + Y^-; \qquad pK = 7.0$$

$$CaY^+ = Ca^{2+} + Y^-; \qquad pK = 2.0$$

$$MgY^+ = Mg^{2+} + Y^-; \qquad pK = 2.0$$

a. Consider no solid phase but all the relevant complexes. Calculate the copper speciation at pH = 7.0 ($Cu_T = 10^{-7}$ M; $Y_T = 10^{-6}$ M). Calculate the copper speciation as a function of pH

b. Consider now that $Cu_T = 10^{-6}$ M and that copper carbonates and/or hydroxide may precipitate (same inorganic and organic species as in part a). Calculate the critical pH('s) of precipitation of the solid(s) in the absence of organic ligand. What solid(s) should form as pH varies from 6 to 12? Sketch a major Cu species diagram on a log Y_T versus pH graph.

CHAPTER SEVEN

OXIDATION-REDUCTION

The geochemical cycles of elements are driven in part by oxidation-reduction reactions: some mineral phases such as metal sulfides are dissolved through oxidation by oxygen, others such as iron oxides through reduction in anoxic environments; conversely, some solutes are ultimately precipitated as reduced sulfides or as oxides. Although significant, this auxiliary role of redox chemistry in the exogenic cycle of elements is not our major focus in this chapter. As unstable chemical entities fueled by a continual diet of decomposing compounds, sustained by a constant energy flux from oxidation reactions, we are motivated by more than pure academic curiosity to study redox chemistry; it is a simple, or not so simple, matter of life and death. Life is by nature a redox process, and a majority of redox processes on the earth are life dependent. From our point of view the subject of the redox cycles of elements thus takes on an importance disproportionate to the relatively small elemental fluxes that are involved.

At the basis of the redox cycles on the earth's surface is one fundamental, nonthermodynamic process, photosynthesis. Using solar energy through specialized pigments and organelles, plants reduce inorganic carbon to organic matter and produce oxygen, thus increasing the Gibbs free energy of the earth's surface:

$$CO_2(g) + H_2O \rightarrow \text{``}CH_2O\text{''} + O_2(g); \quad \Delta G = +114.3 \text{ kcal mol}^{-1} \quad (1)$$

With very few exceptions, this energy-capturing process drives all other redox reactions: organic matter is the ultimate reductant; oxygen, the ultimate oxidant. Powered by photosynthesis, the cycle of organic carbon is the driving belt that propels the other elemental redox cycles. As a first approximation, we can thus view all molecules at the earth's surface as being submitted directly or indirectly to a reducing force (an electron supply) from organic carbon and to an oxidative force (an electron drain) from oxygen. The resulting redox species depend on a local balance between these opposing forces. Fortunately, this balance is rarely an equilibrium state. We are dependent for our survival on the sluggishness of redox reactions.

Besides carbon and oxygen, the elements most involved in redox cycles are those most abundant in living matter: nitrogen, sulfur, and some metals such as iron and manganese. (Phosphorus, which is also abundant in biological material, serves chiefly as a constituent of intracellular energy storage compounds, ATP, NADP, but has no important redox chemistry of its own.) Because life is a matter of stoichiometry as well as one of energetics, the cycles of these elements are then tightly coupled in two different ways. First, the cycles are linked through the energetic relationships imposed by the thermodynamics of the redox reactions. For example, microorganisms must oxidize a certain quantity of organic carbon to obtain the energy necessary to reduce nitrogen. Second, in their ephemeral combination to form living matter, these elements are constrained to cycle within the rather strict stoichiometry of life, as expressed for example, in the Redfield formula, $C_{106}H_{263}O_{110}N_{16}P_1$. These energetic and stoichiometric couplings are ultimately reflected in the chemical composition of aquatic systems.

While equilibrium thermodynamics would seem to have little to do with such highly dynamic processes, it does in fact provide insight into the redox chemistry of aquatic systems in several ways. Continuing in the manner of the previous chapters, we can simply develop partial equilibrium models of natural waters, limiting them to include only the few redox couples that may reasonably be at equilibrium with each other. For example, this approach is helpful in studying the chemistry of trace metals in oxic and reduced systems. For those redox systems that are clearly in a state of disequilibrium, thermodynamics also allows us to calculate the energy required, at the minimum, to maintain that disequilibrium state. For example, we can estimate how much net energy must be fixed by plants to reduce carbon, or what is the maximum energy that can be derived by nonphotosynthetic organisms from any particular oxidation reaction. On the basis of these energetic considerations we can then decide what microbial redox processes are possible. In many cases we are also able to predict in what sequence redox processes will occur in nature since microbes usually exploit first the most energetically favorable reactions. (This amounts to "quasi-kinetic" information on the basis of equilibrium thermodynamics.)

Although the global picture of photosynthesis driving all other redox reactions through production of reduced carbon and oxygen provides a useful conceptual image of the major elemental redox cycles, there are instances where oxidants or reductants originate from other processes. In surface waters a number of redox reactions are driven directly by solar energy through photochemical processes. For example, it is possible for ferric iron in solution to be reduced photochemically to ferrous iron in the absence of reducing agents. Not much is known regarding the importance of these photochemical reactions. The transient formation of highly reactive radicals such as singlet oxygen may dramatically accelerate otherwise sluggish oxidation processes in surface waters.

At the bottom of the oceans one of the most important recent discoveries regarding ocean geochemistry is the role played by geothermal activity at oceanic ridges. Seawater constituents that circulate through ridge "vents" are mixed with reduced compounds from the earth's interior that do not ultimately owe their reduced condition to photosynthesis. These reduced compounds are important in the geochemical balance of many elements in the ocean. Note, however, that the corresponding redox processes—including the survival of the specialized ecosystems that inhabit the vents—are still dependent on photosynthesis since the oxidants that they utilize, O_2, SO_4^{2-}, and Fe(III), are ultimately derived from photosynthetically produced oxygen.

The first section of this chapter is devoted to defining such fundamental concepts as half redox reactions, electron activities, and redox potentials which are often a source of confusion. On the basis of these concepts, redox couples that are present in natural waters are then compared to characterize the redox status of the waters and to estimate the free energies that are involved in maintaining or exploiting existing disequilibria. Because the corresponding reactions are chiefly mediated by microorganisms, this is effectively a study of the energetics of aquatic microbial life. A series of examples of redox equilibrium calculations in complex systems is then presented, including a discussion of pe-pH diagrams. Finally, short elementary introductions to aquatic photochemistry and redox kinetics in natural waters are provided.

1. DEFINITIONS, NOTATIONS, AND CONVENTIONS

1.1 The Electron as a Component

Historically, and etymologically, oxidation reactions are those that involve the combination of oxygen with another element or compound. In the general framework that we have developed so far, we then need one new component to account for these reactions, to introduce O_2 as a reactive aquatic species. If we consider our usual choice of components for aquatic systems, including H_2O and H^+, there are three simple and equivalent possible choices for this new component. The first two are rather obvious; oxygen itself (O_2) and hydrogen (H_2) both provide straightforward stoichiometric expressions for oxygen:

$$O_2 = (O_2)_1$$
$$O_2 = (H_2)_{-2}(H_2O)_2$$

The third choice is that of a component symbolizing electrical charge with no elemental significance. For consistency with what we know of the structure of matter, this last component is taken to have a negative unit charge, is symbolized by e^-, and is referred to as the electron. It provides a formula for oxygen according to

$$O_2 = (H^+)_{-4}(e^-)_{-4}(H_2O)_2.$$

These three choices are strictly equivalent, and each of them has in fact been made at different times in history and in various disciplines for the quantitative study of some particular family of redox reactions. For example, many textbooks still refer to hydrogen rather than electron transfer processes to describe biochemical redox reactions.

Following the modern custom, we choose to use here the electron as our basic redox component, although it may seem perhaps to be the least natural of the three possibilities. With this convention a reductant is a compound that reacts by releasing electrons (an electron donor), and an oxidant is a compound that takes up electrons (an electron acceptor).

Since we do not consider electrons to have an existence of their own (they have been introduced here as purely conceptual components although free hydrated electrons may in fact have a half-life approaching milliseconds under favorable conditions), our convention introduces a whole series of convenient but artificial concepts and quantities in the study of redox processes. In particular, we now have to define reactions involving electrons, called half redox reactions, and their associated thermodynamic constants.

1.2   Half Redox Reactions

Consider the formulae of oxygen and hydrogen as provided from our present choice of components:

$$O_2 = (H^+)_{-4}(e^-)_{-4}(H_2O)_2$$

$$H_2 = (H^+)_2(e^-)_2$$

According to our simple view of the world there are no free electrons in water, and the reactions corresponding to these formulae, the half redox reactions

$$2H_2O = O_2 + 4H^+ + 4e^- \tag{2}$$

$$2H^+ + 2e^- = H_2 \tag{3}$$

can have no tangible chemical reality of their own. While the combination of these two reactions involving no electrons, the complete redox reaction

$$2H_2O = O_2 + 2H_2 \tag{4}$$

can be studied experimentally to calculate a free energy change or an equilibrium constant, the individual half redox reactions cannot. To obtain a thermodynamic description of half redox reactions, we then have to decide arbitrarily on the standard free energy change (or the equilibrium constant) of *one* half redox reaction. By international convention, we take the half reaction between hydrogen ion and hydrogen gas to have a zero standard free energy change:

$$H^+ + e^- = \tfrac{1}{2}H_2(g); \quad \Delta G^0 = 0; \quad K = 1 \tag{5}$$

The standard free energy change (and the equilibrium constant) of any other half redox reaction can then be calculated from the free energy change of complete redox reactions. Most simply, for complete redox reactions involving H^+ as an oxidant and H_2 as a reductant the standard free energy change of the corresponding half redox reaction is equal to that of the complete reaction:

$$\left.\begin{array}{l} \text{Red} + n\text{H}^+ = \text{Ox} + \dfrac{n}{2}\text{H}_2 \\[2mm] \text{Red} = n\text{e}^- + \text{Ox} \end{array}\right\} \text{same } \Delta G^0; \quad \text{same } K \qquad (6)$$
$$\hspace{10.5cm} (7)$$

For example, the oxygen/water half redox reaction has the same standard free energy change and the same equilibrium constant as the reaction of formation of water from oxygen and hydrogen:

$$\tfrac{1}{2}\text{O}_2(\text{g}) + \text{H}_2(\text{g}) = \text{H}_2\text{O}; \quad \Delta G^0 = -56.6 \text{ kcal mol}^{-1}; \quad K = 10^{41.5} \quad (8)$$

therefore

$$\tfrac{1}{2}\text{O}_2(\text{g}) + 2\text{e}^- + 2\text{H}^+ = \text{H}_2\text{O}; \quad \Delta G^0 = -56.6 \text{ kcal mol}^{-1}; \quad K = 10^{41.5} \quad (9)$$

Even if the complete redox reaction with H^+ and H_2 is not conveniently studied experimentally, the thermodynamics of any half redox reaction can be indirectly related to the H^+/H_2 convention through any series of other complete redox reactions involving common half reactions. For example, the reaction of oxidation of glucose by oxygen,

$$\tfrac{1}{6} \text{ glucose} + \text{O}_2(\text{g}) = \text{CO}_2(\text{g}) + \text{H}_2\text{O};$$
$$\Delta G^0 = -114.3 \text{ kcal mol}^{-1}; \quad K = 10^{83.8} \qquad (10)$$

provides the standard free energy change for the glucose/CO_2 half reaction by subtracting the O_2/H_2O half reaction, which is itself obtained from the reaction with hydrogen as shown above:

$$\tfrac{1}{6} \text{ glucose} + \text{H}_2\text{O} = \text{CO}_2(\text{g}) + 4\text{H}^+ + 4\text{e}^-;$$
$$\Delta G^0 = -1.1 \text{ kcal mol}^{-1}; \quad K = 10^{0.8} \qquad (11)$$

1.3 Oxidation State

The "electron levels" of elements in various chemical species are given a formal and conventional value called the *oxidation state* of the element and symbolized by a roman numeral, with a sign, in parentheses following the element: $Fe(+II)$, $Fe(+III)$, $C(O)$, $C(+IV)$, $S(-II)$, and so on (The $+$ sign is often omitted.) Although the assignment of the electrons to the correct element can be difficult and arbitrary in complex molecules, we can define here oxidation states rather simply for most species of interest in aquatic chemistry. Consider as a choice of components, H_2O, H^+, e^-, and the elements themselves, for example, N, S, C, and Fe (but of course *not* O or H). The oxidation state of any element in a compound with oxygen, hydrogen, and electrons is then simply equal to minus the stoichiometric coefficient of the electron, normalized per atom of the element, in the formula of the compound. For a general chemical formula involving the element A,

$$A_aO_oH_h^{n-} = (A)_a(H_2O)_o(H^+)_{-2o+h}(e^-)_{-2o+h+n}$$

this translates into the equation

$$\text{Oxidation state} = \frac{(2o - h - n)}{a} \tag{12}$$

Examples

$N(O)$: N_2

$N(-III)$: NH_3, NH_4^+

$N(+V)$: NO_3^-

$S(-II)$: H_2S, HS^-, S^{2-}

$S(+II)$: $S_2O_3^{2-}$

$S(+VI)$: SO_4^{2-}

$C(-IV)$: CH_4

$C(O)$: C, CH_2O

$C(+IV)$: CO_2, HCO_3^-, CO_3^{2-}

$Fe(+III)$: Fe^{3+}, $Fe(OH)_3$

$Fe(+II)$: Fe^{2+}

$Cr(+III)$: Cr^{3+}, $Cr(OH)_3$

$Cr(+VI)$: CrO_4^{2-}, $Cr_2O_7^{2-}$

The oxidation state of elements does not change due to coordination reactions among metals and ligands. For example, in the ferrous carbonate complex $FeCO_3$, Fe has the oxidation state $(+II)$ and C has the oxidation state $(+IV)$. This leaves the oxidation state of elements in only a few compounds of interest in aquatic chemistry [e.g., cyanide CN^-: $N(-III)$, $C(+II)$] to be assigned by additional rules given in chemistry textbooks.

Any pair of species comprising the same element in different oxidation states constitutes a *redox couple*. A balanced half redox reaction can be written for any redox couple by (1) balancing the stoichiometric coefficients of the element of interest, (2) adding H_2O to balance oxygen, (3) adding H^+ to balance hydrogen, and (4) adding e^- to balance electrical charges. For example, the four steps to balance the $SO_4^{2-}/S_2O_3^{2-}$ reactions are as follows:

1   $2SO_4^{2-} = S_2O_3^{2-}$

2   $2SO_4^{2-} = S_2O_3^{2-} + 5H_2O$

3   $2SO_4^{2-} + 10H^+ = S_2O_3^{2-} + 5H_2O$

4   $2SO_4^{2-} + 10H^+ + 8e^- = S_2O_3^{2-} + 5H_2O$

1.4   Electron Activities and Redox Potentials

On the basis of the convention of an equilibrium constant of unity for the H^+/H_2 couple, we have now defined standard free energies and equilibrium constants for all half redox reactions:

$$Ox + ne^- = Red; \quad \Delta G^0; \quad K \tag{7}$$

In so doing, we have formally and mathematically defined the equilibrium activity of the electron for any redox couple:

$$\{e^-\} = \left[\frac{1}{K}\frac{(Red)}{(Ox)}\right]^{1/n} \tag{13}$$

(The activities of the reductant and the oxidant are replaced here by their free concentrations for simplicity.) This electron activity is defined in any system where the free concentrations (Red) and (Ox) are defined, even though the electron is not considered to be an individual species and thus its concentration is certainly not defined. Electron activities are usually expressed on either of two scales, pe or E_H:

$$pe = -\log\{e^-\} = \frac{1}{n}\left[\log K - \log \frac{(Red)}{(Ox)}\right] \tag{14}$$

$$E_H = \frac{2.3RT}{F} pe \tag{15}$$

therefore

$$E_H = \frac{RT}{F}\frac{1}{n}\left[\ln K - \ln \frac{(Red)}{(Ox)}\right] \tag{16}$$

The parameter pe provides a nondimensional scale (like pH), while E_H, the redox potential, is measured in volts. [F is the Faraday constant (the electric charge of one mole of electrons = 96,500 Coulombs), and 2.3 RT/F has a value of 0.059 V at 25°C.]

One can gain insight into the reason why electron activities are logically expressed on an electric potential scale by considering energies. The progress to the right of the complete redox reaction $Ox_1 + Red_2 = Red_1 + Ox_2$ corresponds to the transfer of n electrons from Red_2 to Ox_1. The energy (ΔG) necessary to transfer this electrical charge of ($-nF$) coulombs can be expressed as the product of the charge and a potential difference ($E_H^1 - E_H^2$):

$$\Delta G = (-nF)(E_H^1 - E_H^2) \tag{17}$$

This energy must also be equal to the free energy change of the reaction:

$$\Delta G = \Delta G^0 + RT \ln \frac{(Red_1)(Ox_2)}{(Ox_1)(Red_2)} \tag{18}$$

The standard free energy ΔG^0 can be expressed as a function of the equilibrium constants of the half reactions:

$$\Delta G^0 = -RT \ln K = -RT \ln \frac{K_1}{K_2} \tag{19}$$

therefore

$$\Delta G = RT \left[\ln K_2 - \ln \frac{(\text{Red}_2)}{(\text{Ox}_2)} \right] - RT \left[\ln K_1 - \ln \frac{(\text{Red}_1)}{(\text{Ox}_1)} \right] \tag{20}$$

A comparison of Equations 17 and 20 leads to the expression of the electrical potential E_H as given in Equation 16 (within a constant which is simply taken to be zero for consistency with $\Delta G = \Delta G^0 = 0$ for the H^+/H_2 reaction when $\{H^+\} = P_{H_2} = 1$).

If we normalize the energy change to one electron transfer and introduce Equation 14 into 20, we obtain the general expressions

$$\frac{\Delta G}{n} = 2.3RT(\text{pe}_2 - \text{pe}_1) \tag{21a}$$

$$= F(E_H^2 - E_H^1) \tag{21b}$$

The difference in electron activities between two half redox reactions is thus directly proportional to the free energy change of the complete redox reaction. In effect pe and E_H define two absolute energy scales (whose origins are fixed by the H^+/H_2 convention), and free energies of complete redox reactions are obtained by simple differences.

If we now consider the *standard* free energy changes of redox reactions (the free energy changes corresponding to unit activities of products and reactants; $\Delta G^0 = -RT \ln K$), we can, for consistency, express the equilibrium constants of half redox reactions as pe^0 or standard potentials E_H^0:

$$\text{pe}^0 = \frac{F}{2.3RT} E_H^0$$

$$= \frac{1}{n} \log K$$

$$= -\frac{1}{n} \frac{\Delta G^0}{2.3RT} \tag{22}$$

At 25°C

$$\text{pe}^0 = 16.9 E_H^0 \quad \text{(in V)} \tag{23a}$$

$$= \frac{1}{n} \log K \tag{23b}$$

$$= -\frac{0.733 \Delta G^0}{n} \quad \text{(in kcal mol}^{-1}) \tag{23c}$$

TABLE 7.1

Half Redox Reactions[1,2,3]

	$pe^0 = \log K$
Hydrogen	
$H^+ + e^- = \frac{1}{2}H_2(g)$	0
Oxygen	
$\frac{1}{2}O_3(g) + H^+ + e^- = \frac{1}{2}O_2(g) + \frac{1}{2}H_2O$	$+35.1$
$\frac{1}{4}O_2(g) + H^+ + e^- = \frac{1}{2}H_2O$	$+20.75$
$\frac{1}{2}H_2O_2 + H^+ + e^- = H_2O$	$+30.0$
(Note also $HO_2^- + H^+ = H_2O$; $\log K = 11.6$)	
Nitrogen	
$NO_3^- + 2H^+ + e^- = \frac{1}{2}N_2O_4(g) + H_2O$	$+13.6$
(Note: $N_2O_4(g) = 2NO_2(g)$; $\log K = -0.47$)	
$\frac{1}{2}NO_3^- + H^+ + e^- = \frac{1}{2}NO_2^- + \frac{1}{2}H_2O$	$+14.15$
(Note $NO_2^- + H^+ = HNO_2$; $\log K = 3.35$)	
$\frac{1}{3}NO_3^- + \frac{4}{3}H^+ + e^- = \frac{1}{3}NO(g) + \frac{2}{3}H_2O$	$+16.15$
$\frac{1}{4}NO_3^- + \frac{5}{4}H^+ + e^- = \frac{1}{8}N_2O(g) + \frac{5}{8}H_2O$	$+18.9$
$\frac{1}{5}NO_3^- + \frac{6}{5}H^+ + e^- = \frac{1}{10}N_2(g) + \frac{3}{5}H_2O$	$+21.05$
$\frac{1}{8}NO_3^- + \frac{5}{4}H^+ + e^- = \frac{1}{8}NH_4^+ + \frac{3}{8}H_2O$	$+14.9$
Sulfur	
$\frac{1}{2}SO_4^{2-} + H^+ + e^- = \frac{1}{2}SO_3^{2-} + \frac{1}{2}H_2O$	-1.65
[Note also $(SO_3^{2-} + H^+ = HSO_3^-$; $\log K \cong 7)$]	
$\frac{1}{4}SO_4^{2-} + \frac{5}{4}H^+ + e^- = \frac{1}{8}S_2O_3^{2-} + \frac{5}{8}H_2O$	$+4.85$
$\frac{1}{6}SO_4^{2-} + \frac{4}{3}H^+ + e^- = \frac{1}{48}S_8^0(s.\ ort.) + \frac{2}{3}H_2O$	$+6.03$
[Note also $\frac{1}{8}S_8^0(s.\ ort.) = \frac{1}{8}S_8^0(s.\ col.)$; $\log K = -0.6$]	
$\frac{3}{19}SO_4^{2-} + \frac{24}{19}H^+ + e^- = \frac{1}{38}S_6^{2-} + \frac{12}{19}H_2O$	$+5.41$
$\frac{5}{32}SO_4^{2-} + \frac{5}{4}H^+ + e^- = \frac{1}{32}S_5^{2-} + \frac{5}{8}H_2O$	$+5.29$
(Note also $S_5^{2-} + H^+ = HS_5^-$; $\log K = 6.1$)	
$\frac{2}{13}SO_4^{2-} + \frac{16}{13}H^+ + e^- = \frac{1}{26}S_4^{2-} + \frac{8}{13}H_2O$	$+5.12$
(Note also $S_4^{2-} + H^+ = HS_4^-$; $\log K = 7.0$)	
$\frac{1}{8}SO_4^{2-} + \frac{5}{4}H^+ + e^- = \frac{1}{8}H_2S(aq) + \frac{1}{2}H_2O$	$+5.13$
(Note also $H_2S(g) = H_2S(aq)$; $\log K_H = 1.0$, and other acid-base, coordination, and precipitation reactions)	
Carbon	
Inorganic	
Carbon monoxide $\quad \frac{1}{2}CO_2(g) + H^+ + e^- = \frac{1}{2}CO(g) + \frac{1}{2}H_2O$	-1.74
Graphite $\quad \frac{1}{4}CO_2(g) + H^+ + e^- = \frac{1}{4}C(s) + \frac{1}{2}H_2O$	$+3.50$
Organic	
C.1	
Formate⁻ $\quad \frac{1}{2}CO_2(g) + \frac{1}{2}H^+ + e^- = \frac{1}{2}HCOO^-$	-5.22
Formaldehyde $\quad \frac{1}{4}CO_2(g) + H^+ + e^- = \frac{1}{4}HCHO(aq) + \frac{1}{4}H_2O$	-1.20

(Continued)

TABLE 7.1 (Continued)

		$pe^0 = \log K$
Methanol	$\frac{1}{6}CO_2(g) + H^+ + e^- = \frac{1}{6}CH_3OH(aq) + \frac{1}{6}H_2O$	$+0.50$
Methane	$\frac{1}{8}CO_2(g) + H^+ + e^- = \frac{1}{8}CH_4(g) + \frac{1}{4}H_2O$	$+2.86$

C.2

Oxalate^{2-}	$CO_2(g) + e^- = \frac{1}{2}(COO^-)_2$	-10.7
Acetate	$\frac{1}{4}CO_2(g) + \frac{7}{8}H^+ + e^- = \frac{1}{8}CH_3COO^- + \frac{1}{4}H_2O$	$+1.27$
Acetaldehyde	$\frac{1}{5}CO_2(g) + H^+ + e^- = \frac{1}{10}CH_3CHO(g) + \frac{3}{10}H_2O$	$+0.99$
Ethanol	$\frac{1}{6}CO_2(g) + H^+ + e^- = \frac{1}{12}CH_3CH_2OH(aq) + \frac{1}{4}H_2O$	$+1.52$
Ethane	$\frac{1}{7}CO_2(g) + H^+ + e^- = \frac{1}{14}C_2H_6(g) + \frac{2}{7}H_2O$	$+2.41$
Ethylene	$\frac{1}{6}CO_2(g) + H^+ + e^- = \frac{1}{12}C_2H_4(g) + \frac{1}{3}H_2O$	$+1.34$
Acetylene	$\frac{1}{5}CO_2(g) + H^+ + e^- = \frac{1}{10}C_2H_2(g) + \frac{2}{5}H_2O$	-0.86

C.3

Pyruvate$^-$	$\frac{3}{10}CO_2(g) + \frac{9}{10}H^+ + e^- = \frac{1}{10}CH_3COCOO + \frac{3}{10}H_2O$	$+0.05$
Lactate$^-$	$\frac{1}{4}CO_2(g) + \frac{11}{12}H^+ + e^- = \frac{1}{12}CH_3CHOHCOO + \frac{1}{4}H_2O$	$+0.68$
Glycerol	$\frac{3}{14}CO_2(g) + H^+ + e^- = \frac{1}{14}CH_2OHCHOHCH_2OH$ $+ \frac{3}{14}H_2O$	$+0.21$
Alanine	$\frac{1}{4}CO_2(g) + \frac{1}{12}NH_4^+ + \frac{11}{12}H^+ + e^-$ $= \frac{1}{12}CH_3CHCOO^-NH_3^+ + \frac{1}{3}H_2O$	$+0.84$

C.4

Succinate^{2-}	$\frac{2}{7}CO_2(g) + \frac{6}{7}H^+ + e^- = \frac{1}{14}(CH_2COO^-)_2 + \frac{2}{7}H_2O$	$+0.77$

C.6

Glucose	$\frac{1}{4}CO_2(g) + H^+ + e^- = \frac{1}{24}C_6H_{12}O_6 + \frac{1}{4}H_2O$	-0.20

Halogens

	$\frac{1}{2}Cl_2(g) + e^- = Cl^-$	$+23.0$
	$\frac{1}{2}HClO + \frac{1}{2}H^+ + e^- = \frac{1}{2}Cl^- + \frac{1}{2}H_2O$	$+25.3$
	(Note also $HClO = H^+ + ClO^-$; $\log K = 7.50$)	
	$\frac{1}{5}IO_3^- + \frac{6}{5}H^+ + e^- = \frac{1}{10}I_2(s) + \frac{3}{5}H_2O$	$+20.1$
	$\frac{1}{2}I_2(s) + e^- = I^-$	$+9.05$

Trace Metals

Cr	$\frac{1}{3}HCrO_4^- + \frac{7}{3}H^+ + e^- = \frac{1}{3}Cr^{3+} + \frac{4}{3}H_2O$	$+20.2$
	(Note $HCrO_4^- = H^+ + \frac{1}{2}Cr_2O_7^{2-} + \frac{1}{2}H_2O$, $\log K = -15$; $HCrO_4^- = H^+ + CrO_4^{2-}$, $\log K = -6.5$; and various Cr(III) precipitation and coordination reactions)	
Mn	$\frac{1}{5}MnO_4^- + \frac{8}{5}H^+ + e^- = \frac{1}{5}Mn^{2+} + \frac{4}{5}H_2O$	$+25.5$
	$\frac{1}{2}MnO_2(s) + 2H^+ + e^- = \frac{1}{2}Mn^{2+} + H_2O$	$+20.8$
Fe	$Fe^{3+} + e^- = Fe^{2+}$	$+13.0$
	$\frac{1}{2}Fe^{2+} + e^- = \frac{1}{2}Fe(s)$	-7.5
	$\frac{1}{2}Fe_3O_4(s) + 4H^+ + e^- = \frac{3}{2}Fe^{2+} + 2H_2O$	$+16.6$
Co	$Co(OH)_3(s) + 3H^+ + e^- = Co^{2+} + 3H_2O$	$+29.5$
	$\frac{1}{2}Co_3O_4(s) + 4H^+ + e^- = \frac{3}{2}Co^{2+} + 2H_2O$	$+31.4$

TABLE 7.1 (Continued)

		$pe^0 = \log K$
Cu	$Cu^{2+} + e^- = Cu^+$	$+2.6$
	$\frac{1}{2}Cu^{2+} + e^- = Cu(s)$	-5.7
Se	$\frac{1}{2}SeO_4^{2-} + 2H^+ + e^- = \frac{1}{2}H_2SeO_3 + \frac{1}{2}H_2O$	$+19.4$
	$\frac{1}{4}H_2SeO_3 + H^+ + e^- = \frac{1}{4}Se(s) + \frac{3}{4}H_2O$	$+12.5$
	$\frac{1}{2}Se(s) + H^+ + e^- = \frac{1}{2}H_2Se$	-6.7
	(Note also $H_2Se = H^+ + HSe^-$, $\log K = -3.9$;	
	$H_2SeO_3 = H^+ + HSeO_3^-$, $\log K = -2.4$;	
	$HSeO_3^- = H^+ + SeO_3^{2-}$, $\log K = -7.9$;	
	$SeO_4^{2-} + H^+ = HSeO_4^-$, $\log K = +1.7$)	
Ag	$AgCl(s) + e^- = Ag(s) + Cl^-$	$+3.76$
	$Ag^+ + e^- = Ag(s)$	-13.5
Hg	$\frac{1}{2}Hg^{2+} + e^- = Hg(l)$	-14.4
	$Hg^{2+} + e^- = \frac{1}{2}Hg_2^{2+}$	$+15.4$
Pb	$\frac{1}{2}PbO_2 + 2H^+ + e^- = \frac{1}{2}Pb^{2+} + H_2O$	$+24.6$
	(Note many other reactions for Mn, Fe, Co, Cu, Se, Ag, Hg, Pb)	

The coexistence of these four different scales (pe^0, E_H^0, K, ΔG^0) to define the thermodynamic characteristics of redox reactions does much to confuse the subject. As shown in Table 7.1, which provides thermodynamic data for several redox reactions of interest in aquatic systems, we shall consistently use pe^0 and pe as our notation, remembering that pe^0 is the log of the equilibrium constant for *the reaction written as a reduction involving one electron*,

$$Ox + e^- = Red$$

and that the ($-\log$ of) electron activity is then given by

$$pe = pe^0 - \log \frac{(Red)}{(Ox)} \tag{24}$$

Equation 16, which defines an electrical potential corresponding to any half redox reaction, is known as the Nernst equation. This potential is in fact measurable, compared to a reference potential defined by a hydrogen gas partial pressure of 1 atmosphere in equilibrium with a solution at pH = 0

$$\{e^-\} = \frac{1}{1} \frac{P_{H_2}^{1/2}}{\{H^+\}} = 1; \quad E_H = 0 \tag{25}$$

This is achieved in electrochemical cells by using the standard hydrogen electrode as reference.

□ □ □

1.5   Similarities and Differences between Redox
and Acid-Base Reactions

It is often noted that there is a parallel between acid-base and redox reactions; in one case H^+ is being exchanged, in the other e^-:

$$\begin{array}{ccccc} HAc & = & H^+ & + & Ac^- \\ \text{acid} (H^+ \text{ donor}) & & \text{proton} & & \text{base} (H^+ \text{ acceptor}) \end{array} \qquad (26)$$

$$\begin{array}{ccccc} Fe^{2+} & = & e^- & + & Fe^{3+} \\ \text{reductant} (e^- \text{ donor}) & & \text{electron} & & \text{oxidant} (e^- \text{ acceptor}) \end{array} \qquad (27)$$

This formal similarity between acid-base and redox processes is conceptually helpful and serves as a thread throughout this chapter, but three major differences have to be kept in mind:

1   H^+, in some hydrated form, exists as a species in water (at least in the sense of balancing electrical charges), while e^- does not.

2   Although acid-base reactions in solution are typically fast, redox reactions are typically slow and mediated by organisms. Total equilibrium for redox processes in natural water results in meaningless—and lifeless—models.

3   Unlike equilibrium constants for acid-base reactions, constants for redox reactions are often extremely large or small. That is to say that the free energies involved are much larger in redox than in acid-base processes and that, at equilibrium, redox reactions proceed to completion in one direction or the other.

These differences between acid-base and redox processes are partly an artifact of the way we have introduced the corresponding concepts and definitions. For convenience we have systematically ignored water as a reactant and failed to differentiate between protons and electrons and their corresponding hydrated species. Reactions that produce or consume protons (per se) are half acid-base reactions, strictly similar to half redox reactions, and they necessitate another half acid-base reaction to yield a complete acid-base reaction. In aqueous systems this matching half acid-base reaction can often be $H_2O + H^+ = H_3O^+$, a reaction that is conceptually and mechanistically important even if in our convention it corresponds to a strict identity. In this sense the definition of pH, like that of pe, is independent of the existence of the species H^+ ($=H_3O^+$) in solution. The concept of pH, and its measurement, depend on the activity ratio for a given acid-base couple, not on the H^+ concentration. pH and pe are intensive thermodynamic quantities not extensive ones. Note also that the energetic difference listed here (3 above) is largely due to our focus on weak acid-base reactions and the fact that many redox reactions involve the transfer of several electrons, while acid-base reactions typically involve the transfer of only one or two protons.

□ □ □

2. COMPARISON AMONG REDOX COUPLES

2.1 pe's of Dominant Redox Couples

Since it is a measure of potential energy, the electron activity that can be calculated for any half redox reaction is a measure of the "reduction power" of the corresponding redox couple in the system of interest. Low pe's $\left[\text{high } \{e^-\}\right]$ correspond to highly reducing species, and high pe's $\left[\text{low } \{e^-\}\right]$ to oxidizing ones. Consider the complete redox reaction:

$$Ox_1 + Red_2 = Red_1 + Ox_2$$

As seen in Equation 21, a necessary and sufficient condition of equilibrium among these two redox couples is given by the equality of their electron activities:

$$\Delta G = 0 = 2.3RT(pe_2 - pe_1) \tag{28}$$

therefore

$$pe_1 = pe_2 \tag{29}$$

Suppose that initially

$$pe_1 > pe_2$$

therefore

$$pe_1 - pe_2 = pe_1^0 - pe_2^0 - \log \frac{(Red_1)(Ox_2)}{(Ox_1)(Red_2)} > 0$$

To achieve equilibrium, the logarithmic term must thus increase; its numerator must increase, and its denominator decrease. The complete redox reaction thus proceeds to the right: Ox_1 oxidizes Red_2 to Ox_2 and Red_2 reduces Ox_1 to Red_1. This may perhaps be seen more readily by considering that the free energy change ΔG of the overall reaction (see Equation 28) is negative. When two redox couples have different electron activities in a given system, the one with the lower pe tends to reduce the other, and vice versa.

If one redox couple is present in much larger concentrations than the other $\left[\text{e.g., }(Ox_1), (Red_1) \gg (Ox_2), (Red_2)\right]$, the corresponding free concentrations of oxidant and reductant (Ox_1) and (Red_1) are unaffected by the advancement of the complete redox reaction toward equilibrium. The equilibrium electron activity is then effectively that of the corresponding dominant redox couple:

$$pe = pe_1 = pe_1^0 - \log \frac{(Red_1)}{(Ox_1)}$$

Redox potentials (electron activities) corresponding to redox couples that are present in relatively large concentrations in natural waters are then of particular interest; they tend to force all other redox couples to the same electron activity; their pe is effectively characteristic of the aquatic system itself and is often (sloppily) referred to as the "pe of the system."

Example 1.   pe of the Organic Matter/CO$_2$ Couple

In much of this chapter it is convenient to use some stoichiometric and thermodynamic description of organic matter as if it were a well-defined unique compound. For this purpose we utilize the symbol "CH$_2$O" with the thermodynamic properties of 1/6 glucose. Using this "average" organic compound permits us to define rough stoichiometric and energetic relationships. (There is of course no substitute for knowing the actual nature of reacting organic species in a particular situation.)

Consider water at neutral pH, in equilibrium with the atmosphere, and containing 10^{-5} M dissolved organic carbon:

$$\tfrac{1}{4}CO_2(g) + H^+ + e^- = \tfrac{1}{4}\text{``CH}_2\text{O''} + \tfrac{1}{4}H_2O; \quad pe^0 = -0.20$$

$$pe = pe^0 - \log \frac{(CH_2O)^{1/4}}{P_{CO_2}^{1/4}\{H^+\}} = -6.83 \tag{30}$$

Note that this pe is rather insensitive to the actual concentration of organic carbon and the partial pressure of CO_2. However, the very low pe of the organic matter/CO_2 couple is not the dominant pe in natural waters because the oxidizing couples O_2/H_2O and SO_4^{2-}/S^{2-} are usually more abundant; yet it may be considered characteristic of intracellular compartments in living organisms. In aquatic systems the biota provides microenvironments that exhibit very high reducing powers and are the loci of most of the reducing activity in the system.

Example 2.   pe of Anoxic Waters: The Sulfate/Sulfide Couple

Anoxic waters are often characterized by the presence of sulfide. When oxygen is exhausted, organisms use sulfate as the electron acceptor to oxidize organic matter. Sulfate being usually abundant in natural waters, the pe of the sulfate/sulfide couple [S(VI)/S(−II)] can be considered as characteristic of anoxic systems. There are of course a number of exceptions to this, and one may find anoxic systems dominated by $N(V)/N(III)$, $N(III)/N(-III)$, $Fe(III)/Fe(II)$, and other redox couples.

Consider, for example, a system at $pH = 8$ where $(HS^-) = 10^{-5}$ M and $(SO_4^{2-}) = 10^{-3}$ M:

$$\tfrac{1}{8}SO_4^{2-} + \tfrac{9}{8}H^+ + e^- = \tfrac{1}{8}HS^- + \tfrac{1}{2}H_2O; \quad pe^0 = 4.25$$

therefore

$$pe = pe^0 - \log \frac{(HS^-)^{1/8}}{(SO_4^{2-})^{1/8}(H^+)^{9/8}} = -4.5 \tag{31}$$

This pe is very insensitive to the actual concentrations of sulfate and sulfide: a decrease of sulfide concentration by four orders of magnitude would only result in a 0.5 increase in pe. The presence of any measurable sulfide in a natural water near neutral pH yields pe's in the range -3.5 to -5.5. Complications introduced by the presence of sulfur species with intermediate oxidation states (between $+VI$ and $-II$) are discussed in Section 4.

Example 3. pe of Oxic Waters: The Oxygen/Water Couple

The major oxidant in oxic systems is oxygen. A straightforward measure of its tendency to oxidize other compounds is given by the pe of the oxygen/water couple.

Consider at $pH = 7$ and $P_{O_2} = 10^{-0.7}$ at

$$\tfrac{1}{4}O_2(g) + H^+ + e^- = \tfrac{1}{2}H_2O; \quad pe^0 = 20.75$$

therefore

$$pe = pe^0 - \log \frac{1}{P_{O_2}^{1/4}(H^+)} = 13.58 \tag{32}$$

Once again, notice that this pe is insensitive to the partial pressure of oxygen; pe's in the range 12 to 14 are characteristic of oxic natural waters.

Example 4. Redox Equilibrium of Iron in Oxic Waters

From the redox potential (pe) of the dominant redox couple in a given aquatic system, one may easily calculate the equilibrium redox speciation of minor species. What is the ratio of $\{Fe^{2+}\}$ to $\{Fe^{3+}\}$ at equilibrium with the oxygen of the atmosphere and water at neutral pH? The pe of the oxygen/water couple has been calculated in Example 3:

$$pe_1 = 13.58$$

$$Fe^{3+} + e^- = Fe^{2+}; \quad pe^0 = 13.0 \tag{33}$$

and

$$pe_2 = 13.0 - \log \frac{(Fe^{2+})}{(Fe^{3+})}$$

Equating the pe's,

$$\frac{(Fe^{2+})}{(Fe^{3+})} = 10^{-0.58} = 0.26 \tag{34}$$

Note that this calculation is only an indirect way to express the mass law of the complete redox reaction ($Fe^{2+} + \frac{1}{4}O_2(g) + H^+ = Fe^{3+} + \frac{1}{2}H_2O$) for specified concentrations of the reactants O_2 and H^+. Note also that it provides the equilibrium ratio of the *free* concentrations of the ferrous and ferric ions, not the ratio of *total* Fe(II) and Fe(III) concentrations.

We shall see later (Section 2.3) how more complicated redox equilibrium problems can be organized in the usual tableau form.

2.2   Redox Reactions as Irreversible Reactions

In Section 2.1 we have noted that the pe of several redox couples is rather insensitive to the actual concentrations of the species involved (except H^+). For example, the presence of any sulfide in a sulfate containing system leads to a very narrow range of redox potentials. Conversely, there can be *no* sulfide at equilibrium with sulfate in an oxygen dominated system:

$$pe = 13.58 \quad \text{(Example 3)}$$

therefore

$$\log \frac{(HS^-)^{1/8}}{(SO_4^{2-})^{1/8}(H^+)^{9/8}} = 4.25 - 13.58 = -9.33 \tag{35}$$

$$(HS^-) = 10^{-74.6}(SO_4^{2-})(H^+)^9$$

Choosing $(SO_4^{2-}) \cong 10^{-2.4} M$ and pH = 7, we obtain the bisulfide concentration:

$$(HS^-) = 10^{-140} M \tag{36}$$

The value of $10^{-140} M$ is a truly meaningless concentration (think how small the magnitudes of Avogadro's number, 6×10^{23}, and the volume of the oceans, 1.4×10^{21} liters, are in comparison).

If we consider the general complete redox reaction, $Red_1 + Ox_2 = Ox_1 + Red_2$, its direction of progress depends *in principle* on the relative concentrations of the various species. If, however, we consider reasonable limits on the values that these concentrations can *actually* take, the direction of progress of the reaction can be a foregone conclusion provided the equilibrium constant is sufficiently large or small. This is the case for many redox reactions in natural systems. For example, the reaction of sulfide with oxygen discussed earlier

$$HS^- + 2O_2(g) = SO_4^{2-} + H^+$$

has an equilibrium constant $K \cong 10^{+132}$.

Consider another example, one that is sadly tangible to Massachusetts car owners:

$$2Fe^0 + \tfrac{3}{2}O_2(g) = Fe_2O_3(s); \quad K = 10^{130} \times \tfrac{2}{3} \tag{37}$$

Under any partial pressure of oxygen greater than 10^{-87} at, elemental iron tends to be oxidized. Again the partial pressure 10^{-130} at is truly meaningless (it only makes conceptual sense), and correct thermodynamic statements can be made in an absolute qualitative way:

"Oxygen oxidizes elemental iron."

"Elemental iron is only stable in *oxygen-free* systems."

"Iron naturally rusts."

Chemically, we can symbolize the reaction as being irreversible.

$$2Fe^0 + \tfrac{3}{2}O_2(g) \rightarrow Fe_2O_3$$

This absolute approach to redox chemistry, which is typified by the question, "What can oxidize or reduce what?" with no regard for the composition of the system, is in some ways contrary to the law of mass action and to the spirit of equilibrium thermodynamics, but it clearly makes practical sense.

The approach can be systematized by comparing equilibrium constants (pe^0) of half redox reactions. The oxidant in reactions with large pe^0 "can" oxidize a reductant at a lower pe^0, and vice versa. To make the comparison more realistic, reactants or products that are involved in these reactions, but are not the oxidant/reductant couples of interest, can be fixed at some typical values (e.g., $(H^+) = 10^{-7} M$, $(HCO_3^-) = 10^{-3} M$, etc.) and included in the constants then denoted pe_w^0.

Consider, for example, the sulfate/sulfide couple:

$$\tfrac{1}{8}SO_4^{2-} + \tfrac{9}{8}H^+ + e^- = \tfrac{1}{8}HS^- + \tfrac{1}{2}H_2O; \quad pe^0 = \log K = 4.25$$

$$pe = pe^0 - \log \frac{(HS^-)^{1/8}}{(SO_4^{2-})^{1/8}(H^+)^{9/8}}$$

$$= 4.25 - \frac{1}{8}\log \frac{(HS^-)}{(SO_4^{2-})} + \frac{9}{8}\log(H^+)$$

We obtain a good approximation (pe_w^0) of this pe for all reasonable values of the sulfide to sulfate ratio and neutral pH by ignoring the second term and setting $(H^+) = 10^{-7} M$:

$$pe_w^0 = 4.25 - (\tfrac{9}{8})7 = -3.63 \tag{38}$$

(More precisely, at pH = 7 we could write the reaction with half of the sulfide as H_2S and half as HS^-.)

Of course, this introduction of pe_w^0 as yet another thermodynamic scale (a fifth one) to describe redox reactions adds to the confusion. Yet it is quite practical for examining the possibility or impossibility of various redox processes.

TABLE 7.2

Effective Equilibrium Constants of Aquatic Redox Couples

	pe^0	pe_w^0 (pH = 7)
$\frac{1}{2}H_2O_2 + H^+ + e^- = H_2O$	+30.0	+23.0
$\frac{1}{4}O_2(g) + H^+ + e^- = \frac{1}{2}H_2O$	+20.75	+13.75
$\frac{1}{5}NO_3^- + \frac{6}{5}H^+ + e^- = \frac{1}{10}N_2(g) + \frac{3}{5}H_2O$	+21.05	+12.65
$\frac{1}{2}NO_3^- + H^+ + e^- = \frac{1}{2}NO_2^- + \frac{1}{2}H_2O$	+14.15	+7.15
$\frac{1}{2}MnO_2(s) + 2H^+ + e^- = \frac{1}{2}Mn^{2+} + H_2O$	+20.8	+6.8
$\frac{1}{8}NO_3^- + \frac{5}{4}H^+ + e^- = \frac{1}{8}NH_4^+ + \frac{3}{8}H_2O$	+14.9	+6.15
$\frac{1}{6}NO_2^- + \frac{4}{3}H^+ + e^- = \frac{1}{6}NH_4^+ + \frac{1}{3}H_2O$	+15.2	+5.8
$\frac{1}{2}O_2(g) + H^+ + e^- = \frac{1}{2}H_2O_2$	+11.5	+4.5
$\frac{1}{6}SO_4^{2-} + \frac{4}{3}H^+ + e^- = \frac{1}{48}S_8(col) + \frac{2}{3}H_2O$	+5.9	−3.4
$\frac{1}{8}SO_4^{2-} + \frac{5}{4}H^+ + e^- = \frac{1}{8}H_2S(g) + \frac{1}{2}H_2O$	+5.25	−3.5
$\frac{1}{8}SO_4^{2-} + \frac{9}{8}H^+ + e^- = \frac{1}{8}HS^- + \frac{1}{2}H_2O$	+4.25	−3.6
$\frac{1}{8}HCO_3^- + \frac{9}{8}H^+ + e^- = \frac{1}{8}CH_4(g) + \frac{3}{8}H_2O$	+3.8	−4.0
$\frac{1}{8}CO_2(g) + H^+ + e^- = \frac{1}{8}CH_4(g) + \frac{1}{4}H_2O$	+2.9	−4.1
$\frac{1}{16}S_8(col) + H^+ + e^- = \frac{1}{2}H_2S(g)$	+3.2	−3.8[a]
$\frac{1}{16}S_8(col) + \frac{1}{2}H^+ + e^- = \frac{1}{2}HS^-$	−0.8	−4.3
$\frac{1}{6}N_2(g) + \frac{4}{3}H^+ + e^- = \frac{1}{3}NH_4^+$	+4.65	−4.7
$Fe(OH)_3(am) + 3H^+ + e^- = Fe^{2+} + 3H_2O$	+16.0	−5.0
$H^+ + e^- = \frac{1}{2}H_2(g)$	0	−7.0
$\frac{1}{4}HCO_3^- + \frac{5}{4}H^+ + e^- = \frac{1}{4}\text{“}CH_2O\text{”} + \frac{1}{2}H_2O$	+1.8	−7.0
$\frac{1}{4}CO_2(g) + H^+ + e^- = \frac{1}{4}\text{“}CH_2O\text{”} + \frac{1}{4}H_2O$	−0.2	−7.2
$\frac{1}{2}HCO_3^- + \frac{3}{2}H^+ + e^- = \frac{1}{2}CO(g) + H_2O$	+2.2	−8.3
$\frac{1}{2}CO_2(g) + H^+ + e^- = \frac{1}{2}CO(g) + \frac{1}{2}H_2O$	−1.7	−8.7

[a] This reaction is listed out of order so as not to separate it from the reaction of formation of the bisulfide ion HS^-.

Consider Table 7.2 which is organized according to pe_w^0's from strong oxidants at the top to strong reductants at the bottom. Looking at such a table one can, for example, decide at a glance that sulfate "can" oxidize organic carbon ("CH_2O") to form carbon dioxide and sulfide but that sulfate "cannot" oxidize nitrite to nitrate.

2.3 Energetics of Microbial Processes

Thermodynamic considerations can do more than just provide criteria for judging the possible reactions in chemical systems. The energies consumed or liberated by changes in chemical composition can actually be estimated. It is in fact remarkable how much insight into the workings of microbial communities, which mediate most redox reactions in aquatic systems, can be gained from purely energetic considerations, often with very limited knowledge of the actual biochemical mechanisms involved.

According to Equation 21 the energy liberated or consumed by any complete redox reaction is directly calculated from the pe^0's of the half reactions:

$$Ox_1 + Red_2 = Ox_2 + Red_1; \quad \log K = pe_1^0 - pe_2^0 \tag{39}$$

therefore

$$\Delta G = -2.3RT(pe_1 - pe_2) = -2.3RT\left[(pe_1^0 - pe_2^0) - \log\frac{(Ox_2)(Red_1)}{(Ox_1)(Red_2)}\right] \tag{40}$$

As discussed in Section 2.2, the concentration-dependent term is often small and can be ignored in such calculations or replaced by a constant term (e.g., $pH = 7$) included in the effective parameter pe_w^0. An approximate value, ΔG_w^0, of the energy liberated or consumed by a complete redox reaction is then obtained from a simple formula:

$$\Delta G_w^0 \cong -2.3RT(pe_{w1}^0 - pe_{w2}^0) \tag{41}$$

therefore

$$\Delta G_w^0 \cong -1.37(pe_{w1}^0 - pe_{w2}^0) \text{ kcal (mol e}^-)^{-1} \tag{42}$$

This last formula has been applied to some familiar microbial processes listed in Table 7.3. (Table 7.2 provides the thermodynamic data used in the calculations.) The reactions of Table 7.3 correspond to very idealized processes, and the calculated energies represent either the maximum energy available from each reaction or the minimum energy necessary to carry out the reaction. Inefficiencies in the energy-transfer mechanisms and various energies necessary for maintenance of cellular integrity, reproduction, or ion transport are not accounted for. For example, about 300 kcal of chemical energy are actually expended by plants to fix one mole of inorganic carbon (even more light energy must be absorbed) and only 40 kcal of useful energy can be obtained from oxidizing 1/6 glucose to CO_2 with oxygen, rather than the standard value of 114 kcal.[4] Energetic efficiencies of the order of 30 to 40% are fairly typical of biochemical processes.

Microorganisms can be classified according to their principal energy source, their principal carbon source, and their electron source[5] (see Figure 7.1). All three can be separate, as is the case for plants which absorb light, fix CO_2, and oxidize water to oxygen; all three can be the same, as exemplified by many bacteria that use organic compounds as a source of energy, carbon, and electrons.

The first section of Table 7.3 shows some reactions that transfer inorganic species into organic compounds. Carbon fixation can be achieved with a variety of electron donors such as water, which is used by plants in aerobic systems, or reduced sulfur or nitrogen compounds, which are used by photosynthetic and chemosynthetic autotrophic bacteria. Incorporation of nitrogen into organic compounds necessitates reduction to the level ammonia, followed by synthesis of amino acids, the latter reaction not being included in the table. Although organic carbon is shown as the electron donor (and the source of energy) for nitrogen reduction, the process can in fact be linked directly to light absorption which then provides some of the necessary energy, while inorganic compounds, including CO_2, contribute some of the electrons for the overall reduction.

Organisms that use light as their energy source (phototrophs) also typically fix inorganic carbon (autotrophs). Phytoplankton and cyanobacteria (blue-

TABLE 7.3

Energetics of Microbial Processes

Fixation of Elements (Organic Synthesis)

Carbon Fixation (Autotrophs)
A $\quad \frac{1}{4}CO_2(g) + \frac{1}{4}H_2O = \frac{1}{4}\text{“}CH_2O\text{”} + \frac{1}{4}O_2(g); \Delta G_w^0 = +28.6 \text{ kcal mol}^{-1}$
B $\quad \frac{1}{4}CO_2(g) + \frac{1}{2}H_2S(g) = \frac{1}{4}\text{“}CH_2O\text{”} + \frac{1}{16}S_8(col) + \frac{1}{4}H_2O; \Delta G_w^0 = +4.7 \text{ kcal mol}^{-1}$
C $\quad \frac{1}{4}CO_2(g) + \frac{1}{6}NH_4^+ + \frac{1}{12}H_2O = \frac{1}{4}\text{“}CH_2O\text{”} + \frac{1}{6}NO_2^- + \frac{1}{3}H^+; \Delta G_w^0 = +17.8 \text{ kcal mol}^{-1}$

Nitrogen Fixation (Nitrogen Fixers)
D $\quad \frac{1}{6}N_2(g) + \frac{1}{3}H^+ + \frac{1}{4}\text{“}CH_2O\text{”} + \frac{1}{4}H_2O = \frac{1}{3}NH_4^+ + \frac{1}{4}CO_2(g); \Delta G_w^0 = -3.5 \text{ kcal mol}^{-1}$

Nitrate Uptake and Reduction
E $\quad \frac{1}{8}NO_3^- + \frac{1}{4}H^+ + \frac{1}{4}\text{“}CH_2O\text{”} = \frac{1}{8}NH_4^+ + \frac{1}{4}CO_2(g) + \frac{1}{8}H_2O; \Delta G_w^0 = -18.3 \text{ kcal mol}^{-1}$

Energy Sources

Light (Phototrophs, Mostly Autotrophs)
Energy per mole of photons of wave length $\lambda(nm) = 25{,}890/\lambda \text{ kcal mol}^{-1}$

Oxidation of Reduced Inorganic Compounds (Chemolithotrophs, Mostly Autotrophs)
F $\quad$ Nitrification
$\quad \frac{1}{6}NH_4^+ + \frac{1}{4}O_2(g) = \frac{1}{6}NO_2^- + \frac{1}{3}H^+ + \frac{1}{6}H_2O; \Delta G_w^0 = -10.8 \text{ kcal mol}^{-1}$
G $\quad \frac{1}{2}NO_2^- + \frac{1}{4}O_2(g) = \frac{1}{2}NO_3^-; \Delta G_w^0 = -9.0 \text{ kcal mol}^{-1}$
H $\quad$ Sulfide oxidation
$\quad \frac{1}{8}H_2S(g) + \frac{1}{4}O_2(g) = \frac{1}{8}SO_4^{2-} + \frac{1}{4}H^+; \Delta G_w^0 = -23.6 \text{ kcal mol}^{-1}$
I $\quad$ Iron oxidation
$\quad Fe^{2+} + \frac{1}{4}O_2(g) + \frac{5}{2}H_2O = Fe(OH)_3(s) + 2H^+; \Delta G_w^0 = -25.7 \text{ kcal mol}^{-1}$
J $\quad$ Hydrogen oxidation
$\quad \frac{1}{2}H_2(g) + \frac{1}{4}O_2(g) = \frac{1}{2}H_2O; \Delta G_w^0 = -28.4 \text{ kcal mol}^{-1}$

Oxidation of Organic Compounds (Chemoorganotrophs, All Heterotrophs)
K $\quad$ Aerobic respiration
$\quad \frac{1}{4}\text{“}CH_2O\text{”} + \frac{1}{4}O_2(g) = \frac{1}{4}CO_2(g) + \frac{1}{4}H_2O; \Delta G_w^0 = -28.6 \text{ kcal mol}^{-1}$
L $\quad$ Denitrification
$\quad \frac{1}{4}\text{“}CH_2O\text{”} + \frac{1}{5}NO_3^- + \frac{1}{5}H^+ = \frac{1}{4}CO_2(g) + \frac{1}{10}N_2(g) + \frac{7}{20}H_2O; \Delta G_w^0 = -27.2 \text{ kcal mol}^{-1}$
M $\quad$ Sulfate reduction
$\quad \frac{1}{4}\text{“}CH_2O\text{”} + \frac{1}{8}SO_4^{2-} + \frac{1}{8}H^+ = \frac{1}{4}CO_2(g) + \frac{1}{8}HS^- + \frac{1}{4}H_2O; \Delta G_w^0 = -4.9 \text{ kcal mol}^{-1}$
N $\quad$ Methane fermentation
$\quad \frac{1}{4}\text{“}CH_2O\text{”} = \frac{1}{8}CO_2(g) + \frac{1}{8}CH_4(g); \Delta G_w^0 = -4.3 \text{ kcal mol}^{-1}$
O $\quad$ Hydrogen fermentation
$\quad \frac{1}{4}\text{“}CH_2O\text{”} + \frac{1}{4}H_2O = \frac{1}{4}CO_2(g) + \frac{1}{2}H_2(g); \Delta G_w^0 = -0.3 \text{ kcal mol}^{-1}$

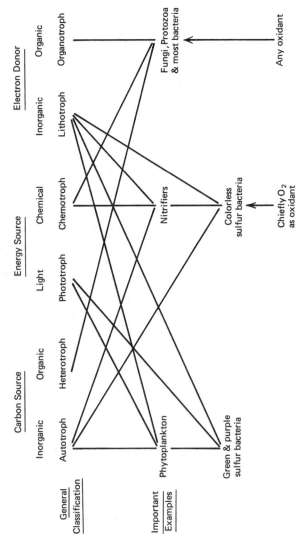

Figure 7.1 Metabolic classification of organisms.

TABLE 7.4

Intermediary Reactions during Methane Fermentation from Glucose[a]

Step in Series	Reaction	ΔG_w^0 kcal	% ΔG_w^0 of Overall Total
1. Partial fermentation of glucose	$0.042C_6H_{12}O_6 + 0.072HCO_3^- = 0.044CH_3COO^- + 0.016CH_3CH_2COO^- + 0.012CH_3CH_2CH_2COO^- + 0.096H_2 + 0.139CO_2 + 0.043H_2O$	-2.74	64.3
2. Butyrate fermentation	$0.012CH_3CH_2CH_2COO^- + 0.012HCO_3^- = 0.024CH_3COO^- + 0.006CH_4 + 0.006CO_2$	-0.075	1.8
3. Propionate fermentation	$0.016CH_3CH_2COO^- + 0.008H_2O = 0.0016CH_3COO^- + 0.012CH_4 + 0.004CO_2$	-0.093	2.2
4. Acetate fermentation	$0.084CH_3COO^- + 0.0084H_2O = 0.084CH_4 + 0.084HCO_3^-$	-0.602	14.1
5. Hydrogen oxidation	$0.096H_2 + 0.024CO_2 = 0.024CH_4 + 0.048H_2O$	-0.75	17.6
Overall	$0.042C_6H_{12}O_2 = 0.126CH_4 + 0.126CO_2$	-4.26	100.0

[a] Source: After McCarty 1972.[6]

green algae) evolve oxygen from water as shown in Reaction A, while green and purple sulfur bacteria, which live in the anaerobic euphotic zone of lakes, take advantage of reduced sulfur compounds (Reaction B).

Bacteria that derive their energy from oxidizing reduced inorganic compounds (chemotrophs) are usually aerobes, using oxygen to carry out the oxidation reaction, although in some cases nitrate can be substituted as the oxidant. They are chiefly autotrophs and use the same inorganic compounds as reductants for their energy source and as electron donors for fixing CO_2 (lithotrophs, e.g., Reactions F and C). Although few in number, these bacteria are important in the geochemical cycle of elements, particularly nitrogen and sulfur. Nitrifying bacteria oxidize ammonium to nitrite (F) and nitrate (G), leading to the regeneration of dinitrogen, $N_2(g)$, through denitrification (Reaction L). Chemosynthetic bacteria that oxidize sulfide with oxygen are called colorless sulfur bacteria to distinguish them from the photosynthetic green and purple sulfur bacteria. They can use a variety of reduced sulfur compounds (H_2S, $S_2O_3^{2-}$, S_8^0) and may produce a number of intermediary redox sulfur species. Other chemosynthetic bacteria oxidize iron, manganese, and hydrogen.

By far the largest number of microorganisms use organic compounds as their energy (chemotrophs), carbon (heterotrophs), and electron (organotrophs) sources. These organisms include fungi, protozoa, and most species of bacteria. Almost any oxidant (e.g., oxygen, nitrate, or sulfate; Reactions K, L, M in Table 7.3) can be used for these chemoorganotrophs to oxidize carbon. In fermentation processes (Reactions N and O) organic carbon itself or water can serve as the oxidant. The oxidation of organic carbon rarely proceeds directly to the final end products shown in Table 7.3. In some cases the oxidant is reduced to an intermediary species such as N_2O or S_8^0. More important, intermediary organic oxidation products are usually formed and utilized by specialized microorganisms. Table 7.4 presents examples of complete redox reactions for intermediary carbon compounds, acetate, propionate, and butyrate which are formed during the anaerobic fermentation of glucose to methane.[6] Each of the reactions in the table is carried out by different bacterial species.

As seen in Table 7.4, the oxidation of the various organic substrates yields widely different free energies, so the "average" organic formula "CH_2O" would not provide an adequate general representation of methane fermentation. This is true of all microbial processes; for example, the oxidation of glucose by any of the electron acceptors in Table 7.1 yields about 13 kcal more per carbon than a similar oxidation of acetate. This difference is of little significance (12%) when oxygen is the electron acceptor. However, when sulfate is the oxidant, this difference corresponds to a relatively large decrease (69%) in the energy made available to the organism.

This is shown for several organic compounds and several potential oxidants in Figure 7.2: the free energies released by all organic compounds (normalized to one electron) are within a few percent of each other when the oxidant is strong [i.e., O_2 yields $\cong 28$ kcal (mol e^-)$^{-1}$], but their relative values decrease markedly, from formate to methane, when the oxidant is weak (i.e., SO_4^{2-}). The

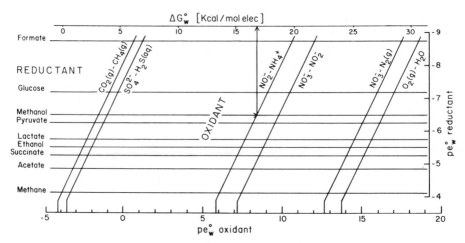

Figure 7.2 Free energy available from the oxidation of various organic substrates with various electron acceptors. The free energy released by a complete redox reaction (per mole of electrons transferred) is obtained from the intersection of the horizontal line corresponding to the reductant with the oblique line corresponding to the oxidant, as illustrated for the oxidation of methanol by nitrite. Data from Table 7.1. Adapted from McCarty, 1972.[6]

use of an average organic compound "CH_2O" = 1/6 glucose is thus appropriate when dealing with well-oxidized systems; it can lead to relatively large errors in calculating free energies in reduced systems. Compare, for example, the energies released in the fermentation of methane from acetate and from glucose (Figure 7.2). In the first case the energy released is only about 1 kcal $(mol\ e^-)^{-1}$, in the second it is about four times larger.

Although thermodynamic calculations can only provide a very rough idea of microbial energetics (and can be quite misleading), they do give insight into the workings of aquatic communities without requiring detailed understanding of the physiological processes. An example of such calculations concerns the energetic problem faced by various species of nitrifiers.

Example 5.   Inorganic Carbon Fixation by Nitrifiers

Consider the chemoautotrophic bacteria of the genera Nitrosomonas and Nitrobacter who are specialized in oxidizing ammonium to nitrite and nitrite to nitrate, respectively. From the data of Table 7.1 (and Table 7.3) we can write the appropriate oxidation and carbon fixation reactions:

Nitrosomonas

Ammonium oxidation

$$\frac{1}{6}NH_4^+ + \frac{1}{4}O_2(g) = \frac{1}{6}NO_2^- + \frac{1}{3}H^+ + \frac{1}{6}H_2O$$
$$\Delta G_w^0 = -10.8 \text{ kcal mol}^{-1}$$
$$\Delta G_1 = \text{energy available per N oxidized}$$
$$= 10.8 \times 6 = 65 \text{ kcal} \tag{43}$$

Carbon fixation with
NH$_4^+$ as reductant

$$\tfrac{1}{4}CO_2 + \tfrac{1}{6}NH_4^+ + \tfrac{1}{12}H_2O$$
$$= \tfrac{1}{4}\text{``CH}_2\text{O''} + \tfrac{1}{6}NO_2^- + \tfrac{1}{3}H^+$$
$$\Delta G_w^0 = +17.8 \text{ kcal mol}^{-1} \tag{44}$$
$$\Delta G_2 = \text{energy necessary per inorganic C fixed}$$
$$= 17.8 \times 4 = 71 \text{ kcal}$$
$$\Delta G_2/\Delta G_1 \cong 1.1$$

Nitrobacter

Nitrite oxidation

$$\tfrac{1}{2}NO_2^- + \tfrac{1}{4}O_2(g) = \tfrac{1}{2}NO_3^-$$
$$\Delta G_w^0 = -9.0 \text{ kcal mol}^{-1} \tag{45}$$
$$\Delta G_1 = 9 \times 2 = 18 \text{ kcal}$$

Carbon fixation with
NO$_2^-$ as reductant

$$\tfrac{1}{4}CO_2 + \tfrac{1}{2}NO_2^- + \tfrac{1}{4}H_2O$$
$$= \tfrac{1}{4}\text{``CH}_2\text{O''} + \tfrac{1}{2}NO_3^-$$
$$\Delta G_w^0 = +19.6 \text{ kcal mol}^{-1} \tag{46}$$
$$\Delta G_2 = 19.6 \times 4 = 78 \text{ kcal}$$
$$\Delta G_2/\Delta G_1 \cong 4.3$$

In fact is found experimentally that various strains of Nitrosomonas oxidize approximately 14 to 70 mol of ammonium for every carbon assimilated: in Nitrobacter this ratio varies from 76 to 135.[7] For both types of organisms the optimal energetic efficiency thus appears to be on the order of 5 to 10%, perhaps not an unexpected value considering the coupling of two biochemical pathways, each with a total energetic efficiency on the order of 30% or less.

On the basis of simple energetic calculations it is also possible to estimate the relative biomass yield from alternative heterotrophic processes or to decide which of two competing processes is more energetically favorable and hence more ecologically probable.

Example 6. Relative Biomass Yields from Aerobic and Denitrifying Bacteria

Consider two heterotrophic microbial populations, one composed of aerobes and the other one of denitrifiers, utilizing the same organic carbon source, "CH$_2$O." It is found from experiments that 80% of the organic carbon source is accounted for in the aerobe biomass. What is the expected yield for the denitrifiers? Are denitrifiers likely to outcompete aerobes in a mixed culture?

We assume that 20% of the organic carbon is utilized by the aerobes for various energetic sustenance and growth requirements: for every four organic carbon assimilated by the aerobes, one is burned yielding 114.3 kcal. Assuming that the denitrifiers have approximately the same energetic requirement and the same efficiency, for every four organic carbon they assimilate they must expend 114.3 kcal and thus oxidize $114.3/(27.2 \times 4) = 1.05$ mol of organic carbon. The relative yield of the denitrifiers is thus $4/5.05 \cong 79\%$, hardly distinguishable from the 80% of the aerobes. Energetically, both microbial populations are roughly equivalent and equally likely to succeed. This is because oxygen and nitrate are approximately equally good oxidants.

Note, however, that this result is pH dependent. For every increase by one in pH units above 7.0, the energy yield for the denitrifiers decreases by $1.36 \times 4/5 \cong 1.1$ kcal mol^{-1} of carbon oxidized, and aerobic bacteria are thus increasingly favored energetically at high pH. Conversely, denitrifiers can actually have an energetic advantage at low pH (< 3.5).

Example 7. Competition among Sulfate Reducers and Methane Fermenters

The importance of remembering that there is a concentration-dependent term in the free energy expression is exemplified by the question of competition between sulfate reducing and methanogenic bacteria.[8] From the data of Table 7.3 the sulfate reducers seem to obtain more energy per unit of substrate oxidized than the methane fermenters $[\Delta G_w^0 = -4.9 \text{ vs.} -4.3 \text{ kcal (mol e}^-)^-]$. It is thus often assumed that the sulfate reducers should outcompete the methane fermenters as long as sulfate is available in the system. This result is of course unaffected by the exact nature of the organic substrate as long as suitable organisms of both types can use it. However, the two calculated energies are fairly similar, and it is necessary to consider more precisely the actual chemistry of the system. Both processes yield the same energy when the reaction of oxidation of methane by sulfate is at equilibrium (subtract one reaction from the other; at equilibrium, $\Delta G = 0$. and thus $\Delta G_1 - \Delta G_2 = 0$):

$$\tfrac{1}{8}CH_4(g) + \tfrac{1}{8}SO_4^{2-} + \tfrac{1}{8}H^+ = \tfrac{1}{8}HS^- + \tfrac{1}{4}H_2O + \tfrac{1}{8}CO_2(g) \qquad (47)$$

and from the appropriate half reactions in Table 7.1

$$\log K = 4.25 - 2.86 = 1.39$$

The equilibrium relation

$$\frac{(HS^-)P_{CO_2}}{P_{CH_4}(SO_4^{2-})(H^+)} = 10^{11.1}$$

describes then the conditions under which both types of organisms can derive the same energy from the same organic substrate [e.g., $(HS^-) \cong 10^{-3} M$, $P_{CO_2} = 10^{-2}$ at, $P_{CH_4} = 10^{-6}$ at, $(SO_4^{2-}) = 10^{-2} M$, $(H^+) = 10^{-8} M$]. Low sulfate, methane, and hydrogen ion concentrations concurrent with high sulfide and carbon dioxide concentrations will tend to favor methanogenesis over sulfate reduction, and vice versa. Under some conditions organisms can in fact derive energy from Reaction 47, oxidizing methane with sulfate; under other conditions the reaction is energy consuming.

There are many examples of ecological succession among microbes that can be rationalized, if not explained, on the basis of the redox functions of the various organisms: stepwise oxidation of reduced compounds promotes the sequential growth of specialized genera, as does the gradual exhaustion of the various oxidants. As a case in point it is of course tempting to consider the stepwise formation of ethanol and acetic acid from fructose, which often seems so important in academic circles. Perhaps more geochemically relevant, however, is the sequence of events leading to the formation of anoxic hypolimnetic waters: assemblages of aerobic organisms, denitrifiers, sulfate reducers, and methanogenic bacteria flourish one at a time as O_2, NO_3^-, and SO_4^{2-} are sequentially exhausted from the system. Such a sequence is directly predicted by the relative energies of the corresponding oxidation reactions; as shown in Table 7.2 the pe_w^0's decrease in the order O_2/H_2O, NO_3^-/N_2, SO_4^{2-}/HS^-, and CH_2O/CH_4. Note, however, that the reason why the various organisms do not bloom simultaneously on the same organic substrate is not simply a matter of relative energetic advantage. In many cases there are specific biochemical processes that inhibit the growth of some classes of organisms. For example, sulfate reduction is biochemically impossible in the presence of even traces of oxygen.

From a macroscopic point of view, the energy fixed into reduced carbon by the initial process of photosynthesis is gradually utilized by the living biota in a multitude of serial and parallel interlocking processes. Step by step the energy is degraded into heat, except for a small amount of storage as refractory organic compounds and reduced minerals.

From a microscopic point of view, the energetic couplings among the various redox reactions are carried out by organisms using a few intracellular energy-transfer compounds. Several of these compounds, NADP, NAD, and ferredoxin, transfer energy by undergoing changes in oxidation state and coupling with other appropriate redox couples to carry out complete redox reactions (see Table 7.5). The principal compound for energy transfer in cells, ATP, does not undergo a redox reaction, however. Instead of an electron exchange, ATP transfers energy by exchanging phosphate groups with metabolic intermediates. In addition to the energetic scale provided by half redox reactions, biochemists thus also define an energetic scale by classifying "half" phosphorylation-dephosphorylation reactions in which inorganic phosphate is taken as a reactant.

A somewhat more mechanistic and biochemical approach to microbially mediated redox processes can be taken by coupling aquatic redox couples with the proper intracellular redox couples. For example, the light and dark reactions of photosynthesis can thus be separated:

Light reaction
$$2H_2O + 2NADP^+ \rightarrow O_2 + 2H^+ + 2NADPH$$
$$\Delta G_w^0 = +105.5 \text{ kcal mol}^{-1} \tag{48}$$

Dark reaction
$$2H^+ + 2NADPH + CO_2 \rightarrow 2NADP^+ + H_2O + CH_2O$$
$$\Delta G_w^0 = +9.3 \text{ kcal mol}^{-1} \tag{49}$$

TABLE 7.5

Some Cellular Energy-Transfer Reactions[9,10]

Half Redox Reactions (Reduction)	pe_w^0
$NAD^+ + 2H^+ + 2e^- = NADH^+ + H^+$	-5.4
$NADP^+ + 2H^+ + 2e^- = NADPH + H^+$	-5.5
2 ferredoxin(Ox) $+ 2e^- = 2$ ferredoxin(Red)	-7.1
ubiquinone $+ 2H^+ + 2e^- =$ ubiquinol	$+1.7$
2 cytochrome C(Ox) $+ 2e^- = 2$ cytochrome C(Red)	$+4.3$
Half Phosphate Exchange Reactions (Hydrolysis)	ΔG_w^0
(Pi = inorganic phosphate)	
phosphoenol pyruvate = pyruvate + Pi	-14.8
phosphocreatinine = creatinine + Pi	-10.3
acetylphosphate = acetate + Pi	-10.1
adenosine triphosphate (ATP) = ADP + Pi	
$37°C$, pH = 7.0, excess Mg^{2+}	-7.3
$25°C$, pH = 7.4, $10^{-3}\ M\ Mg^{2+}$	-8.8
$25°C$, pH = 7.4, no Mg^{2+}	-9.6
adenosine diphosphate (ADP) = AMP + Pi	-7.3
glucose-1-phosphate = glucose + Pi	-5.0
glucose-6-phosphate = glucose + Pi	-3.3
glycerol-1-phosphate = glycerol + Pi	-2.2

The stoichiometry of these reactions is not exact, and both reactions involve ATP. Altogether the reduction of one carbon is found to necessitate 2.2NADPH and 3ATP, plus 2ATP for auxiliary processes. In bacterial photosynthetic reactions where a reductant other than water is used as the electron donor, the light reaction requires much less energy:

$$2H_2S + 2NADP^+ \rightarrow 2S^0 + 2H^+ + 2NADPH; \quad \Delta G_w^0 = +9.3 \text{ kcal mol}^{-1}$$

but the dark reaction is of course the same. Note that energetic calculations involving intracellular compounds are very much affected by the precise chemical composition of intracellular compartments. ATP, NADPH, and NADH, for example, have weak acid-base properties and coordinate strongly with major cations (Table 7.5). Maintenance of H^+, Ca^{2+}, and Mg^{2+} gradients can be used to favor particular reactions, and ion "pumps" are an integral part of the energy-transfer machinery of cells. Note also that the formation of high-energy compounds such as NADPH and ATP during the light reaction can drive energy-consuming reactions other than the reduction of inorganic carbon. Rather than depending exclusively on the utilization (respiration) of carbohydrates, many cellular processes in plants are driven in whole or in part by direct coupling to the light reaction. This is, for example, the case for nitrate reduction, whose energy is derived only in part from dark respiration, the symbolism of Table 7.3 notwithstanding.

3.   PARTIAL REDOX EQUILIBRIUM CALCULATIONS: PE CALCULATION IN COMPLEX SYSTEMS

Although many redox reactions in natural waters do not reach equilibrium, there are situations where several redox couples are roughly at equilibrium with each other. To perform the corresponding equilibrium calculations, it is not always possible to consider that the pe is fixed by the "major" redox couple and simply to equate the pe of the "minor" couples as shown in Example 4. If the redox potential (pe) of the system results from a balance between several redox couples with similar concentrations, the objective of the calculation is in fact to obtain the equilibrium pe by considering simultaneously the various redox equilibria. We wish to examine here how this is done using the general methodology of Chapter 3, constructing tableaux and choosing appropriate components. For this purpose we use as an example the common situation of oxygen depletion in hypolimnetic waters.

3.1   Definition of the Problem and Choice of Components

Example 8.   Anoxia in the Hypolimnion of a Lake

Consider the composition of the hypolimnion of a lake that becomes anoxic due to the decomposition of 0.5 *mM* of organic carbon:

Recipe
$$(SO_4)_T = 2 \times 10^{-4} \ M$$
$$[Fe(III)]_T = 10^{-5} \ M$$
$$\left.\begin{array}{l} Alk = 5 \times 10^{-4} \ M \\ (CO_3)_T = 10^{-3} \ M \end{array}\right\} \ \text{before any redox reaction pH} = 6.3$$
$$(CH_2O)_T = 5 \times 10^{-4} \ M$$
$$(O_2)_T = 3 \times 10^{-4} \ M$$

As discussed previously, there is little point in considering the CH_2O/CO_2 and O_2/H_2O equilibria. If thermodynamics has its way, the reaction

$$CH_2O + O_2 \rightarrow CO_2 + H_2O \tag{50}$$

must exhaust the oxygen of the system, and the reaction

$$CH_2O + \tfrac{1}{2}SO_4^{2-} + H^+ \rightarrow CO_2 + \tfrac{1}{2}H_2S + H_2O \tag{51}$$

must proceed until the organic substrate is eliminated (we do not consider here intermediary carbon or sulfur redox species). These reactions are important because they define the mole balance constraints on the system, but they do not result in interesting equilibrium relationships. Were we to write the mass law equations for the corresponding half redox reactions, we would end up calculating ridiculously small concentrations for CH_2O and O_2.

As is typical of such systems, the redox reactions for which we want to consider redox equilibrium are the $S(+IV)/S(-II)$ and the $Fe(III)/Fe(II)$ couples:

$$\tfrac{1}{8}SO_4^{2-} + \tfrac{5}{4}H^+ + e^- = \tfrac{1}{8}H_2S + \tfrac{1}{2}H_2O; \quad pe^0 = 5.12 \tag{52}$$

$$Fe^{3+} + e^- = Fe^{2+}; \qquad\qquad\qquad pe^0 = 13.0 \tag{53}$$

The list of interesting species included:

1 H^+ and OH^-.
2 The usual carbonate species: $H_2CO_3^*$, HCO_3^-, CO_3^{2-}.
3 SO_4^{2-} *and* H_2S, HS^-, S^{2-}.
4 Fe^{3+}, $Fe(OH)^{2+}$, $Fe(OH)_2^+$, $Fe(OH)_4^-$, *and* Fe^{2+}.

There is also a possibility of precipitating $Fe(OH)_3(s)$, $FeCO_3(s)$, $Fe(OH)_2(s)$, or $FeS(s)$. Let us consider no solid initially.

Note. We have not include O_2 as a species because this system contains an excess of reduced compounds [compare $(O_2)_T$ and $(CH_2O)_T$], so O_2 will be effectively exhausted. If, however, we had an excess of oxygen, it would be critical to include O_2 as a species.

From the earlier discussion in this chapter we include the electron among our choice of components for such a system. Without a good basis on which to decide what the principal components are, let us choose arbitrarily as a component set: H^+, e^-, $H_2CO_3^*$, SO_4^{2-}, Fe^{2+} (see Tableau 7.1).

TABLEAU 7.1

	H^+	e^-	$H_2CO_3^*$	SO_4^{2-}	Fe^{2+}	
Species						
H^+	1					
OH^-	-1					
$H_2CO_3^*$			1			
HCO_3^-	-1		1			
CO_3^{2-}	-2		1			
SO_4^{2-}				1		
H_2S	10	8		1		
HS^-	9	8		1		
S^{2-}	8	8		1		
Fe^{3+}		-1			1	
$Fe(OH)_x^{(3-x)+}$	$-x$	-1			1	
Fe^{2+}					1	
Recipe						
$(SO_4)_T$				1		$2 \times 10^{-4} \ M$
$[Fe(III)]_T$		-1			1	$10^{-5} \ M$
Alk	-1					$5 \times 10^{-4} \ M$
$(CO_3)_T$			1			$10^{-3} \ M$
$(CH_2O)_T$	$+4$	$+4$	1			$5 \times 10^{-4} \ M$
$(O_2)_T$	-4	-4				$3 \times 10^{-4} \ M$

The mole balance equations for H^+ and e^- are written

$$
\begin{aligned}
TOTH &= (H^+) - (OH^-) - (HCO_3^-) - 2(CO_3^{2-}) + 10(H_2S) + 9(HS^-) \\
&\quad + 8(S^{2-}) - x[Fe(OH)_x^{(3-x)+}] \\
&= -Alk + 4(CH_2O)_T - 4(O_2)_T = 3 \times 10^{-4}\ M
\end{aligned}
\tag{54}
$$

$$
\begin{aligned}
TOTe &= 8(H_2S) + 8(HS^-) + 8(S^{2-}) - (Fe^{3+}) - [Fe(OH)_x^{(3-x)+}] \\
&= -[Fe(III)]_T + 4(CH_2O)_T - 4(O_2)_T = 7.9 \times 10^{-4}\ M
\end{aligned}
\tag{55}
$$

On inspection, it is clear that these two equations are not the most convenient or elegant ones that can be written. For example, taking their difference leads to an equation that does not contain the high stoichiometric coefficients for all the sulfide species nor the $[4(CH_2O)_T - 4(O_2)_T]$ term which dominates in both equations. This new equation would be obtained directly if H_2S, rather than e^-, were chosen as a component. In effect, what we are discovering here is an extension of our previous rules for choosing principal components: *instead of the electron, the principal components of the system include both the reduced and the oxidized species of the major redox couple.* The choice of the electron as a component, though conceptually pleasing and yielding straightforward stoichiometric expressions for all species, is numerically awkward. When the species of the major redox couple are chosen as components, the mole balance equations can often be greatly simplified, with one concentration being in large excess of all others.

3.2   pe Calculation: Solution of Example 8

As argued in Section 3.1, our choice of components should include $S(+VI)$ and $S(-II)$ species. Since the oxidation of organic matter given in Reactions 50 and 51 is acid producing (CO_2 and H_2S), we expect the equilibrium pH to be below the value of 6.3 which prevailed before any redox reaction. Thus H_2S and $H_2CO_3^*$ are the principal components for sulfide and carbonate. Choosing arbitrarily Fe^{2+} as our iron component then yields a complete component set: H^+, $H_2CO_3^*$, SO_4^{2-}, H_2S, and Fe^{2+}; see Tableau 7.2. [The stoichiometric coefficients are obtained by considering that $e^- = (H_2S)_{1/8}(SO_4^{2-})_{-1/8}(H^+)_{-5/4}.$]

$$
\begin{aligned}
TOTH &= (H^+) - (OH^-) - (HCO_3^-) - 2(CO_3^{2-}) - (HS^-) - 2(S^{2-}) \\
&\quad + \tfrac{5}{4}(Fe^{3+}) + (\tfrac{5}{4} - x)[Fe(OH)_x^{(3-x)+}] \\
&= + \tfrac{5}{4}[Fe(III)]_T - Alk - (CH_2O)_T + (O_2)_T \\
&= -6.88 \times 10^{-4}\ M
\end{aligned}
\tag{56}
$$

$$
\begin{aligned}
TOTH_2CO_3 &= (H_2CO_3^*) + (HCO_3^-) + (CO_3^{2-}) = (CO_3)_T + (CH_2O)_T \\
&= 1.5 \times 10^{-3}\ M
\end{aligned}
\tag{57}
$$

TABLEAU 7.2

	H^+	$H_2CO_3^*$	SO_4^{2-}	H_2S	Fe^{2+}
Species					
H^+	1				
OH^-	-1				
$H_2CO_3^*$		1			
HCO_3^-	-1	1			
CO_3^{2-}	-2	1			
SO_4^{2-}			1		
H_2S				1	
HS^-	-1			1	
S^{2-}	-2			1	
Fe^{3+}	$+\frac{5}{4}$		$+\frac{1}{8}$	$-\frac{1}{8}$	1
$Fe(OH)_x^{(3-x)+}$	$+\frac{5}{4} - x$		$+\frac{1}{8}$	$-\frac{1}{8}$	1
Fe^{2+}					1
Recipe					
$(SO_4^{2-})_T$			1		$2 \times 10^{-4}\ M$
$[Fe(III)]_T$	$+\frac{5}{4}$		$+\frac{1}{8}$	$-\frac{1}{8}$	$1 \quad 10^{-5}\ M$
Alk	-1				$5 \times 10^{-4}\ M$
$(CO_3)_T$		$+1$			$10^{-3}\ M$
$(CH_2O)_T$	-1	$+1$	$-\frac{1}{2}$	$+\frac{1}{2}$	$5 \times 10^{-4}\ M$
$(O_2)_T$	$+1$		$+\frac{1}{2}$	$-\frac{1}{2}$	$3 \times 10^{-4}\ M$

$$TOTSO_4 = (SO_4^{2-}) + \tfrac{1}{8}(Fe^{3+}) + \tfrac{1}{8}\left[Fe(OH)_x^{(3-x)+}\right]$$
$$= (SO_4^{2-})_T + \tfrac{1}{8}\left[Fe(III)\right]_T - \tfrac{1}{2}(CH_2O)_T + \tfrac{1}{2}(O_2)_T$$
$$= 1.01 \times 10^{-4}\ M \tag{58}$$

$$TOTH_2S = (H_2S) + (HS^-) + (S^{2-}) - \tfrac{1}{8}(Fe^{3+}) - \tfrac{1}{8}\left[Fe(OH)_x^{(3-x)+}\right]$$
$$= -\tfrac{1}{8}\left[Fe(III)\right]_T + \tfrac{1}{2}(CH_2O)_T - \tfrac{1}{2}(O_2)_T$$
$$= 0.99 \times 10^{-4}\ M \tag{59}$$

$$TOTFe = (Fe^{3+}) + \left[Fe(OH)_x^{(3-x)+}\right] + (Fe^{2+}) = \left[Fe(III)\right]_T$$
$$= 10^{-5}\ M \tag{60}$$

All Fe terms are negligible in a first approximation in each but the last equation. The first equation is then the usual alkalinity expression, and the others are straightforward expressions of conservation for CO_3^{2-}, SO_4^{2-}, S^{2-}, and Fe.

Since $TOTH_2CO_3$ is in excess of $TOTH_2S$ by a factor of 15, the sulfide terms in the $TOTH$ equation are relatively small, and the pH is roughly obtained in the usual manner for a pure carbonate system, by comparing Equations 56 and

57. [Since C_T is a little more than twice the alkalinity, the pH should be slightly below 6.3.]

$$TOTH \cong -(HCO_3^-) = -6.9 \times 10^{-4} = 10^{-3.15}$$

$$TOTH_2CO_3 \cong (H_2CO_3^*) + (HCO_3^-) = 1.5 \times 10^{-3}$$

therefore

$$(H^+) = \frac{10^{-6.3}(H_2CO_3^*)}{(HCO_3^-)} \cong \frac{10^{-6.3}(1.5 \times 10^{-3} - 6.9 \times 10^{-4})}{(6.9 \times 10^{-4})}$$

$$pH = 6.23$$

$$TOTSO_4 \cong (SO_4^{2-}) = 10^{-4} \ M$$

$$TOTH_2S \cong (H_2S) + (HS^-) = (H_2S)[1 + 10^{-7} \times 10^{6.23}] = 10^{-4} \ M$$

$$(H_2S) = 10^{-4.07} \ M$$

$$(HS^-) = 10^{-4.84} \ M$$

Other minor carbon and sulfur species can then be obtained in a straightforward fashion.

To solve Equation 60, we need to write the constants corresponding to the Fe(III) species in the tableau. This is of course strictly equivalent to equating the pe of the Fe^{3+}/Fe^{2+} couple to that of SO_4^{2-}/H_2S, which is the pe of the system:

$$\frac{SO_4^{2-}}{H_2S}: \qquad pe = 5.12 - \tfrac{5}{4} pH + \tfrac{1}{8} \log \frac{(SO_4^{2-})}{(H_2S)} = -2.66$$

$$\frac{Fe^{3+}}{Fe^{2+}}: \qquad -2.66 = 13.0 + \log \frac{(Fe^{3+})}{(Fe^{2+})}$$

therefore

$$(Fe^{2+}) = 10^{+15.66}(Fe^{3+})$$

At a pH of 6.2 none of the hydrolysis species of Fe(III) have a sufficiently high formation constant to make them dominant over Fe^{2+} (see Chapter 5):

$$(FeOH^{2+}) = 10^{11.8} \times 10^{-7.8}(Fe^{3+}) = 10^{-11.66}(Fe^{2+}) \ll (Fe^{2+})$$

$$(Fe(OH)_2^+) = 10^{22.3} \times 10^{-15.6}(Fe^{3+}) = 10^{-8.96}(Fe^{2+}) \ll (Fe^{2+}))$$

$$(Fe(OH)_4^-) = 10^{34.4} \times 10^{-31.2}(Fe^{3+}) = 10^{-12.46}(Fe^{2+}) \ll (Fe^{2+})$$

Even the solid $am \cdot Fe(OH)_3(s)$ cannot form

$$(Fe^{3+}) = 10^{-15.66}(Fe^{2+}) \le 10^{-20.66}$$

therefore

$$(Fe^{3+})(OH^-)^3 \le 10^{-20.66} \times 10^{-23.6} \ll K_s = 10^{-38.8}$$

(Fe^{2+}) is thus the dominant term in Equation 60:

$$TOTFe \cong (Fe^{2+}) = 10^{-5} \ M$$

We still have to verify if $FeCO_3(s)$ or $FeS(s)$ precipitate:

$$(Fe^{2+})(CO_3^{2-}) = 10^{-5} \times \frac{10^{-10.3}}{10^{-6.23}} \times 10^{-3.15} = 10^{-12.22} < K_s = 10^{-10.7} \, M$$

$$(Fe^{2+})(S^{2-}) = 10^{-5} \times \frac{10^{-20.9} \times 10^{-4.06}}{10^{-12.46}} = 10^{-17.50} > K_s = 10^{-18.1} \, M$$

So in fact $FeS(s)$ does precipitate. This affects the sulfide minimally since the total sulfide is 10 times in excess of the iron, and it affects the alkalinity and the pH even less:

$$(Fe^{2+}) = \frac{10^{-18.1}}{(S^{2-})} = \frac{10^{-18.1} \times 10^{-12.46}}{10^{-20.9} \times 10^{-4.06}}$$

$$= 10^{-5.6} \, M$$

Three-quarters of the iron is precipitated as FeS.

The calculations could be redone by choosing FeS as a component instead of Fe^{2+}, but the results would be practically unchanged. The composition of this hypolimnetic water is thus characterized by the following chemistry:

$$pH = 6.23; \qquad pe = -2.66; \qquad (O_2) = 0 \, M$$

$$(H_2CO_3^*) = 10^{-3.08} \, M; \quad (HCO_3^-) = 10^{-3.15} \, M$$

$$(H_2S) = 10^{-4.06} \, M; \qquad (HS^-) = 10^{-4.86} \, M$$

$$(SO_4^{2-}) = 10^{-4.0} \, M$$

$$(Fe^{2+}) = 10^{-5.6} \, M; \qquad (FeS \cdot s) = 10^{-5.1} \, M$$

3.3   Redox in Wastewater Disposal Fields

Other interesting examples of pe calculations are found in aquatic systems where anoxic and oxic waters are mixed. A case in point is sewage or sludge discharge into fresh or marine waters where the total oxygen consumption of the waste organic matter is usually described by a familiar empirical parameter, the "biological oxygen demand" (BOD). For engineering design purposes the rate of oxygen consumption is then approximated by a first-order rate constant (also determined by empirical testing) that aggregates all oxidative biological processes. Slightly more sophisticated description of such waste disposal situations can include several oxidation kinetic parameters to represent the growth of subpopulations of aerobic microbes. In particular, one may include parameters for the "nitrogenous BOD" to reflect the oxidation of reduced nitrogen compounds (nitrification) which normally follows the oxidation of the non-refractory organic carbon ("carbonaceous BOD"). One complication in such empirical description of biological oxidation processes is the difficulty in matching the microbial community in the BOD test inoculum with that of the receiving waters.[5] Other kinetic parameters for oxygen addition by diffusion through the air-water interface and by algal photosynthesis can be superimposed on the oxygen-consumption equation and included in a hydrodynamic transport model of the system. Because waste disposal systems are specifically designed to avoid anoxic conditions (and the BOD test is an aerobic test), their redox chemistry is in fact rarely interesting to the aquatic chemist: as seen in Example 3, the pe of such oxic systems is more sensitive to the pH than to the partial pressure of oxygen as long as oxygen is present in measurable concentration.

□ □ □

3.4   Redox in Hydrothermal Oceanic Vents

Leaving the murky waters of sewage disposal fields for the bottom of the blue ocean, a system that does provide an interesting case of redox calculations is that of the mixing of hydrothermal fluid from the ridge vents with deep ocean water. Although some of the important chemical processes in such a system are controlled by temperature effects ("pure" hydrothermal fluid is estimated to have a temperature of $350°C$), it is instructive to study a simple model system, assuming uniform low temperature (our usual $25°C$, of course) for the mix.

Example 9.   Mixture of Hydrothermal Fluid and Seawater*

Concentrations for the various constituents of the two *end members* of the vent mixture, seawater and hydrothermal fluid, are given in Table 7.6. To keep the problem manageable, let us fix the ionic strength (say, $I = 0.5 M$) and ignore the chemistry of Na^+, Mg^{2+}, and Ca^{2+} (including the possible precipitation of $CaCO_3$), K^+, Cl^-, and H_2SiO_3 (which is considered a good conservative tracer in this system). We define h as the proportion of hydrothermal water in the mix and consider the composition of the system as h varies from 1 (hydrothermal fluid) to 0 (seawater). Because of temperature effects, all calculations for $h > 0.1$ should be taken only as very rough indications of the possible chemistry of the system. Our principal goals are to obtain the pH and the pe of the system as a function of h and to calculate the equilibrium chemistry of sulfur, iron, and manganese.

Before embarking on detailed calculations we should note that at high values of h, the pH and the pe of the system are low because of the negative alkalinity—mineral acidity—of the hydrothermal fluid and its high sulfide content. Under these conditions H^+, $H_2CO_3^*$, H_2S, Fe^{2+}, and Mn^{2+} are the major species and the principal components in the system. As h decreases, the pH should increase

* Specific conditions for this example were provided by R. E. McDuff.

TABLE 7.6

Concentrations of Major Constituents in Seawater and Hydrothermal Fluid in Oceanic Vents[a,b]

		Seawater, $1-h$ (2°C)	Hydrothermal Fluid, h (350°C)
Major cations	Na^+	466	527
	Mg^{2+}	52.7	0
	Ca^{2+}	10.3	25
	K^+	10.0	15
Major anions	Cl^-	542	594
	SO_4^{2-}	28.9	0
	Alk	2.45	−2.0
	C_T	2.38	9.0
Leached species	H_2SiO_3	0.16	18.0
Reduced species	H_2S	0	8.0
	Fe(II)	0	0.3
	Mn(II)	0	2
Oxygen	O_2	0.12	0

[a] Note: All concentrations are given in mM.
[b] The composition of the hydrothermal fluid was obtained from extrapolation of early data with lower temperature samples. More direct data suggest a higher alkalinity (ca 0 to -0.5 mM). This affects the results of the example at low dilutions and together with the neglected temperature effects is responsible for obvious discrepancies with more recent analytical information.

significantly, due to the increasing alkalinity of the system, much before the pe increases to oxic values: for example for $h = 0.1$ the alkalinity of the mix is close to that of seawater, while the sulfide concentration from the hydrothermal fluid is still high (10^{-3} M) and much in excess of the oxygen from seawater (10^{-4} M). At some point, as the pH increases, iron and manganese may precipitate, perhaps as sulfides or as carbonates or hydroxides depending on the relative increases in S^{2-}, CO_3^{2-}, and OH^-. For small values of h the oxygen of seawater becomes important, while the sulfide is diluted and the system becomes oxic. Iron and manganese solids may then dissolve due to dilution and/or oxidation of sulfide before their ultimate precipitation as hydroxides or oxides. Let us then focus on the intermediate situation and hypothesize with hindsight that FeS is precipitated but not MnS. (Previous calculations showed MnS not to precipitate for any value of the dilution factor h, as is in fact known experimentally.) Our goals are to calculate the point of precipitation and dissolution of FeS and to follow the evolution of pH and pe as a function of h.

pH   Consider first low dilution factors ($h \simeq 1$), and assume that the pH is below 6. The principal components of the system are then: H^+, $H_2CO_3^*$, SO_4^{2-}, H_2S, FeS, and Mn^{2+} (see Tableau 7.3). The mole balance equations are written:

TABLEAU 7.3

	H^+	H_2CO_3	SO_4^{2-}	H_2S	FeS	Mn^{2+}	
Species							
H^+	1						
OH^-	-1						
$H_2CO_3^*$		1					
HCO_3^-	-1	1					
CO_3^{2-}	-2	1					
O_2	1		$\frac{1}{2}$	$-\frac{1}{2}$			
SO_4^{2-}			1				
H_2S				1			
HS^-	-1			1			
S^{2-}	-2			1			
Fe^{2+}	2			-1	1		
Fe^{3+}	$\frac{13}{4}$		$\frac{1}{8}$	$-\frac{9}{8}$	1		
$Fe(OH)_x^{(3-x)+}$	$\frac{13}{4}-x$		$\frac{1}{8}$	$-\frac{9}{8}$	1		
Mn^{2+}						1	
$FeS\cdot s$				1			
Hydrothermal fluid							
Alk	-1						-2×10^{-3}
$(CO_3)_T$		1					9×10^{-3}
$(H_2S)_T$				1			8×10^{-3}
$(Fe\cdot II)_T$	2			-1	1		3×10^{-4}
$(Mn\cdot II)_T$						1	2×10^{-3}
Seawater							
Alk	-1						2.45×10^{-3}
$(O_2)_T$	1		$\frac{1}{2}$	$-\frac{1}{2}$			1.2×10^{-4}
$(CO_3)_T$		1					2.38×10^{-3}
$(SO_4)_T$			1				2.89×10^{-2}

$$TOTH = (H^+) - (OH^-) - (HCO_3^-) - 2(CO_3^{2-}) + (O_2) - (HS^-)$$
$$- 2(S^{2-}) + 2(Fe^{2+}) + \tfrac{13}{4}(Fe^{3+})$$
$$+ (\tfrac{13}{4} - x)\left[Fe(OH)_x^{(3-x)+}\right]$$
$$= h(2 + 0.6)10^{-3} + (1 - h)(-2.45 + 0.12)10^{-3}$$
$$= (4.93h - 2.33)10^{-3} \tag{61a}$$

$$TOTH_2CO_3 = (H_2CO_3^*) + (HCO_3^-) + (CO_3^{2-})$$
$$= 9h \times 10^{-3} + (1 - h)2.38 \times 10^{-3}$$
$$= (6.62h + 2.38)10^{-3} \tag{62a}$$

$$TOTSO_4 = \tfrac{1}{2}(O_2) + (SO_4^{2-}) + \tfrac{1}{8}(Fe^{3+}) + \tfrac{1}{8}\left[Fe(OH)_x^{(3-x)+}\right]$$
$$= (1 - h)(0.6 \times 10^{-4} + 2.89 \times 10^{-2})$$
$$= (2.9 - 2.9h)10^{-2} \qquad\qquad (63a)$$
$$TOTH_2S = -\tfrac{1}{2}(O_2) + (H_2S) + (HS^-) + (S^{2-}) - (Fe^{2+}) - \tfrac{9}{8}(Fe^{3+})$$
$$- \tfrac{9}{8}\left[Fe(OH)_x^{(3-x)+}\right]$$
$$= h(8 - 0.3)10^{-3} - (1 - h)0.6 \times 10^{-4}$$
$$= (7.76h - 0.06)10^{-3} \qquad\qquad (64a)$$
$$TOTFeS = (FeS \cdot s) + (Fe^{2+}) + (Fe^{3+}) + \left[Fe(OH)_x^{(3-x)+}\right]$$
$$= 3h \times 10^{-4} \qquad\qquad (65a)$$
$$TOTMn = (Mn^{2+})$$
$$= 2h \times 10^{-3} \qquad\qquad (66)$$

Neglecting the iron and oxygen terms in the first four equations, using the usual approximations for a low pH system, and expressing the appropriate mass laws (with apparent seawater constants), one can simplify the equations to

$$TOTH = (H^+) - (HCO_3^-) - (HS^-) = (4.93h - 2.33)10^{-3} \quad (61b)$$
$$TOTH_2CO_3 = (H_2CO_3^*) + (HCO_3^-) = (HCO_3^-)[1 + 10^{6.0}(H^+)]$$
$$= (6.62h + 2.38)10^{-3} \qquad\qquad (62b)$$
$$TOTSO_4 = (SO_4^{2-}) = (2.9 - 2.9h)10^{-2} \qquad\qquad (63b)$$
$$TOTH_2S = (H_2S) + (HS^-) = (HS^-)[1 + 10^{6.9}(H^+)]$$
$$= (7.76h - 0.06)10^{-3} \qquad\qquad (64b)$$

Substitution of Equations 62b and 64b into 61b provides an implicit expression of (H^+) as a function of h:

$$(H^+) - \frac{(6.62h + 2.38)10^{-3}}{1 + 10^{6.0}(H^+)} - \frac{(7.76h - 0.06)10^{-3}}{1 + 10^{6.9}(H^+)} = (4.93h - 2.33)10^{-3} \quad (67a)$$

This expression which is plotted as line A in Figure 7.3 is valid as long as the solid FeS (and not MnS) is precipitated and the oxygen concentration is small compared to that of sulfide. The relationship between (O_2) and (H_2S) is given by combination of the appropriate redox reactions:

$$\tfrac{1}{4}O_2(aq) + e^- + H^+ = \tfrac{1}{2}H_2O; \qquad\qquad pe^0 = 21.45 \qquad (68)$$
$$\tfrac{1}{8}SO_4^{2-} + e^- + \tfrac{5}{4}H^+ = \tfrac{1}{8}H_2S(aq) + \tfrac{1}{2}H_2O; \quad pe^0 = 5.12 \qquad (69)$$

therefore

$$O_2(aq) + \tfrac{1}{2}H_2S(aq) = \tfrac{1}{2}SO_4^{2-} + H^+; \quad \log K = +65.3 \qquad (70)$$
$$(O_2) = 10^{-65.3}(H^+)(SO_4^{2-})^{1/2}(H_2S)^{-1/2} \qquad\qquad (71)$$

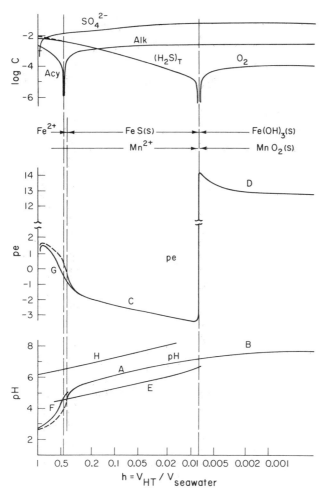

Figure 7.3 Variations of the calculated equilibrium composition of hydrothermal fluid as a function of its dilution (h) into seawater (Example 9). The letters labeling the graphs in the lower part of the figure correspond to equations in the text: $G, C, D = $ pe; $F, A, B = $ pH; $E = $ pH of saturation of FeS(s); and $H = $ pH of saturation of MnS(s). Although temperature and kinetic effects unaccounted for in the calculations are important in the real system, this example illustrates the coupling between redox, acid-base, and precipitation-dissolution reactions in natural waters.

In the range $0.1 > h > 0.01$, pH $\cong 6.5$ (see Figure 7.3), and $(SO_4^{2-}) \cong 10^{-1.5}$ (see Equation 63b):

$$(O_2)(H_2S)^{1/2} \cong 10^{-73} \qquad (72)$$

This very small value of the product of (O_2) and (H_2S) concentrations makes it possible for only one of them to be important at any time (this is of course a general result, see Section 2.2). In other words (O_2) is the dominant term in Equation 64a for any negative value of the right-hand side, and (H_2S) is the dominant term for any positive value:

$$TOTH_2S = -\tfrac{1}{2}(O_2) + (H_2S) + (HS^-) = (7.76h - 0.06)10^{-3} \qquad (64c)$$

The value of the dilution parameter yielding $TOTH_2S = 0$,

$$h = \frac{0.06}{7.76} \cong 0.008 \qquad (73)$$

corresponds thus to a critical dilution of the hydrothermal fluid, at which the system undergoes a sudden transition from anoxic to oxic conditions.

Under oxic conditions ($h < 0.008$), the pH is given from the $TOTH$ equation using H^+, HCO_3^-, O_2, SO_4^{2-}, $Fe(OH)_3(s)$, and Mn^{2+} as the principal components [guessing that $Fe(OH)_3$ precipitates; see Tableau 7.4]:

$$\begin{aligned} TOTH &= (H^+) - (OH^-) + (H_2CO_3^*) - (CO_3^{2-}) + 2(H_2S) + (HS^-) \\ &\quad + 2(Fe^{2+}) + 3(Fe^{3+}) + (3 - x)\left[Fe(OH)_x^{(3-x)+}\right] \\ &= h(2 + 9 + 16 + 0.6)10^{-3} + (1 - h)(-2.45 + 2.38)10^{-3} \\ &= (27.67h - 0.07)10^{-3} \end{aligned} \qquad (74a)$$

TABLEAU 7.4

	H^+	HCO_3^-	O_2	SO_4^{2-}	$Fe(OH)_3$	Mn^{2+}	
Species							
H^+	1						
OH^-	-1						
$H_2CO_3^*$	1	1					
HCO_3^-	0	1					
CO_3^{2-}	-1	1					
O_2			1				
SO_4^{2-}				1			
H_2S	2		-2	1			
HS^-	1		-2	1			
S^{2-}			-2	1			
Fe^{2+}	2		$\frac{1}{4}$		1		
Fe^{3+}	3		$\frac{9}{4}$	-1	1		
$Fe(OH)_x^{(3-x)+}$	$3 - x$		$\frac{9}{4}$	-1	1		
Mn^{2+}						1	
$Fe(OH)_3 \cdot s$					1		
Hydrothermal fluid							
Alk	-1						-2×10^{-3}
$(CO_3)_T$	1	1					9×10^{-3}
$(H_2S)_T$	2		-2	1			8×10^{-3}
$(Fe \cdot II)_T$	2		$-\frac{1}{4}$		1		3×10^{-4}
$(Mn \cdot II)_T$						1	2×10^{-3}
Seawater							
Alk	-1						2.45×10^{-3}
$(O_2)_T$			1				1.2×10^{-4}
$(CO_3)_T$	1	1					2.38×10^{-3}
$(SO_4)_T$				1			2.89×10^{-2}

The carbonate species concentrations are much larger than the others, leading to the simplified equation:

$$TOTH = (H_2CO_3^*) - (CO_3^{2-}) = (27.67h - 0.07)10^{-3} \qquad (74b)$$

[This equation can also be obtained from the previous simplified mole balances by the linear transformation: $TOTH' = TOTH + TOTH_2CO_3 + 2 \times TOTH_2S$.]

The carbonate mole balance given by Equation 62a remains unchanged:

$$TOTHCO_3 = TOTH_2CO_3 = (H_2CO_3^*) + (HCO_3^-) + (CO_3^{2-})$$

$$= (6.62h + 2.38)10^{-3}$$

Introducing the appropriate mass laws into Equation 62a and 74b, and combining the equations yields an expression for (H^+):

$$\frac{10^{6.0}(H^+) - 10^{-8.9}/(H^+)}{1 + 10^{6.0}(H^+) + 10^{-8.9}/(H^+)}(6.62h + 2.38) = 27.67h - 0.07 \qquad (75)$$

This implicit function (H^+) of h is plotted as line B on Figure 7.3. For $h = 0.008$ the pH change due to sulfide and iron oxidation is so small (ca. 0.03) that line B is indistinguishable from line A. At higher dilutions, the possible precipitation of MnO_2 would decrease the pH, but this effect is negligible due to the very small value of h (and hence the great dilution of Mn), as will be seen later. Note that the final seawater pH calculated in this example (7.45) is too low because we have neglected such species as borate which are important in the acid-base chemistry of seawater.

pe In the anoxic region the pe is given by the sulfate-sulfide ratio according to Reaction 69:

$$pe = 5.12 - \tfrac{5}{4}\,pH + \tfrac{1}{8}\log\frac{(SO_4^{2-})}{(H_2S)} \qquad (76)$$

Substituting Equations 63b and 64b yields the expression

$$pe = 5.12 - \tfrac{5}{4}\,pH + \tfrac{1}{8}\log(1 - h)10^{-1.54} - \tfrac{1}{8}\log\frac{7.76h - 0.06}{1 + 10^{-6.9}/(H^+)}10^{-3}$$

therefore

$$pe = 5.19 - \tfrac{5}{4}\,pH + \tfrac{1}{8}\log\frac{(1 - h)[1 + 10^{-6.9}/(H^+)]}{h - 0.008} \qquad (77a)$$

(see line C in Figure 7.3). In the oxic region the oxygen concentration controls the pe according to Reaction 68:

$$pe = 21.45 - pH + \tfrac{1}{4}\log(O_2)$$

The aqueous oxygen concentration is obtained by neglecting the sulfide species in Equation 64c (which results from Tableau 7.4 as well):

$$pe = 20.7 - pH + \tfrac{1}{4}\log(0.12 - 15.52h) \qquad (78)$$

(see line D in Figure 7.3). Lines C and D exhibit a sharp discontinuity at $h = 0.008$, corresponding to a sudden change from a control of the redox by sulfide to a control by oxygen.

Fe   The solubility of ferrous sulfide is given by the reaction

$$Fe^{2+} + H_2S = FeS(s) + 2H^+; \quad \log K = -2.8 \tag{79}$$

therefore

$$(Fe^{2+}) \leq 10^{2.8}(H^+)^2(H_2S)^{-1} \tag{80}$$

Solving Equation 64b for (H_2S) and equating (Fe^{2+}) to the total iron concentration (Equation 65a) produces an equation for the pH of FeS precipitation:

$$(Fe^{2+}) = 10^{2.8}(H^+)^2(H_2S)^{-1} = Fe_T \tag{81}$$

therefore

$$\frac{10^{2.8}(H^+)^2[1 + 10^{-6.9}/(H^+)]}{(7.76h - 0.06)10^{-3}} = 3h \times 10^{-4} \tag{82}$$

For small values of h the pH is low, and (HS^-) can be neglected $[10^{-6.9}/(H^+) \ll 1]$, leading to the simple equation

$$pH = 4.66 - \tfrac{1}{2} \log (7.76h^2 - 0.06h) \tag{83}$$

This equation is plotted as line E on Figure 7.3; its intersection with the pH graph provides the conditions for precipitation:

$$h_{crit} \cong 0.42; \quad pH \cong 4.64$$

Because of the rapid pH change in that region, near the CO_2 equivalence point, this result is not very sensitive to the solubility product of FeS which might be affected by temperature or the exactitude of the approximations that we have performed.

Below the critical dilution value ($h > 0.42$), Fe^{2+} is the principal iron species and should be used as a component. Three mole balance equations are affected by this change (see Tableau 7.3), and they can be written without developing a whole new tableau:

1   (FeS·s) is of course eliminated from the iron mole balance Equation 65, but the right-hand side of the equation remains the same, leading to the simplification

$$TOTFe = (Fe^{2+}) = 3h \times 10^{-4} \tag{65b}$$

2   (Fe^{2+}) is eliminated from the $TOTH$ equation, and the right-hand side is decreased by $6h \times 10^{-4}$. The previous simplified expression is changed little:

$$TOTH = (H^+) - (HCO_3^-) - (HS^-) = (4.33h - 2.33)10^{-3} \tag{61c}$$

3   In the same way the sulfide mole balance is only slightly changed:

$$TOTH_2S = (H_2S) + (HS^-) = (8.06h - 0.06)10^{-3} \tag{64c}$$

Equation 67a, which describes the variation in pH as a function of h (line A), must thus be modified for high values of h ($h > 0.42$). The appropriate terms in Equation 67a can be replaced by the right-hand side of Equations 61c and 64c. In addition, since the pH is low in that region, the terms corresponding to the HCO_3^- and HS^- concentrations can be neglected:

$$(H^+) - \frac{(6.62h + 2.38)10^{-3}}{10^{6.0}(H^+)} - \frac{(8.06h - 0.06)10^{-3}}{10^{6.9}(H^+)} = (4.33h - 2.33)10^{-3} \quad (67b)$$

(see line F on Figure 7.3). In the same way the pe is affected by FeS dissolution, and Equation 77a must be replaced by

$$pe = 5.25 - \tfrac{5}{4} pH + \tfrac{1}{8} \log \frac{2.9 - 2.9h}{8.06h - 0.06} \quad (77b)$$

(see line G on Figure 7.3). Both lines F and G provide only small corrections to the original pH and pe graphs, A and C.

As can be seen on Figure 7.3, the line of FeS precipitation-dissolution (line E) does not cross the pH graph again at low values of h. This indicates that FeS does not dissolve by simple dilution or acid-base processes. We thus expect the dissolution of FeS to depend on $Fe(OH)_3$ precipitation as the hydrothermal iron is oxidized by the oxygen of seawater. Although the exact equations for the point of transition between FeS and $Fe(OH)_3$ are very cumbersome to develop, it is straightforward to verify that $Fe(OH)_3$ does not precipitate below the critical dilution for the anoxic to oxic transition ($h \cong 0.008$) and is precipitated above it. Thus FeS dissolves, and $Fe(OH)_3$ precipitates at the sharp jump in pe when the oxygen of seawater titrates the hydrothermal sulfide.

Mn We can now verify that MnS does not precipitate as hypothesized. Similar to the case for iron, the solubility of manganous sulfide is obtained by equating the appropriate solubility product and total manganese concentrations:

$$Mn^{2+} + H_2S = MnS(s) + 2H^+; \quad \log K = -7.4 \quad (84)$$

$$(Mn^{2+}) = 10^{7.4}(H^+)^2(H_2S)^{-1} = (Mn \cdot II)_T \quad (85)$$

therefore

$$\frac{10^{7.4}(H^+)^2}{(7.76h - 0.06)10^{-3}} = 2h \times 10^{-3} \quad (86)$$

$$pH = 6.55 - \tfrac{1}{2} \log(5.76h^2 - 0.06h)$$

This equation is plotted as line H on Figure 7.3. The line does not intersect the pH graph, so MnS does not precipitate.

At the point where the system becomes oxic ($h < 0.008$), the manganese concentration is still very high and the oxide MnO_2 is undoubtedly the stable form of the metal. The precipitation of MnO_2 is slow at the pH of seawater, however. Kinetic hindrance of the precipitation should allow for a significant dissolved manganese enrichment until Mn^{2+} is eventually scavenged into ferromanganese nodules.

Summary As shown in Figure 7.3, the chemistry of the mixture of hydrothermal fluid and seawater can be divided into roughly three domains, depending on the dilution factor. The two critical dilution factors separating these three domains are given approximately by the conditions Alk $= 0$ ($h \cong 0.47$) and $(O_2)_T \cong 2(S \cdot -II)_T$ ($h \cong 0.008$). Below the first of these critical dilution factors, both the acid-base and the redox chemistry of the mix are dominated by the hydrothermal fluid: the alkalinity of the water is negative, the system is reduced, and the principal iron and manganese species are Fe^{2+} and Mn^{2+}.

In the intermediate domain ($0.47 > h > 0.008$) the pH of the mix is approximately neutral and dominated by the seawater carbonate system; the pe is still very negative and controlled by the sulfide of the hydrothermal fluid. Note the paradoxical, and typical, decrease in pe as the hydrothermal fluid is initially diluted with seawater. At dilutions too low for the seawater oxygen to oxidize the hydrothermal sulfide, the increase in pH leads to an increase in $\{S^{2-}\}$ and thus also $\{e^-\}$. In this intermediate dilution domain, FeS is precipitating but not MnS which is too soluble.

At the second critical dilution factor the oxygen of seawater completely oxidizes the hydrothermal sulfide, and the chemistry of the mix becomes similar to that of seawater. This transition is very sudden, the activity of the electron decreasing by some 17 orders of magnitude, with a very small change in the dilution factor. At this transition FeS dissolves, and $Fe(OH)_3$ precipitates while MnO_2 becomes saturated. The much sharper changes in chemistry at the anoxic-oxic transition ($h \cong 0.008$) than at the zero alkalinity transition ($h \cong 0.47$) are a good illustration of the differences between redox and acid-base processes in aquatic systems.

In reality the FeS appears to precipitate at a lower dilution of the hydrothermal fluid in oceanic vents and the oxic-anoxic transition is not as sharp as suggested by the equilibrium calculations. This underscores the importance of temperature and kinetic effects which are difficult to examine quantitatively at this point. Nonetheless this example provides a general sense for the processes at play in hydrothermal vents and it illustrates the methodology for equilibrium calculations involving coupled acid-base, precipitation-dissolution, and redox reactions.

□ □ □

4. pe-pH DIAGRAMS

A common and convenient way to explore the redox chemistry of systems of known composition is to develop "pe-pH diagrams" (also known as E_H-pH or Pourbaix diagrams) which show the major species of the various components as a function of both pH and pe. Such diagrams display the pe-pH domains of stability of possible solids and the regions of dominance of soluble species.

Given a half redox reaction involving H^+, the relative concentrations of the oxidant and the reductant are readily given as a function of pH and pe:

$$Ox + ne^- + mH^+ = Red; \quad pe^0 \tag{87}$$

therefore

$$pe = pe^0 - \frac{1}{n} \log \frac{(Red)}{(Ox)} + \frac{m}{n} \log \{H^+\} \tag{88}$$

$$n(pe - pe^0) + mpH = \log \frac{(Ox)}{(Red)} \tag{89}$$

When the expression on the left-hand side is greater than zero, the reductant is the more important species, and vice versa. The dividing line $[(Red) = (Ox)]$ is obtained by equating the left-hand side to zero:

$$n(pe - pe^0) + mpH = 0 \tag{90}$$

This expression defines a straight line of slope m/n on a pe-pH diagram. With pe as the ordinate, the oxidant is the dominant species above the line, the reductant below it. The difficulty in developing a pe-pH diagram lies in choosing the correct redox couple for which to write a half redox reaction. The reductant and the oxidant have to be the major species of the component under study, in the pe-pH domains on each side of the dividing line. For example, depending on whether the pH is below or above 7, the $S(+VI)/S(-II)$ couple should be represented by the SO_4^{2-}/H_2S or the SO_4^{2-}/HS^- reaction. Precipitation of sulfide by an excess of Fe(II) would make SO_4^{2-}/FeS the reaction to be studied. Short of repetitive trial and error procedures, the only solution is to develop through experience an intuition for the probable major couples.

4.1 Example 9. The Water, O_2, H_2 System

The half redox reactions that express the oxidation of water to oxygen gas and its reduction to hydrogen gas provide two lines on a pe-pH diagram:

$$\tfrac{1}{4}O_2(g) + H^+ + e^- = \tfrac{1}{2}H_2O; \quad pe^0 = 20.75 \tag{91}$$

therefore

$$pe + pH = 20.75 + \tfrac{1}{4} \log P_{O_2} \tag{92}$$

$$H^+ + e^- = \tfrac{1}{2}H_2(g); \quad pe^0 = 0 \tag{93}$$

$$pe + pH = -\tfrac{1}{2} \log P_{H_2} \tag{94}$$

Certainly one atmosphere is an upper limit on the partial pressure of oxygen or hydrogen in natural waters and the equations

$$pe + pH = 20.75 \tag{95}$$

$$pe + pH = 0 \tag{96}$$

provide, respectively, upper and lower limits on a pe-pH diagram for natural waters (see Figure 7.4). Above the upper line water becomes an effective reductant (producing oxygen); below the lower line water is an effective oxidant (producing hydrogen). Figure 7.4a shows how these limits vary with the partial pressures of the gases.

The possible oxidizing role of photochemically produced ozone and peroxide in natural waters is illustrated in Figure 7.4b which provides the pe-pH diagrams corresponding to the reactions:

$$\tfrac{1}{2}O_3(g) + H^+ + e^- = \tfrac{1}{2}O_2(g) + \tfrac{1}{2}H_2O; \quad pe^0 = 35.1 \qquad (97)$$

$$\tfrac{1}{2}O_2(g) + H^+ + e^- = \tfrac{1}{2}H_2O_2; \qquad\qquad pe^0 = 11.5 \qquad (98)$$

$$\tfrac{1}{2}H_2O_2 + H^+ + e^- = H_2O; \qquad\qquad pe^0 = 30.0 \qquad (99)$$

(and also the acid-base reaction $H_2O_2 = H^+ + HO_2^-$; $pK = 11.6$).

Ozone and peroxide are much stronger oxidants than oxygen in water, producing, respectively, oxygen and water as the reduced species. For peroxide this may seem paradoxical since H_2O_2 is a product of the partial reduction of oxygen itself: the acquisition of the first two electrons (per mole of oxygen), from O_2 to H_2O_2, is less energetically favorable than the acquisition of the

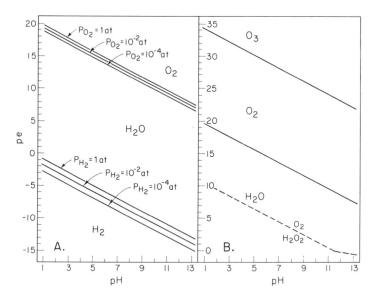

Figure 7.4 pe-pH diagrams for the O_2—H_2O—H_2 system (A) and the O_3—O_2—H_2O_2—H_2O system (B). In panel A the limits of the domain of H_2O stability are shown as a function of P_{O_2} and P_{H_2}. Within this domain O_2 acts as an oxidant (which it does not in the O_2 domain), and H_2 as a reductant (which it does not in the H_2 domain). The domain of stability of ozone, O_3, is shown in panel B. Hydrogen peroxide is never stable in the presence of oxygen and water; the dashed line in panel B indicates the region above which H_2O_2 is a better oxidant than O_2 in water.

next two electrons, from H_2O_2 to H_2O. As a result there is no region of Figure 7.4b where H_2O is the dominant species. The reaction

$$H_2O_2 = H_2O + \tfrac{1}{2}O_2(g)$$

has a negative standard free energy,

$$\Delta G^0 = -\frac{(30 - 11.5)}{0.733} = -25.2 \text{ kcal mol}^{-1}$$

and tends to proceed to the right for measurable concentrations of the reactants. The line separating the regions where O_2 and H_2O_2 are the second most important species in the H_2O domain (Reaction 98) is shown on the graph. Above this line H_2O_2 is a better oxidant than O_2, and vice versa below the line. From a purely energetic point of view we can thus expect peroxide to be a more reactive oxidant than oxygen itself in oxic waters, and the concentration of H_2O_2 in natural waters is accordingly low. The complete reduction of oxygen to water thus provides the major oxidizing reaction in natural waters and the O_2/H_2O couple effectively determines the "the pe" of oxic waters—even if in some situations (such as the surface of platinum electrodes) the acquisition of the last two electrons, from H_2O_2 to H_2O, is relatively slow.[11]

4.2   The Sulfate/Sulfide System

When oxygen is exhausted in the process of oxidizing organic matter (anoxic waters), sulfate becomes the primary oxidant. Although there exist important intermediary redox sulfur compounds (e.g., S^0), the reaction usually proceeds all the way to the formation of sulfides and can be written

$$\tfrac{1}{8}SO_4^{2-} + \tfrac{10}{8}H^+ + e^- = \tfrac{1}{8}H_2S + \tfrac{1}{2}H_2O; \quad pe^0 = 5.1 \qquad (100)$$

A simplified sulfur system thus includes only sulfate and sulfide species.

Example 10.   pe-pH Diagram for the S(VI)/S(−II) System

In addition to Reaction 100 the pertinent reactions are

$$HSO_4^- = H^+ + SO_4^{2-}; \quad pK = 2.0 \qquad (101)$$

$$H_2S = H^+ + HS^-; \quad pK = 7.0 \qquad (102)$$

$$HS^- = H^+ + S^{2-}; \quad pK = 13.9 \qquad (103)$$

Reactions 101, 102, and 103 provide three vertical lines separating the various acid-base species on the pe-pH diagram (see Figure 7.5):

1   $HSO_4^- - SO_4^{2-};$ pH = 2.0
2   $H_2S - HS^-;$ pH = 7.0
3   $HS^- - S^{2-};$ pH = 13.9

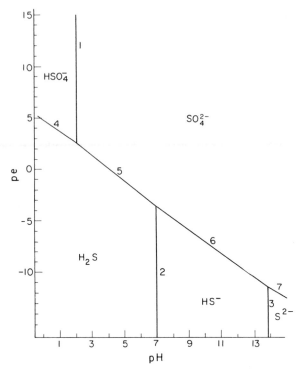

Figure 7.5 pe-pH diagram for the sulfate-sulfide system. See text for explanation of the numbered boundaries.

The separation of the sulfate and sulfide species then consists of four segments:

4 $HSO_4^{2-} - H_2S$ for pH < 2.0

5 $SO_4^{2-} - H_2S$ for 2.0 < pH < 7.0

6 $SO_4^{2-} - HS^-$ for 7.0 < pH < 13.9

7 $SO_4^{2-} - S^{2-}$ for 13.9 < pH

The corresponding reactions and pe-pH equations are obtained by combinations of Reactions 100, 101, 102, and 103, and expressions of mass laws

4 $\frac{1}{8}HSO_4^{2-} + \frac{9}{8}H^+ + e^- = \frac{1}{8}H_2S + \frac{1}{2}H_2O$; $pe^0 = 4.85$

$\qquad 8pe + 9pH = 38.8$ for $(HSO_4^-) = (H_2S)$ $\hfill (104)$

5 $\frac{1}{8}SO_4^{2-} + \frac{10}{8}H^+ + e^- = \frac{1}{8}H_2S + \frac{1}{2}H_2O$; $pe^0 = 5.1$

$\qquad 8pe + 10pH = 40.8$ for $(SO_4^{2-}) = (H_2S)$ $\hfill (105)$

6 $\frac{1}{8}SO_4^{2-} + \frac{9}{8}H^+ + e^- = \frac{1}{8}HS^- + \frac{1}{2}H_2O$; $pe^0 = 4.23$

$\qquad 8pe + 9pH = 33.8$ for $(SO_4^{2-}) = (HS^-)$ $\hfill (106)$

7 $\frac{1}{8}SO_4^{2-} + H^+ + e^- = \frac{1}{8}S^{2-} + \frac{1}{2}H_2O$; $pe^0 = 2.49$

$\qquad pe + pH = 2.5$ for $(SO_4^{2-}) = (S^{2-})$ $\hfill (107)$

The various equations are plotted in Figure 7.5.

In each region of Figure 7.5 the free concentration of any sulfur species can be obtained from the total sulfur concentration by considering it to be equal to the concentration of the appropriate major species and expressing the corresponding mass law. For example, the free sulfide concentration (S^{2-}), which is important for assessing the solubility of metal sulfides (see Example 11), is given by

$$pS^{2-} = -19.9 + pS_T + 8pe + 8pH \quad \text{(in the SO}_4^{2-} \text{ region)} \tag{108}$$

$$pS^{2-} = 20.9 + pS_T - 2pH \quad \text{(in the H}_2\text{S region)} \tag{109}$$

$$pS^{2-} = 13.9 + pS_T - pH \quad \text{(in the HS}^- \text{ region)} \tag{110}$$

If we consider the presence of hydrogen sulfide gas, the relevant redox reaction is obtained by combination of Reactions 100 and 111:

$$H_2S(g) = H_2S; \quad pK_H = 1.0 \tag{111}$$

therefore

$$\tfrac{1}{8}SO_4^{2-} + \tfrac{10}{8}H^+ + e^- = \tfrac{1}{8}H_2S(g) + \tfrac{1}{2}H_2O; \quad pe^0 = 5.23$$
$$\log(SO_4^{2-}) = -41.8 + 8pe + 10pH + \log(P_{H_2S}) \tag{112}$$

This equation gives the sulfate concentration in redox equilibrium with any partial pressure of hydrogen sulfide.

Figure 7.5 displays the characteristic features of a pe-pH diagram. Most noteworthy is the fact that the regions of dominance of various species meet at common corners; typically three boundaries intersect at the same point. This is because the equations of the boundaries (which are logarithmic expressions of mass laws) are linear combinations of each other. Very simply, the point where $(SO_4^{2-}) = (H_2S)$ and $(SO_4^{2-}) = (HS^-)$ is by necessity a point where $(H_2S) = (HS^-)$. This feature of pe-pH diagrams is useful for detecting errors in the construction of the graphs which it facilitates greatly. In redox systems that include high stoichiometric coefficients, this unique property of pe-pH diagrams is conserved only when the graph is strictly a dominance diagram, when the boundaries represent the points where half the element of interest is in each of the major species on each side of the line. To obtain the equations of these boundaries, the concentrations of the species themselves must be divided by the stoichiometric coefficient of the element, for example, $(HS^-) = \tfrac{1}{2}S_T$, $(S_2O_3^{2-}) = \tfrac{1}{4}S_T$, and $(S_6^{2-}) = \tfrac{1}{12}S_T$. Note also that in this case the boundaries around the domain of a solid species correspond to the points where half of the solid is already precipitated, and not the onset of precipitation. The diagram provides the region of dominance of the solid, not its domain of stability. (The two are usually separated by only a fraction of a log unit.)

In the process of reducing sulfate or oxidizing sulfide, a number of compounds with intermediate oxidation state can be formed. As seen in Table 7.1, these include sulfite (SO_3^{2-}), thiosulfate $(S_2O_3^{2-})$, polysulfides (S_n^{2-}), and solid sulfur in colloidal $(S_8^0 \cdot \text{col})$ or orthorombic $(S_8^0 \cdot \text{ort})$ form. Although the production of these intermediary sulfur compounds is largely under biological control, partial chemical oxidation of sulfide by oxygen leads to the formation of elemental sulfur, polysulfides, and thiosulfate. Direct reaction between sulfide and bacterially produced elemental sulfur also leads to the formation of polysulfides.[12]

The domains of dominance of these various intermediary sulfur species are illustrated in the pe-pH diagram of Figure 7.6, where $S_T = 2 \times 10^{-2} M$ and the possible formation of sulfate is not considered; (SO_4^{2-} would dominate a large part of the graph, obliterating the sulfite and most of the thiosulfate domains). Note that the less stable colloidal form of elemental sulfur is considered and that S_6^{2-} is the only polysulfide ever to dominate sulfur speciation.

The equations of the various boundaries in Figure 7.6 are obtained by

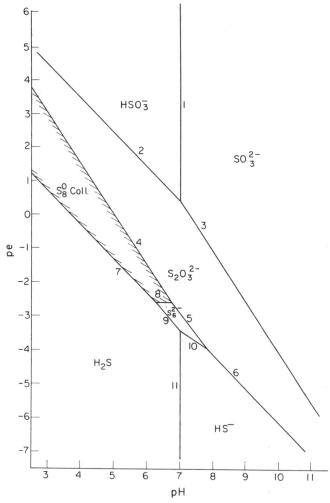

Figure 7.6 pe-pH diagram for sulfur: $S_T = 2 \times 10^{-2} M$; no sulfate. See text for explanation of the numbered boundaries.

combination of the appropriate half redox (and acid-base) reactions of Table 7.1:

1 $HSO_3^- - SO_3^{2-}$
$$HSO_3^- = H^+ + SO_3^{2-}; \quad \log K = -7.0 \tag{113}$$
$$pH = 7.0 + 2.0 - 2.0 = 7.0$$

2 $HSO_3^- - S_2O_3^{2-}$
$$\tfrac{1}{2}HSO_3^- + H^+ + e^- = \tfrac{1}{4}S_2O_3^{2-} + \tfrac{3}{4}H_2O; \quad pe^0 = 7.85 \tag{114}$$
$$pH + pe = 7.85 + \tfrac{1}{4}(2.3) - \tfrac{1}{2}(2.0) = 7.43$$

3 $SO_3^{2-} - S_2O_3^{2-}$
$$\tfrac{1}{2}SO_3^{2-} + \tfrac{3}{2}H^+ + e^- = \tfrac{1}{4}S_2O_3^{2-} + \tfrac{3}{4}H_2O; \quad pe^0 = 11.35 \tag{115}$$
$$\tfrac{3}{2}pH + pe = 11.35 + \tfrac{1}{4}(2.3) - \tfrac{1}{2}(2.0) = 10.93$$

4 $S_2O_3^{2-} - S_8^0$
$$\tfrac{1}{4}S_2O_3^{2} + \tfrac{3}{2}H^+ + e^- = \tfrac{1}{16}S_8^0(col) + \tfrac{3}{4}H_2O; \quad pe^0 = 8.09 \tag{116}$$
$$\tfrac{3}{2}pH + pe = 8.09 - \tfrac{1}{4}(2.3) = 7.52$$

5 $S_2O_3^{2-} - S_6^{2-}$
$$\tfrac{3}{14}S_2O_3^{2-} + \tfrac{9}{7}H^+ + e^- = \tfrac{3}{14}H_2O + \tfrac{1}{14}S_6^{2-}; \quad pe^0 = 6.37 \tag{117}$$
$$\tfrac{9}{7}pH + pe = 6.37 + \tfrac{1}{14}(2.78) - \tfrac{3}{14}(2.3) = 6.08$$

6 $S_2O_3^{2-} - HS^-$
$$\tfrac{1}{8}S_2O_3^{2-} + H^+ + e^- = \tfrac{1}{4}HS^- + \tfrac{3}{8}H_2O; \quad pe^0 = 3.66 \tag{118}$$
$$pH + pe = 3.66 + \tfrac{1}{4}(2.0) - \tfrac{1}{8}(2.3) = 3.87$$

7 $S_8^0 - H_2S$
$$\tfrac{1}{16}S_8^0 + H^+ + e^- = \tfrac{1}{2}H_2S; \quad pe^0 = 2.73 \tag{119}$$
$$pH + pe = 2.73 + \tfrac{1}{2}(2.0) = 3.73$$

8 $S_8^0 - S_6^{2-}$
$$\tfrac{3}{8}S_8^0 + e^- = \tfrac{1}{2}S_6^{2-}; \quad pe^0 = -3.97 \tag{120}$$
$$pe = -3.97 + \tfrac{1}{2}(2.78) = -2.58$$

9 $S_6^{2-} - H_2S$
$$\tfrac{1}{10}S_6^{2-} + \tfrac{6}{5}H^+ + e^- = \tfrac{3}{5}H_2S; \quad pe^0 = 4.07 \tag{121}$$
$$\tfrac{6}{5}pH + pe = 4.07 + \tfrac{3}{5}(2.0) - \tfrac{1}{10}(2.78) = 4.99$$

10 $S_6^{2-} - HS^-$
$$\tfrac{1}{10}S_6^{2-} + \tfrac{3}{5}H^+ + e^- = \tfrac{3}{5}HS^-; \quad pe^0 = -0.13 \tag{122}$$
$$\tfrac{3}{5}pH + pe = 0.13 + \tfrac{3}{5}(2.0) - \tfrac{1}{10}(2.78) = 0.79$$

11 $H_2S - HS^-$
$$H_2S = HS^- + H^+; \quad \log K = -7.0 \tag{123}$$
$$pH = 7.0 + 2.0 - 2.0 = 7.0$$

4.3   Metals in Carbonate and Sulfur-Bearing Waters

As seen in Chapter 5, many metals form carbonate, sulfide, and hydroxide complexes and solids in natural waters. Without considering the redox chemistry of the metal itself, the speciation of a metal such as cadmium will then depend not only on the pH but also on the pe of the water due to the redox chemistry of sulfur.

Example 11.   pe-pH Diagram for Dominant Cadmium Species

Assume, for example,

$$(CO_3)_T = 10^{-2} \, M$$

$$(S)_T = 10^{-4} \, M$$

$$(Cd)_T = 2 \times 10^{-8} \, M$$

The major species of interest are Cd^{2+}, $CdCO_3(s)$, $Cd(OH)_2(s)$, and $CdS(s)$ ($pK_s = 13.7$, 14.3, and 27.0 respectively). For simplicity, we consider only sulfate and sulfide as sulfur species (see Example 10) and denote the total concentrations by C_T, S_T, and Cd_T. In the high pe region where sulfide is not important, the vertical separations between the Cd^{2+}, $CdCO_3(s)$, and $Cd(OH)_2$ regions are obtained in the manner studied in Chapter 5:

1   $Cd^{2+} - CdCO_3(s)$. The boundary is in the region where HCO_3^- is the major carbonate species:

$$pCd^{2+} + pCO_3^{2-} = 13.7$$

$$pCO_3^{2-} = 10.3 + pHCO_3^- - pH$$

therefore

$$pCd^{2+} + pHCO_3^- = 3.4 + pH$$

Equating $pCd^{2+} = p(Cd_T/2) = 8$ and $pHCO_3^- = pC_T = 2$ yields pH = 6.6.

2   $CdCO_3(s) - Cd(OH)_2(s)$:

$$pCd^{2+} + pCO_3^{2-} = 13.7$$

$$pCd^{2+} + 2pOH^- = 14.3$$

$$pH + pOH^- = 14.0$$

therefore

$$pCO_3^{2-} + 2pH = 27.4$$

The boundary is in the region where CO_3^{2-} is the major carbonate species: $pCO_3^{2-} = pC_T = 2$ yields pH = 12.7.

It remains to define the boundaries between Cd^{2+}, $CdCO_3(s)$, $Cd(OH)_2(s)$, and

CdS(s). These are obtained readily by considering the expressions of free sulfide concentrations in the various regions (see Example 10):

3 $Cd^{2+} - CdS(s)$ (in the SO_4^{2-} region):

$$pCd^{2+} + pS^{2-} = 27.0$$

$$pS^{2-} = -19.9 + pS_T + 8pe + 8pH$$

therefore

$$pCd^{2+} + pS_T + 8pe + 8pH = 46.9$$

Equating $pCd^{2+} = p(Cd_T/2) = 8$ and $pS_T = 4$ yields $pe + pH = 4.36$.

4 $Cd^{2+} - CdS$ (in the H_2S region):

$$pCd^{2+} + pS^{2-} = 27.0$$

$$pS^{2-} = 20.9 + pS_T - 2pH$$

therefore

$$pCd^{2+} + pS_T - 2pH = 6.1$$

Equating $pCd^{2+} = p(Cd_T/2) = 8$ and $pS_T = 4$ yields $pH = 2.95$.

5 $CdCO_3(s) - CdS(s)$ (in the HCO_3^- and SO_4^{2-} region):

$$pCd^{2+} + pCO_3^{2-} = 13.7$$

$$pCd^{2+} + pS^{2-} = 27.0$$

$$pCO_3^{2-} = 10.3 + pHCO_3^- - pH$$

$$pS^{2-} = -19.9 + pS_T + 8pe + 8pH$$

therefore

$$pS_T - pHCO_3^- + 8pe + 9pH = 43.5$$

Equating $pHCO_3^- = pC_T = 2$ and $pS_T = 4$ yields $8pe + 9pH = 41.5$.

6 $CdCO_3(s) = CdS(s)$ (in the CO_3^{2-} and SO_4^{2-} region). Taking $pCO_3^{2-} = pC_T = 2$ in the above equations yields $pe + pH = 3.9$.

7 $Cd(OH)_2(s) - CdS(s)$ (in the SO_4^{2-} region):

$$pCd^{2+} + 2pOH^- = 14.3$$

$$pCd^{2+} + pS^{2-} = 27.0$$

$$pS^{2-} = -19.9 + pS_T + 8pe + 8pH$$

$$pH + pOH^- = 14.0$$

$$pS_T + 8pe + 10pH = 60.6$$

Equating $pS_T = 4$ yields $4pe + 5pH = 28.3$.

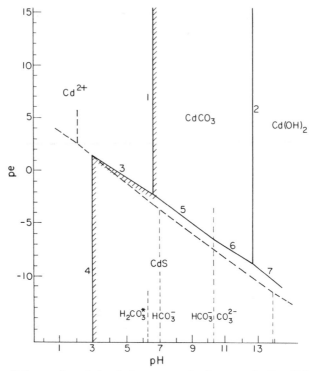

Figure 7.7 pe-pH diagram for cadmium in the presence of carbonate and sulfur: $(CO_3)_T = 10^{-2} M$; $(S)_T = 10^{-4} M$; $(Cd)_T = 2 \times 10^{-8} M$ (Example 11). Dashed lines are those of Fig. 7.5.

The boundary lines (1) to (7) are plotted on the pe-pH diagram of Figure 7.7 which shows the domains of dominance of the various cadmium species.

This figure exhibits the characteristic features of many pe-pH diagrams for dominant metal species. At high pe's the metal is controlled from low to high pH by the free metal ion, the carbonate, and the hydroxide solids. As the system becomes more reducing in the neutral pH range, the sulfide solid precipitates and replaces the free metal ion and the carbonate solid as the major species, before sulfide itself becomes dominant over sulfate. At low pe, if the pH becomes low enough, the solid sulfide dissolves simply due to the decrease in free sulfide ion activity. With increasing pH, depending on the exact value of the pe, the carbonate and hydroxide solid can precipitate. In the diagram of Figure 7.6, however, the direct transition from sulfide to hydroxide solid occurs at too low a pe to be relevant to aquatic systems.

Although the equations are rarely difficult, the construction of such pe-pH diagrams for metal species can become quite time-consuming. This is especially so when the metal itself is subject to redox reactions such as:

1 Formation of oxides:

$$Fe_3O_4, \ MnO_2, \ PbO_2, \ etc.$$

2   Equilibrium among various soluble oxidation states:

$$Fe(II)/Fe(III), Cu(II)/Cu(I), etc.$$

3   Formation of the elemental state:

$$Cu^0, Hg^0, Ag^0, etc.$$

For example, iron and manganese oxides, Fe_3O_4 and MnO_2, typically dominate the high pe and pH region of the diagrams, and elemental copper is stable in a region below the sulfide precipitation domain. In multimetal diagrams, even greater complications are introduced by the formation of mixed metal phases. At low pe's, in a system containing copper, iron, and sulfur, such solids as $CuFeS_2$ and Cu_3FeS_4 may control the speciation of the metals in large regions of the graph.

5.   PHOTOCHEMICAL REACTIONS

So far in this chapter our attention has been focused principally on slow redox processes, usually mediated by microorganisms and often located near the bottom of the water column where organic debris accumulate. These redox processes depend ultimately on the biological utilization of solar energy by photosynthesis. It is becoming increasingly clear that solar energy also generates a host of rapid redox reactions in surface waters through strictly abiotic processes. Photochemical reactions produce highly reactive radicals and unstable redox species which may be important in some elemental cycles in the euphotic zone of natural waters. Although not all photochemical reactions are redox reactions, in aqueous solution most are, and the subject of aquatic photochemistry is conveniently approached in the framework of this chapter. Unlike the related topics of photosynthesis and atmospheric photochemistry, it is a subject in its infancy, and only an elementary introduction is provided here. The usual difficulties encountered in aquatic chemistry—very low concentrations, multiplicity of chemical constituents, presence of suspended colloidal phases—are dramatically enhanced when studying photochemical processes, which are characterized by short-lived chemical species and are not easily amenable to the phenomenological approaches that dominate natural water chemistry.[13-15] (The field of photochemistry itself focuses more on mechanisms at the molecular level than most other areas of chemistry.)

5.1   General Principles

Photochemical processes include three separate steps: absorption of light by a molecule (a chromophore), primary reactions usually within the molecule, and secondary reactions involving other molecules. The fundamental principle of photochemistry is that only absorbed light can lead to chemical changes. The absorption spectrum of a molecule thus defines its ability to reach electronically excited states, whose initial energies correspond to that of the absorbed photons. This energy is given by the Planck relation

$$E = Nh\nu = \frac{Nhc}{\lambda} \tag{124}$$

where E is the energy absorbed per mole, N is Avogadro's number (6.02×10^{23}), h is Planck's constant (1.58×10^{-37} kcal s), c is the velocity of light (3.00×10^8 m s^{-1}), and ν and λ are, respectively, the frequency and the wavelength of the absorbed radiation. With λ in nanometers the expression can be written:

$$E = \frac{28,600}{\lambda} \text{ kcal mol}^{-1} \tag{125}$$

A molecule that absorbs at 520 nm must thus reach an excited state with an energy in excess of 50 kcal mol^{-1} above the ground state, some large fraction of that energy being potentially available for photochemical reactions. Such high energy is reached by a negligible fraction of the molecules through purely thermal processes ($n_1/n_0 = e^{-(E_1 - E_0)/RT} \cong e^{-50/0.58} \cong 3.6 \times 10^{-38}$ at room temperature according to the Boltzmann distribution law), and it is also comparable to typical bond and activation energies. This relatively large energy available from light absorption is a major reason why many chemical reactions that are impossible thermally are easily achieved photochemically.

Once the light has been absorbed, and the molecule is in an electronically excited state, the energy can be dissipated in several ways, only one of which is a primary photochemical reaction; for example,

$$NO_2^- + h\nu \rightarrow {}^*NO_2^- \tag{126}$$

$${}^*NO_2^- \rightarrow NO + O^- \tag{127}$$

Other energy-dissipating processes include luminescence (fluorescence and phosphorescence), heat production through vibrational relaxation processes, and energy transfer to other molecules (which may also initiate photochemical reactions). The products of primary photochemical reactions are usually very reactive and lead to secondary chemical reactions. For example, the species NO and O$^-$ are free radicals (i.e., they have an unpaired electron) that react rapidly with other water constituents. Within nanoseconds, O$^-$ reacts with water itself to yield an OH radical:

$$O^- + H_2O \rightarrow OH + OH^- \tag{128}$$

In turn, OH reacts rapidly (in microseconds in seawater) with other constituents such as bromide or dissolved organic matter:

$$OH + Br^- \rightarrow Br + OH^- \tag{129}$$

$$OH + DOC \rightarrow {}^*DOC + H_2O \tag{130}$$

These reactions illustrate three important characteristics of many photochemical reactions in water: they are principally one-electron-transfer redox reactions (in this case through free radical formation and dissipation), they involve transient high-energy compounds that react rapidly, and they often lead to sequential reactions that may ultimately affect molecules with no light absorption capability of their own.

The efficiency with which the absorbed light energy is utilized in photochemical processes is measured by the quantum yield, defined as the number of molecules of reactant consumed per photon of light absorbed. For primary reactions the quantum yield (primary quantum yield) is at the most equal to unity and is usually less than one since processes other than chemical change may dissipate absorbed light quanta. Because chain reactions can lead to more molecules of products than the initial number of reactant molecules, overall quantum yields can theoretically be greater than one.

5.2 Photochemistry in Aquatic Systems

The incident solar radiation reaching the earth's surface (Figure 7.8) has a characteristic spectrum encompassing near uv (290 to 400 nm), visible (400 to 700 nm), and infrared (>700 nm) wavelengths. The infrared, which accounts for some 50% of the light energy, is absorbed and dissipated as heat in the surface waters (<1 m). Some 90% of the rest penetrates into the water (10% is lost in reflection and backscattering), following a classical exponential attenuation curve according to Beer-Lambert's law. The maximum penetration of this photon flux varies from a fraction of a meter in very turbid and colored water to some 100 m in clear ocean water. Since there are no drastic changes in the near uv-visible light spectrum as a function of depth, the whole euphotic zone can be subjected to photochemical processes, initiated by absorption of active wavelengths below 700 nm. As can be seen in Figure 7.8b, many bond energies are smaller than the available light quanta. Photochemical reactions in water are thus not limited to a thin layer at the surface, although it is true that the surface microlayer which is rich in chromophores and receives the most reactive uv photons is a prime target for photochemical processes.

The three principal aquatic constituents responsible for the absorption of light are (1) water itself, whose absorption is weak and poorly characterized but is certainly important in clear water bodies; (2) dissolved organic matter, principally the ill-defined colored humic material since identified aquatic organic solutes do not absorb in the relevant wavelength range (this may be partly tautological since photoreactive solutes produced in surface water would be rapidly destroyed by the very large flux of light energy); and (3) suspended particles, dead or alive, whose photoreactivity—aside from photosynthesis—is simply unknown and outside the realm of classical photochemistry. As is shown by the great transparency of many saline waters, the inorganic solutes contribute little to the overall light absorption. Note, however, that while absorption by trace elements, particularly transition trace metals, may play an insignificant role in overall light attenuation, it may be very important in the chemistry of these elements.

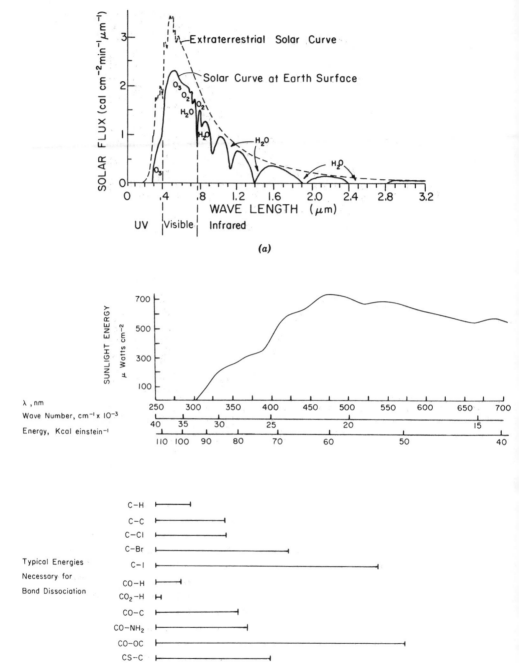

Figure 7.8 (a) Solar spectrum at the earth's surface. From Gates, 1962.[16] (b) Comparison of the uv-visible solar flux with typical bond-breaking energies. From Zika, 1981.[14]

Little is known about specific photochemical reactions in natural waters. Because of inherent experimental difficulties very few systems have been studied, and those that have been studied represent the more tractable, not necessarily the most important, ones. For example, the photolysis of nitrite which is well documented probably represents only a minor sink for NO_2^- in seawater. Possibly more significant are the free radicals that are generated in the process (e.g., OH) which may lead ultimately to important redox reactions. Another and more complex example is the formation of hydrogen peroxide according to the overall schematic reaction

$$SW + h\nu \rightarrow HOOH \tag{131}$$

(SW = seawater with all its constituents). This process probably also involves free radicals

$$SW + O_2 + h\nu \rightarrow SW\text{cation}^+ + O_2^- \tag{132}$$

$$H^+ + 2O_2^- \rightarrow O_2 + OOH^- \tag{133}$$

$$H^+ + OOH^- \rightarrow HOOH \tag{134}$$

While the study of anoxic euphotic zones may prove particularly intriguing to photochemists as it is to microbiologists, the presence of oxygen in most euphotic waters is likely to dominate the photochemistry of aquatic systems. Because of its energetic characteristics and its abundance, oxygen should be the recipient of most energy-transfer deactivation processes in water, leading to the reactive singlet oxygen (an excited electronic state of molecular O_2). Dissolved oxygen is probably also a major reactant with photochemically produced radicals, forming strong redox reagents (e.g., superoxide radical O_2^-, hydrogen peroxide H_2O_2, and organic peroxides ROO^-) as well as partially oxidized and highly oxidizable new radicals. Photochemical reactions in the presence of oxygen are thus expected to be ultimately oxidizing, and one of their principal impacts in aquatic systems is probably the formation and degradation of refractory organic compounds. The fate of both natural humic and anthropogenic refractory compounds in surface waters is certainly controlled in part, if not in whole, by photochemical processes.[17]

Although they are not the primary chromophores, the abundance of such inorganic species as bromide in seawater may well be important in its overall photochemical response and differentiate it from that of freshwater systems. Formation of short-lived radicals by photolysis may be ineffective in bringing significant chemical changes if these radicals react rapidly and indiscriminately with a variety of substrates. Longer-lived radicals such as Br, Cl, and CO_3^- would be expected to react much more selectively with specific dissolved compounds, leading to a focusing of their chemical impact. For example, one may speculate that the relatively high transparency of productive seawater could be linked to its saltiness.

While the major fraction of photochemical reactions in aquatic systems must be ultimately oxidative, the more subtle reductive role of these reactions may also be important in the geochemical cycle of trace elements. If a reactant is oxidized in a photochemical processes, another must be reduced (for example, NO_2^- is reduced to NO in the photolysis of nitrite). Although oxygen is often the ultimate electron acceptor, the formation of intermediate reduced species is of potential interest. Most simple among such reduced species is the hydrated electron itself which is probably formed by photochemically induced charge transfer from phenolic groups to water:

$$\text{phenolate} + hv = \text{phenoxyl radical}^+ + e_{aq}^- \qquad (135)$$

The subsequent formation of superoxide radical

$$e_{aq}^- + O_2 \rightarrow O_2^- \qquad (136)$$

then leads to hydrogen peroxide as mentioned earlier.

Transition metals that have two stable oxidation states one electron apart and form organic complexes in natural waters, even relatively unstables ones, are also likely targets for photoreduction, for example,

$$Cu(II)Y + hv \rightarrow Cu^+ + \text{oxidized } Y \qquad (137)$$

$$Fe(III)Y + hv \rightarrow Fe^{2+} + \text{oxidized } Y \qquad (138)$$

(Note that photooxidation of the reduced metal is also likely to occur, though probably with lower quantum efficiency than the reduction, e.g., $Fe^{2+} + hv \rightarrow Fe^{3+} + e_{aq}^-$.) Such reactions may be important in providing a source of reduced metals and in decomposing complexing organic ligands in natural waters. Although many reactions of this type are well documented in laboratory systems, they are difficult to demonstrate in the field where the metal concentrations are very low and reoxidation by oxygen is often rapid.

The best-documented case of geochemically relevant photoreduction of a metal is that of iron in low pH waters where the reoxidation of Fe(II) is slow.[18,19] This is shown in Figure 7.9 which illustrates the diurnal and seasonal variations in the photochemical formation of epilimnetic Fe(II). Such photoreduction may also be taking place in surface waters at higher pH where the oxidation rate of ferrous iron is fast, thus maintaining Fe(II) at very low concentration. A low steady-state concentration of photochemically produced Fe^{2+} has been shown to be important for iron uptake by phytoplankton in culture;[20] the same may be true in nature. Also the reoxidation of Fe(II) to Fe(III) and its precipitation as a fresh hydrous oxide may maintain iron in a highly disperse colloidal form with high adsorption capacity and affinity for aquatic solutes.

Although there is some evidence for Cu(I) production in surface water, manganese may in fact provide the most convenient demonstration of the importance of photoreduction of metals in aquatic systems. This is because the oxidation rate of Mn^{2+} is relatively slow under most conditions of interest, and because the two principal oxidation states of manganese in aquatic systems, Mn (II) and Mn(IV), are easily identified as the soluble and insoluble fractions of the metals, Mn^{2+} and MnO_2.

The shape of manganese concentration profiles as a function of depth in the oceans are indeed consistent with a photoreduction mechanism. The profiles exhibit a pronounced Mn(II) maximum, and Mn(IV) minimum, at the surface where the photon flux is highest.[21,22] (A secondary Mn(II) maximum and Mn(IV) minimum are also often observed at the oxygen minimum where the reoxidation rate should be slowest.) There have been only preliminary laboratory studies of manganese photoreduction, but the phenomenon can be clearly observed as a solubilization of MnO_2 suspension in seawater exposed to light.[22] Like that of Fe(III), the reduction rate is markedly increased by the presence of humic compounds. Although the mechanism of manganese photoreduction is unknown, it is likely to depend on the presence of the +III oxidation state of the metal, and one should recall that the formal oxidation state of Mn in aquatic oxides is in fact between 2.5 and 4.

Overall it appears that photochemical processes are important in the oxidation of organic matter, leading to the formation and partial destruction of refractory humic compounds.[23] This process is catalyzed by trace metals (particularly Fe) which are concomitantly reduced. Depending on the reoxidation rate, thermodynamically unstable reduced metallic species may thus be maintained at significant concentrations in euphotic waters. This interaction between light, trace metals and trace organic compounds (one could add particles) is probably a major controlling factor in the geochemistry and geobiology of surface waters. It seems that, unlike photosynthesis, photochemical reactions in water may only accelerate thermodynamically favorable processes that would otherwise depend on slower thermal production of free radicals. For example, in the context of a partial equilibrium system that excludes O_2 from consideration, the concomittant reduction of iron and manganese and oxidation

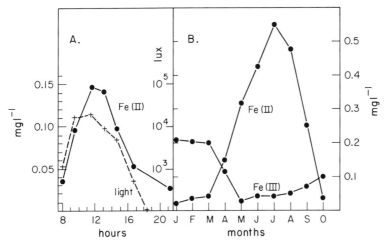

Figure 7.9 Daily (A) and seasonal (B) accumulation of photochemically produced ferrous ion in a well-oxygenated acid lake. Adapted from Collienne, 1983.[19]

of organic matter are indeed thermodynamically favorable, and in preliminary experiments dark metal reduction rates are observed to be nonnegligeable.

6. REDOX KINETICS

Much like precipitation-dissolution reactions, and unlike acid-base or co-ordination reactions, redox reactions in natural waters are often sufficiently slow (and sufficiently rapid) that a kinetic description would be desirable in most cases. As repeated throughout this chapter, aquatic redox reactions are most often mediated by microorganisms. In some simple cases it is possible to provide simplified aggregate mathematical descriptions of these microbial redox processes, as exemplified by the classical BOD-DO equations mentioned in Section 3. From a more chemical point of view, however—and apart from photochemical processes which we are still too ignorant to describe kinetically—there are but a few redox reactions in aquatic systems that are controlled by purely chemical mechanisms and are amenable to kinetic description by fundamental rate laws. Among these reactions that occur chemically, at least in part, one can list the oxidation of Fe(II), Mn(II), and to a lesser extent S($-$II) by oxygen and the oxidative dissolution of sulfide solids.

Even for those relatively few redox reactions that are known to occur chemically, the applicability of rate laws determined in simple laboratory systems can be questionable. For example, catalysis by trace elements and compounds, by light, and by solid surfaces can greatly increase the effective rate constants in nature. It is only by comparing predicted rates to the rates measured *in situ* that one can gain confidence in the appropriateness of any given kinetic equation. Nonetheless, such experimentally derived equations provide first estimates of characteristic half-lives and should define lower limits for rate constants.

The best-studied examples of redox kinetics particularly relevant to natural waters are those of the oxygenation of Fe(II) and Mn(II).[24-29] For iron, the reaction is second order in (OH^-) at high pH and zero order at low pH (<4).[24] First-order kinetics with respect to Fe(II) and P_{O_2} leads to the rate law (at high pH):

$$\frac{d(\text{Fe} \cdot \text{II})}{dt} = -k(\text{Fe} \cdot \text{II})P_{O_2}(OH^-)^2 \tag{139}$$

where $k \cong 2 \times 10^{13} \, M^{-2} \, \text{at}^{-1} \, \text{min}^{-1}$ at 25°C and $I = 5 \times 10^{-2} \, M$, corresponding to a 10-min half-life for Fe(II) when the pH is neutral and the partial pressure of oxygen is atmospheric.

A more general rate law can be written as a function of the activities of Fe^{2+}, O_2, and OH^-:[25]

$$\frac{d(\text{Fe} \cdot \text{II})}{dt} = -k'\{Fe^{2+}\}\{O_2 \cdot \text{aq}\}\{OH^-\}^2 \tag{140}$$

where the value of the rate constant, $k' = 1.2 \times 10^{17} \, M^{-3} \, min^{-1}$, fits approximately all experimental data in inorganic systems. In the range of interest the principal effect of temperature on the oxidation rate at a given pH is accounted for by the changes in the solubility of oxygen and in the ion product of water, resulting in a half-life increase by a factor of almost 10 from 25°C to 5°C. In seawater the effective rate constant is found to be approximately 30 times smaller ($\tau_{1/2} \cong 3$ min at pH = 8.0)[26] than in freshwaters. With the activity of the ferrous ion in the rate expression, roughly half of this retardation is expected through pure ionic strength effects and the rest is due to complexation of Fe^{2+} by Cl^- and SO_4^{2-} (see Table 6.1).

Equation 140 accounts for the effects of temperature, ionic strength, and inorganic complexation in controlling the kinetics of Fe(II) oxidation but not those of (organic) complexing agents. For example, some organic chelators (e.g., EDTA) can dramatically enhance the oxidation kinetics, while some others can effectively stabilize Fe(II) (e.g., tannate). In general, strong Fe(III) complexing agents accelerate the oxidation, while strong Fe(II) complexing agents retard it. The issue is complicated by the role of organic complexing agents in promoting photoreduction and photooxidation of iron, both in solution and in particulate phases.

In simple systems at pH > 7 an autocatalytic effect by the precipitated hydrous ferric oxide—in this case a microcystalline lepidocrocite, γ-FeOOH—is observed.[25] The oxidation kinetics can then be described by the rate law:

$$\frac{d(Fe \cdot II)}{dt} = -[k_1 + k_2(Fe \cdot III)](Fe \cdot II) \tag{141}$$

where k_1 (in min^{-1}) represents the previous homogeneous rate constant and the apparent heterogeneous rate constant, k_2, is on the order of 10^2 to $10^3 \, M^{-1}$ min^{-1}. Such an autocatalytic effect can decrease the half-life of ferrous iron by as much as a factor of two at pH = 7.2 in high ionic strength systems.

The same general rate laws that are applicable to the oxygenation of iron also seem to be applicable to manganese, with two major quantitative differences: the oxygenation of Mn^{2+} in solution is much slower than that of Fe^{2+}, and the catalytic effect of precipitated oxides, in particular MnO_2 and FeOOH, is much more important.[27-29] In the rate law

$$\frac{d(Mn \cdot II)}{dt} = -[k_1 + k_2(solid)](Mn \cdot II) \tag{142}$$

the constants have the approximate values

$$k_1 \cong 10^{-15} \frac{(O_2)}{(H^+)^2} \, d^{-1}$$

$$k_2 \cong 5 \times 10^{-10} \frac{(O_2)}{(H^+)^2} \, M \, d^{-1} \quad \text{(when the solid is colloidal } MnO_2\text{)}$$

$$k_2 \cong 2 \times 10^{-10} \frac{(O_2)}{(H^+)^2} \, M \, d^{-1} \quad \text{(when the solid is } \gamma\text{-FeOOH)}$$

In seawater these values lead to half-lifes of several years for the homogeneous oxidation reaction and of a few weeks for the surface catalyzed process. (Estimated residence times on the order of two days have been inputed to bacterial catalysis.[30]) A mechanistic interpretation of the catalysis of oxygenation by oxide solids is dependent upon proper quantification of the adsorption of the soluble species (Mn^{2+} or Fe^{2+}) at the oxide surface. A possible interpretation of the variable stoichiometry of manganese dioxide formed by Mn(II) oxygenation in aquatic systems ($MnO_{1.3}$ to $MnO_{1.9}$) is to view the solid as a solid solution of MnO_2 and $Mn(OH)_2$, the relative proportions of each depending on the relative rates of oxidation and hydrolysis of the adsorbed Mn^{2+} at the solid surface.

REFERENCES

1   L. G. Sillen and A. E. Martell, *Stability Constants of Metal Ion Complexes*, Special Publication 17, Chemical Society, London, 1964.
2   W. M. Latimer, *The Oxidation States of the Elements and Their Potentials in Aqueous Solutions*, 2nd ed., Prentice-Hall, Englewood Cliffs, N.J., 1952.
3   J. Boulègue and G. Michard, *J. Francais d'Hydrologie*, **9**, 27–35 (1978).
4   J. A. Raven, "Energetics and Transport in Marine Plants," Lecture Notes, Marine Biological Laboratories, Woods Hole, Ma., 1982.
5   A. F. Gaudy and E. T. Gaudy, *Microbiology for Environmental Scientists and Engineers*, McGraw-Hill, New York, 1980.
6   P. L. McCarty, in *Water Pollution Microbiology*, R. Mitchell, Ed., Wiley, New York, 1972.
7   M. Alexander, *Introduction to Soil Microbiology*, 2nd ed., Wiley, New York, 1965.
8   R. W. Howarth, "The Role of Sulfur in Salt Marsh Metabolism," Ph.D. Dissertation, Woods Hole Oceanographic Institution and Massachusetts Institute of Technology Joint Program in Biological Oceanography, 1979.
9   A. L. Lehninger, *Biochemistry*, 2nd ed., Worth, 1975.
10   *CRC Handbook of Biochemistry*, 2nd ed. H. A. Sober, ed. Chemical Rubber Co. Cleveland, 1970.
11   W. Stumm and J. J. Morgan, *Aquatic Chemistry*, 2nd ed., Wiley, New York, 1981, pp. 461–462.
12   J. Boulègue and G. Michard, in *Chemical Modeling in Aqueous Systems*, E. A. Jenne, Ed., ACS Symposium Series 93, ACS, Washington, D.C., 1979.
13   O. C. Zafiriou, *Mar. Chem.*, **5**, 497–522 (1977).
14   R. G. Zika, in *Marine Organic Chemistry*, E. K. Duursma and R. Dawson, Eds., Elsevier, Amsterdam, 1981.
15   O. C. Zafiriou, in *Chemical Oceanography*, vol. 8, J. P. Riley and R. Chester, Eds., 2nd ed., Academic, New York, in press.
16   D. M. Gates, *Energy Exchange in the Biosphere*, Harper & Row Biological Monographs, New York, 1962.
17   R. G. Zepp and D. M. Cline, *Env. Sci. Technol.*, **11**, 359–366 (1977).
18   J. W. McMahon, *Limnol. Oceanogr.*, **14**, 357–367 (1969).
19   R. Collienne, *Limnol. Oceanogr.*, **28**, 83–100 (1983).
20   M. A. Anderson and F. M. M. Morel, *Limnol. Oceanogr.*, **27**, 789–813 (1982).
21   J. H. Martin and G. A. Knauer, *Earth. Planet. Sci. Letters*, **51**, 266–274 (1980).
22   W. G. Sunda, S. A. Huntsman, and G. R. Harvey, *Nature*, in press.
23   C. J. Miles and P. L. Brezonick, *Environ. Sci. Technol.*, **15**, 1089–1095 (1981).
24   P. C. Singer and W. Stumm, *Science*, **167**, 3921 (1970).
25   W. Sung and J. J. Morgan, *Environ. Sci. Technol.*, **14**, 561–568 (1980).
26   D. A. Kester, R. H. Byrne, and Y. J. Liang, in *Marine Chemistry in the Coastal Environment*, T. M. Church, Ed., ACS Symposium Series 18, ACS, Washington, D.C., 1975.

27   J. J. Morgan, in *Principles and Applications of Water Chemistry*, S. D. Faust and J. V. Hunter, Eds., Wiley, New York, 1981.

28   W. Sung and J. J. Morgan, *Geochim. Cosmochim. Acta*, **45**, 2377–2383 (1981).

29   R. W. Coughlin and I. Matsui, *J. Catal.*, **41**, 108–123 (1976).

30   S. Emerson, S. Kalhorn, M. B. Tebo, K. H. Nealson, and R. A. Rosson, *Geochim. Cosmochim. Acta*, **46**, 1073–1079 (1982).

PROBLEMS

7.1   Given $(Mn)_T = (Co)_T = 10^{-6} M$, what are the principal oxidation states of manganese and cobalt in oxic waters at pH $= 7$? What concentration of a complexing ligand ($K_{Mn}^{app} = 10^{11}$; $K_{Co}^{app} = 10^{12}$) would be necessary to keep Mn(II) and Co(II) in solution? Same questions at pH $= 9$.

7.2   Denitrification can lead to the formation of NO(g) or $N_2O(g)$ instead of $N_2(g)$. At pH $= 7$, under what partial pressure of NO or N_2O in the atmosphere are these processes favorable compared to N_2 evolution? At pH $= 10$? At pH $= 5$?

7.3   Suppose you were a chemoautotroph trying to make a living out of oxidizing Fe(II) to Fe(III) with oxygen, in a solution saturated with $FeCO_3(s)$ (pH $= 8$, $C_T = 10^{-3} M$) and am·$Fe(OH)_3(s)$. Assuming 10% overall efficiency, how much Fe do you need to oxidize in order to fix one mole of inorganic carbon. (Assume that Fe^{2+} is the reductant in C fixation.) Discuss the effects of pH and Fe(II) or Fe(III) chelation on your survival.

7.4   During a storm a reduced swamp doubles its volume and reaches complete acid-base and redox equilibrium. (Assume organic carbon formation and utilization to be negligible.) Given the initial conditions pH $= 8.0$, Alk $= 10^{-3} M$, $(SO_4)_T = 10^{-3} M$, and $(H_2S)_T = 4 \times 10^{-4} M$, and assuming rainwater to be slightly polluted (Alk $= -10^{-4} M$) and saturated with atmospheric CO_2 and $O_2 [(H_2CO_3^*) \cong 10^{-5} M; (O_2 \cdot aq) \cong 3 \times 10^{-4} M]$, what is the final composition (Alk, pH, pe) of the swamp?

7.5   With the data of Tables 7.1 and 6.1 prepare a pe-pH diagram for chromium. What are the principal stable forms of Cr in natural waters?

7.6   a. Given the data of Tables 7.1, 6.1, and 5.3, prepare a pe-pH diagram for iron in an oxidized carbonated system $[(CO_3)_T = 10^{-3} M$; pe $> 0]$; major species to consider are $Fe(OH)_x^{(3-x)+}$, α-$Fe_2O_3(s)$, $Fe_3O_4(s)$, Fe^{2+}, $FeCO_3(s)$, and $Fe(OH)_2(s)$.

   b. Prepare a pe-pH diagram for iron in reduced systems (pe < 0), given $S_T = 10^{-4} M$, $(CO_3)_T = 10^{-3} M$. Do not consider elemental sulfur, sulfite, thiosulfate, polysulfides, or pyrite; major species to consider are Fe^{2+}, FeS(s), $FeCO_3(s)$, and $Fe(OH)_2(s)$.

   c. Complete the diagram by determining the boundaries between parts a and b.

7.7 An anoxic freshwater is characterized by

$$Na_T = Ca_T = Mg_T = K_T = Cl_T = (SO_4)_T = C_T = (S \cdot -II)_T = 10^{-3} \ M$$

a. Given $pH = 8.0$, calculate the alkalinity, the carbonate, and the sulfide species and the partial pressures of CO_2 and H_2S.

b. Calculate the pe and the partial pressure of O_2.

c. Is methane fermentation likely to take place in this system?

d. Write the complete redox reactions for methane fermentation of methanol (CH_3OH) and formaldehyde (CH_2O). What are the qualitative effects of such fermentation on alkalinity and pH?

e. Using the buffer capacity of the system (considered closed and isolated from gas phases), calculate the pH change due to fermentation of $10^{-4} \ M \ CH_2O$.

f. Considering that $10^{-3} \ M$ of O_2 are introduced into the system and react rapidly with the sulfide, calculate the new composition of the system: pe, pH, alkalinity, carbonate, and sulfide species (no gas exchange).

CHAPTER EIGHT

REACTIONS ON
SOLID SURFACES

As noted in Chapter 5, some 80% of the material transported by rivers and discharged into the oceans is in the form of suspended particles, and only some 20% in the form of dissolved species. Although we have emphasized the role of acid-base and precipitation-dissolution reactions as controlling factors in the exogenic cycle, it is in fact the reaction of solutes with solid surfaces that are thought to control the geochemistry of most trace elements. For example, simple chemical models for the residence time of trace metals in the ocean are based on particle sedimentation rates and on the partitioning of species between soluble and particulate fractions.[1,2] (Other, biological, models emphasize the uptake of trace elements by living microorganisms.) In most cases this partitioning can be shown to be controlled not by the precipitation of pure solid phases but by the adsorption of solute species on the surface of suspended solids of different chemical composition. The extent of this adsorption is in fact practically the *only chemical parameter* entering such geochemical models.

The topic of solute adsorption on suspended solids in aquatic systems is one that has undergone important conceptual evolution over the past decade. While physical processes, particularly the electrostatic interactions between surfaces and solutes, used to receive most of the attention, the chemical reactions between solutes and functional groups at surfaces are now the principal focus.[3,4] As argued by Stumm and Morgan (1981),[5] we are witnessing in effect a return to the earlier chemical theory of surface processes. Such evolution of our scientific paradigms compounds the inherent difficulty and the wide scope of issues related to surface phenomena, making it the most difficult topic of aquatic chemistry to present in a reasonably pedagogical yet rigorous manner. It is significant that one speaks of adsorption *theories* and *models* rather than principles; there is still some divergence of views on this topic and not all adsorption processes can be accounted for in a unique and elegant conceptual framework. As a result the examples of this chapter are closely tied to experimental data for the purpose of delineating the foundations and the limits of the theories. Only a very introductory presentation is attempted here, and the reader is reminded that practically all aspects of colloid and interface science are relevant to the study of aquatic chemistry. Specialized texts and journal articles in this field should be the basis of a more advanced topical study.

By considering that the reactions between solutes and functional groups on solid surfaces are effectively coordination reactions that obey mass law equations, the concepts and the mathematics of complexation processes become applicable to adsorption. It is a convenient way to introduce the topic, focusing on the nature and coordinative properties of the surface groups found on aquatic particles. The importance of adsorption in the speciation of trace aquatic constituents is illustrated in numerical examples showing how both metals and ligands—sometimes interacting with each other—in fresh or salt water can be controlled by adsorption on particles.

A major difference between complexation reactions among (small) solutes and coordination reactions on solid surfaces stems from the long-range nature of electrostatic interactions and the physical proximity of adsorption sites. We have seen how long-range electrostatic forces affect the interactions among solutes in high ionic strength media and how these effects can be accounted for by nonideal corrections applied to the free energies of the reactions (or their equilibrium constants). For adsorption reactions the role of the electrostatic interactions between the adsorbing solutes and the charged surface groups is so important and the charge of the surface depends so much on the composition of the solution that the total free energy of coordination has to be separated into a purely chemical term and an electrostatic term. There lies the principal difficulty in applying complexation concepts to adsorption processes; no universal equilibrium constant can be simply defined and various adsorption models differ principally by the manner in which the electrostatic interaction term is estimated. This subject is addressed here, starting with a study of the charging mechanism for aquatic particles and the presentation of the classical Gouy-Chapman theory of the double layer. Stern's extension of the theory to include purely chemical interactions and account for the size of adsorbing ions then serves as the basis of a simple one-layer adsorption model which is applied to some experimental data.

Finally, a few related topics such as ion exchange, adsorption-desorption kinetics, and particle coagulation are introduced briefly.

1. AQUATIC PARTICLES (NATURAL HYDROSOLS)

Suspended particles in natural waters comprise both inorganic solids and dead or live organic matter, the proportion of each varying widely in time and place.

Inorganic solids account for most of the total suspended matter transported by rivers and consist principally of alumino-silicates derived from physical erosion and partial weathering of continental rock. They constitute the red clays that dominate the composition of much of the ocean sediments. Other inorganic solids such as oxides and carbonates, (e.g., FeO_x, MnO_x, $CaCO_3$, and even SiO_2) are for a large part precipitated *in situ* and account for most of the total mass sedimenting on the ocean floor. However, only a fraction of this mass is accumulated in the deep-sea sediments due to partial redissolution of the major "organic" constituents, SiO_2 and $CaCO_3$.* In the euphotic zone of lakes and oceans, organisms—principally bacteria, micro-algae, and zooplankton—contribute typically a major portion (>90%) of the total mass of suspended particles, most of it (ca. 75%) in the form of dead remains and fragments.[6,7] In the deep ocean where suspended solids are chiefly inorganic, the major part of the suspended organic matter originates from bacteria and zooplankton.

The concentration range of suspended particles in natural waters spans some seven orders of magnitude from 0.01 mg liter^{-1} in the deep ocean to 50,000 mg liter^{-1} in particularly turbid rivers and estuaries. Concentrations of a few milligrams per liter are characteristic of productive surface waters, while eutrophic lakes may reach up to a few hundred milligrams per liter of suspended solids.

The size range of aquatic particles is also very large, from 10^{-9} m for colloidal oxides and humic substances to 10^{-2} m for large zooplankton or fecal aggregates. (We are not considering here a wide assortment of important, large particles such as sargassum weed, large animals, icebergs, ships, flotsam, tar balls, and bottles with desperate messages.) The particle size distribution usually follows a power law function of the form[8]

$$n(r) = \frac{dN}{dr} = Ar^{-p} \tag{1}$$

where A and p are constants and dN is the number of particles per fluid volume with radii between r and $r + dr$; that is, the total number of particles is given by the integral

$$N = \int_{r_{min}}^{r_{max}} n\, dr$$

In aquatic systems the exponent p of such a power law ranges from 2 to 5, most of the data being well approximated by an exponent close to 4 in the size range 1 to 100 μm (Table 8.1). This size distribution is observed at all depths in the

---

* Recall that oceanographers use the adjective "organic" in a peculiar way when referring to sediments, designating in this way material derived from organisms, including SiO_2 and $CaCO_3$. Recall also that in the same context the adjective detrital refers not to fecal or decomposing organic matter but to solid material of continental origin, mostly alumino-silicates.

oceans, even though the total particle concentration decreases sharply with depth in the top mixed layer (Figure 8.1). A fourth power distribution can be predicted by considering a steady state between aggregation of small particles due to fluid shear forces and settling of larger aggregates[10] (see Section 7).

As a result of this particle size distribution, the particles of radius r contribute proportionally to r^{1-p} to the total number of particles (per liter). In the same way, they contribute proportionally to r^{3-p} and r^{4-p} to the total area and the total volume of suspended solids, respectively. For a typical exponent of 4, the small particles thus contribute most of the area available for

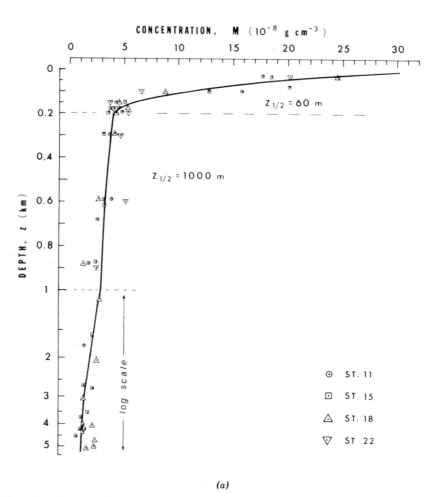

(a)

Figure 8.1 Variation of particle concentration (*a*) and of particle size distribution (*b*) as a function of depth in the Pacific Ocean. Note the constancy of the slope of the logarithmic size distribution (the exponent of the power law) with depth as the particle concentration varies by a factor of more than 10. (Newer data indicate lower deep ocean concentrations, on the order of 10^{-9} g cm^{-3}.) From Lerman, 1979.[9]

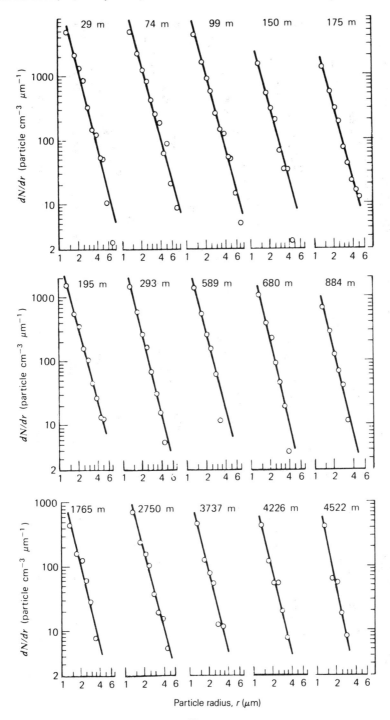

Particle radius, r (μm)

(*b*)

TABLE 8.1

Particle Size Distributions in Aquatic Systems[3,8]

	Coefficient of Power Law Distribution
Lakes	
Lake Zurich, Switzerland	1.7–2.4
Lake Murten, Switzerland	1.6–2.2
Deer Creek Reservation, Utah	3.4
Rivers	
American, California	3.4
San Joaquin, California	4.2
Sacramento, California	3.8
Oceans	
Indian	4.5
(Forams, diatoms; surface)	
Western Mediterranean	4.0
Equatorial Atlantic	4.0
East Atlantic	4.2
(Calcareous material)	
Soils	
Soils in granitic region	4.0
Calcareous sediments in West Equatorial Pacific	4.0
Wastewaters	
Digested sludge	4.2
Activated sludge	3.3
Primary sludge	4.3

adsorption (proportionally to r^{-1}), while the particle volume and mass are distributed more or less uniformly over the various size classes.

2. COORDINATIVE PROPERTIES OF SURFACES

As we shall see in this section, the surfaces of aquatic particles contain functional groups whose acid-base and other coordinative properties are similar to (and can be correlated with) those of their counterparts in soluble compounds. What then makes the coordinative behavior of surfaces different and so difficult to study in comparison with inorganic and organic complexing weak acids such as those we examined in Chapter 6? The main answer to this question lies simply in the geometric restrictions imposed by the solid nature of surfaces. Before examining the surface chemistry of specific types of solids, let us briefly mention the effects of these geometric restrictions that are common to all solids.

2.1 General Considerations

First of all, and as mentioned earlier, the obligatory proximity of the various functional groups on a surface makes them susceptible to long-range coulombic (i.e., electrostatic) interactions from other neighboring groups. (Recall that this is also the case for polyelectrolytes such as humic acids.) Consider, for example, a surface composed of identical negatively charged weak acid groups. Upon acidimetric titration the tendency of these groups to coordinate a proton decreases as the increasingly positive charge of the overall surface (due to the presence of the coordinated H^+) contributes a decreasing attractive electrostatic energy of bonding to each remaining acid group. The net result is a continuously decreasing equilibrium constant for the protonation reaction, a situation that renders difficult both the analysis of experimental data and the calculation of model systems.

The geometry of solid-solution interfaces creates other complications. The necessary exclusion of water molecules from the solid volume may lead to a greater degree of dehydration for coordinated species than in similar soluble complexes. Due to these and other steric effects, and to the influence of neighboring atoms, chemically similar groups in different geometrical configurations may exhibit very different affinities for solutes. Superimposed on the continuously varying coulombic effect, the chemically reactive groups at solid surfaces may then also exhibit an *intrinsic* variability in their coordinative properties.

Finally, solid surfaces in solution exhibit adsorptive properties in the absence of any specific chemical affinity for solutes. Obviously, electrostatically charged surfaces—whatever the origin of the charge—must attract ions of opposite charge. In addition, the mere exclusion of water molecules from the solid volume makes it energetically favorable for hydrophobic molecules to be situated at the solid-water interface. (The total free energy of the system includes a term for the surface tension which is decreased by organic adsorption.) These two adsorptive processes based on purely electrostatic and hydrophobic effects are those that have been given the most complete theoretical treatments, chiefly according to physical rather than chemical principles. As a result they are difficult to integrate into a general analytic description of the coordinative properties of solid surfaces in solution.

2.2 Adsorption Isotherms

The adsorption of a solute C on a solid X is traditionally quantified by an *adsorption density* parameter Γ (mol g^{-1}) representing the number of moles of C adsorbed per unit mass of X. Other dimensions such as $g\ g^{-1}$, mol mol^{-1}, and mol m^{-2} are of course possible and equivalent. Various theoretical and empirical mathematical expressions for Γ as a function of medium chemistry, chiefly the concentration of C, can be formulated. They are known as adsorption isotherms because they are strictly applicable to systems at constant temperature—a condition much more critical in the gaseous systems for which the formulae were originally developed.

Consider first the simplest case in which the adsorption process can be represented as a coordination reaction with $1:1$ stoichiometry

$$X + C = XC \tag{2}$$

where X is now taken to represent an adsorptive surface site on the solid. If a constant standard free energy of adsorption, ΔG_{ads}^0, can be assigned to this reaction, the mass law is written

$$\frac{\{XC\}}{\{X\}\{C\}} = K_{ads} = \exp\left(-\frac{\Delta G_{ads}^0}{RT}\right) \tag{3}$$

Making the further assumptions that (1) the activities of the surface species are proportional to their concentration (a key assumption throughout this chapter),* and (2) the number of adsorption sites X_T is fixed, this mass law can be combined with the mole balance for surface sites

$$TOTX = (X) + (XC) = X_T \tag{4}$$

to yield the expression

$$(XC) = X_T \frac{K_{ads}(C)}{1 + K_{ads}(C)} \tag{5}$$

Defining $\Gamma = (XC)/\text{solid mass}$ and $\Gamma_{max} = X_T/\text{solid mass}$, and substituting in Equation 5 yield the well-known expression of the *Langmuir isotherm*:

$$\Gamma = \Gamma_{max} \frac{K_{ads}(C)}{1 + K_{ads}(C)} \tag{6}$$

This hyperbolic expression is widely used to interpret adsorption data that demonstrate a saturation in surface coverage (Figure 8.2).

Although the great majority of adsorption data do exhibit a decrease in differential adsorption at increasing solute concentrations, not all show a clear maximum as does the Langmuir isotherm. To generalize the issue, let us write the adsorption reaction as a simple change in the nature of the species C.

$$C = C_{ads}; \quad \Delta G^{0\prime}$$

---

* Surface species concentrations can be expressed in moles per liter of solution, per gram of solid, per square meter of solid surface, or per mole of solid. For convenience we choose here the first of these possibilities (mol liter^{-1}) which provides a consistent set of units for all aqueous and adsorbed species.

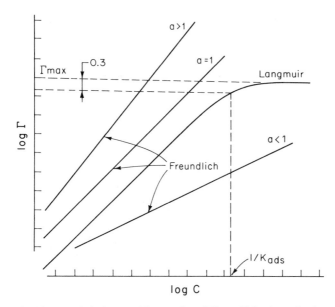

Figure 8.2 Characteristic shapes of Langmuir and Freundlich adsorption isotherms.

therefore

$$\Delta G = \Delta G^{0'} + RT \ln \frac{(C_{\text{ads}})}{(C)} \tag{7}$$

In order for the ratio of surface to bulk concentrations, $(C_{\text{ads}})/(C)$, to decrease at increasing concentrations, the adsorption reaction must somehow become less favorable, that is, $\Delta G^{0'}$ must increase. If we compare Equation 8 with that corresponding to the Langmuir equation,

$$\Delta G = \Delta G^{0} + RT \ln \frac{(XC)}{(X)(C)} \tag{8}$$

we see that the Langmuir equation provides for an increase in the effective free energy of adsorption $\Delta G^{0'}$ through the decrease in the activity of free surface sites (X) as the surface becomes saturated:

$$\Delta G^{0'} = \Delta G^{0} - RT \ln (X) = \Delta G^{0} - RT \ln [X_{T} - (C_{\text{ads}})] \tag{9}$$

As (C_{ads}) approaches X_{T}, $\Delta G^{0'}$ goes to infinity, and further adsorption becomes energetically impossible. There are obviously many alternative formulations to account for an increase in $\Delta G^{0'}$ as the adsorption reaction proceeds. Some are purely empirical, and some based on theories; all result in an isotherm expression—typically attached to a Germanic name—relating the concentration sorbed at the surface to that in the bulk solution.

Perhaps the simplest approach is to consider that the free energy of adsorption increases linearly with the (log of the) concentration of adsorbate. It is then not necessary to consider that the solid surface ever becomes saturated:

$$\Delta G^{0\prime} = \Delta G^0 + \alpha RT \ln(C_{ads}) \qquad (10)$$

At equilibrium

$$\Delta G = 0 = \Delta G^0 + (1 + \alpha)RT \ln(C_{ads}) - RT \ln(C)$$

therefore

$$\ln(C_{ads}) = -\frac{\Delta G^0}{(1 + \alpha)RT} + \frac{1}{1 + \alpha} \ln(C) \qquad (11)$$

Defining $a = 1/(1 + \alpha)$ and $\ln A = -a\Delta G^0/RT$, we obtain the exponential expression

$$(C_{ads}) = AC^a \qquad (12)$$

Dividing by the solid mass to obtain the adsorption density, this expression is that well-known as the Küster or Freundlich isotherm

$$\Gamma = A'C^a \qquad (13)$$

which accomodates many data sets over some range of concentrations.

Of the many other simple formulations that correspond to an increase in the free energy of adsorption with surface coverage, there is particular interest in those that yield a continuum between adsorption and precipitation ($\Delta G^{0\prime}$ varies from ΔG^0_{ads} to ΔG^0_{ppn}). The famous Brunauer, Emmett, and Teller (BET) isotherm for gases corresponds to such a formulation; it considers sorption in multiple layers as providing a continuum between monolayer gas adsorption on a solid and ultimate condensation of the gas.[11] Similar formulations for solute adsorption are possible and warranted by the available data. For example, one can consider, at the surface of a solid, the formation of a solid solution (see Section 6.2) whose composition varies continuously between that of the original solid (the adsorbent) and a pure precipitate of the adsorbate. Such a model may prove useful in describing adsorption at high surface coverage.

2.3   The Complexation Model for Adsorption

The basic adsorption model that we shall use throughout this chapter is a straightforward generalization of the Langmuir formulation: instead of considering that only one 1:1 surface complex can be formed, we allow for the presence of any number of such surface complexes with any reasonable stoichiometry. As in the Langmuir derivation, the fundamental assumption (hence the word "model") is that one may apply mass laws to the corresponding complexation reactions, using activities of surface species that are proportional to their concentrations.

In other words, we consider that aquatic surfaces contain functional groups with well-defined coordinative properties (surface sites), that for each type of surface site a total concentration X_T can be defined, and that a standard free energy of adsorption ΔG^0_{ads} can be assigned to each adsorption reaction. In order to account separately for the electrostatic adsorption term, the free energy of adsorption can be taken as the sum of an intrinsic and a coulombic free energy:

$$\Delta G^0_{ads} = \Delta G^0_{int} + \Delta G^0_{coul} \tag{14}$$

The corresponding adsorption constant is of course given as the product of two constants

$$K_{ads} = K_{int} \times K_{coul} \tag{15}$$

with the usual formula for each constant

$$K = \exp\left(-\frac{\Delta G^0}{RT}\right)$$

For the adsorption reaction

$$X + C = XC$$

the mass law is then written

$$\frac{(XC)}{(X)(C)} = K_{ads} = K_{int} \times K_{coul} \tag{16}$$

In the case of a polydentate surface complex the formulation of the mass law presents a minor difficulty. Consider, for example, a general adsorption reaction in which a solute C reacts with x surface sites X:

$$xX + C = X_xC \tag{17}$$

One may be tempted to write the mass law:

$$\frac{(X_xC)}{(X)^x(C)} = K_{ads} \tag{18}$$

However, the exponent x in this expression has doubtful validity. At a naive level it expresses that the probability of finding x sites together for C to react with is proportional to the xth power of the site concentrations. While this is valid for "independent" molecules in solution, it is certainly not true of fixed sites on a surface. A more satisfying approach is to consider that C is reacting with a single multidentate surface species X_x. At low surface coverage the concentration of such a species is simply given by $(X)/x$. As long as x is not too large (typically $x \leq 3$) and the surface is not close to saturation, an exponent of 1 is more satisfactory in the mass law expression:

$$\frac{(X_xC)}{(C)(X)/x} = K \tag{19}$$

The coefficient $1/x$ can of course be included in the equilibrium constant (as are the activity coefficients) leading to the expression:

$$\frac{(X_xC)}{(X)(C)} = K_{ads} = K_{int} \times K_{coul} \tag{20}$$

To describe adsorption processes in aquatic systems on the basis of the surface complexation model, it is then necessary and sufficient to define the stoichiometry and the intrinsic constants of the various surface species and to provide an expression for the coulombic term. To make matters simple, we shall consider initially only the intrinsic adsorption term, effectively ignoring the variable electrostatic interaction. This is equivalent to studying the adsorption process under conditions where the surface is (approximately) uncharged or the coulombic term is taken to be constant and included in the overall adsorption constant. The energetics and the stoichiometry of the adsorption process can then be studied without the complication of unspecific long-range coulombic interactions.

As we have noted for other chemical processes, over a certain range of conditions only one reaction often dominates the chemical behavior of a complex system. When this is true for adsorption processes—a situation often encountered in dilute solutions at fixed pH—the surface complexation model simply degenerates to the Langmuir formulation, a simple and satisfactory result in view of the wide applicability of the Langmuir isotherm.

3. ADSORPTION OF METALS AND LIGANDS ON AQUATIC PARTICLES

3.1 Oxide Surfaces

The bulk of the recent literature on adsorption in aquatic systems deals with metal oxides, principally those of silicon, aluminum, iron, and manganese. At the surface of a solid not all the coordination possibilities of the metal or the ligand atoms can be satisfied. If the metal is in excess of the ligand, the surface must exhibit metal-like coordination properties; conversely, if the ligand is in excess, the surface must behave like a ligand toward solutes at the interface. For dispersed suspensions of an oxide in water, both situations are true simultaneously because the medium itself is effectively an infinite source of exchangeable ligands. The metal atoms closest to the surface can always bind the appropriate number of hydroxyl groups from the solution to fulfill their coordination requirements (Figure 8.3). The surface can thus be seen either as ligand rich, if one considers the surface layer of hydroxyl ions as belonging to the solid, or metal rich, if the surface hydroxyl ions are seen as adsorbed solution species. Indeed, metal oxides are observed to adsorb both metals and ligands from solution.

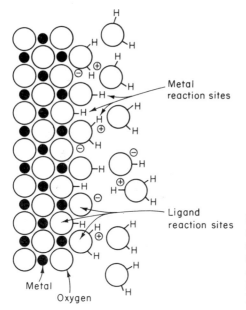

Metal reaction sites

Ligand reaction sites

Metal

Oxygen

Figure 8.3 A schematic model of the surface of metal oxides illustrating the sites of metal and ligand reactions. The surface oxygen atoms, hydroxyl ions, and water molecules act as exchangeable ligands, giving the surface both ligand-like and metal-like properties. Modified from Schindler, 1981.[12]

The most commonly accepted symbolism is to represent surface groups of metal oxides by a general hydrolyzed species XOH and acid-base reactions are written by considering the loss or gain of a proton rather than that of a hydroxyl ion, as is more usual for the soluble hydroxide species

$$XOH = H^+ + XO^-$$

$$XOH + H^+ = XOH_2^+$$

Replacing H^+ by other metals provides the symbolism for metal adsorption reactions, while ligand adsorption is represented by an exchange with the hydroxyl groups as illustrated in Figure 8.4. Several types of species may be formed in this manner, including ion pairs with major bulk electrolytes, surface chelates of metal ions (since the surface is clearly multidentate), and ternary surface complexes in which metal-ligand combinations are adsorbed at the surfaces.

Note. One should be careful to distinguish between the physical reality of these various surface species—a reality that is very difficult to ascertain experimentally—and their appropriateness in representing experimental data in the framework of a particular adsorption model. For example, one may argue whether or not ion pairs such as $XO \cdot Na$ or $XOH_2 \cdot Cl$ do or do not actually exist—and if they do exist, whether they are inner or outer sphere complexes as suggested by their name. The issue is more conceptual than experimental. What is an undeniable experimental fact is that the nature and the concentration of the background electrolyte affect the adsorption behavior of oxides, including their acid-base properties. As is the case for solutions, but without much of the

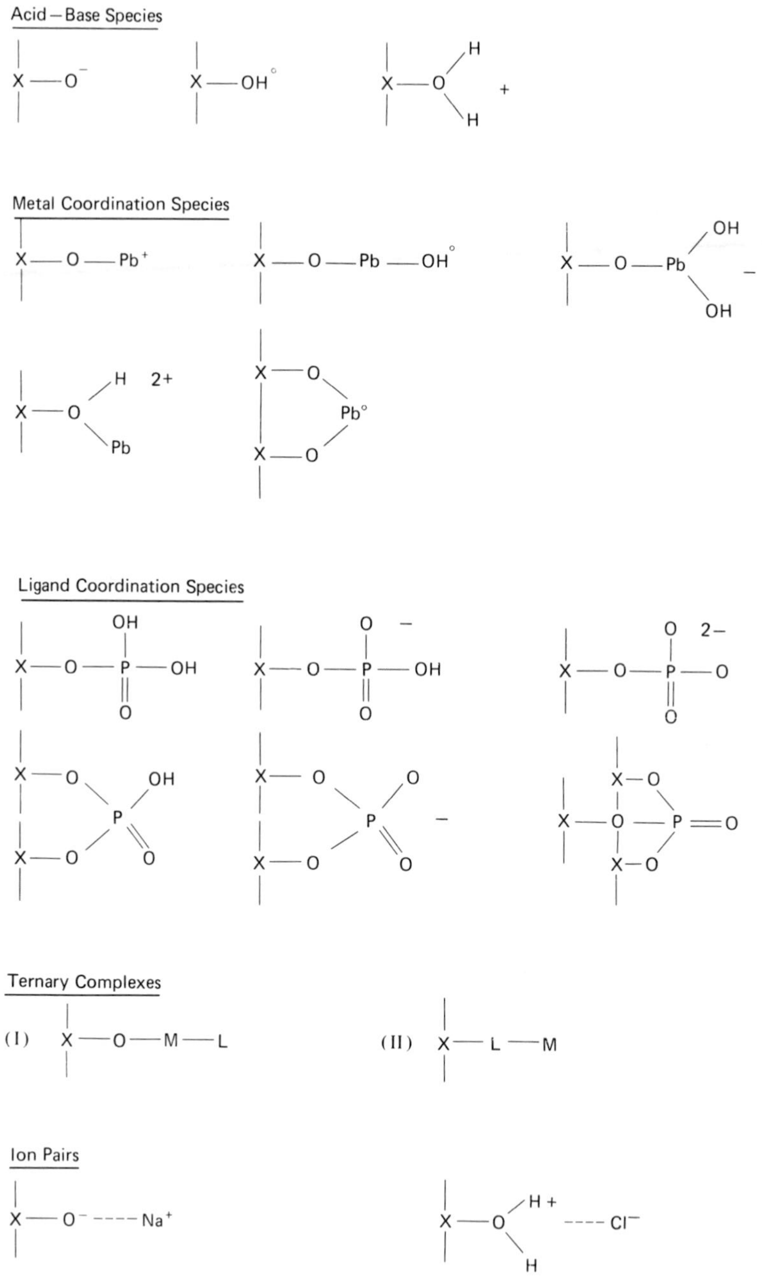

Figure 8.4 Some possible coordination species that may form between solutes and an oxide surface. Direct evidence for the existence of a particular surface species is generally lacking, and "reasonable" stoichiometries are thus chosen simply to fit available adsorption data.

experimental backing, one may wish to account for such effects by either ion pair formation or activity coefficient corrections (included in an effective equilibrium constant). From a practical point of view the advantage of more universal applicability of the ion pair model is largely offset by its greater complexity and its greater data requirement compared with the effective constant approach. The degree of hydration of sorbed ions has of course great theoretical importance for our understanding the nature of the interactions between solid surfaces, solutes, and water.

In terms of the choice of specific surface species such as XOM^+, $(XO)_2M^0$, or $XOMOH^0$, the main rationale for choosing one rather than another is to match experimental data describing the effects of pH and sorbate concentrations on adsorption. Since these are also the principal parameters affecting surface charge, the choice of the species and the values of the corresponding intrinsic constants are largely dependent on the way the coulombic interaction is evaluated. This is discussed in Section 5. To illustrate the effect of sorbate concentration and pH on adsorption, let us now consider the case of adsorption of a metal on a metal oxide.

Example 1. Adsorption of Lead on Alumina

Consider a solution containing a background electrolyte (say 0.1 M NaCl) and ad hoc concentrations of strong acid and base to adjust the pH in the range 4 to 7.

The solution contains a suspension of alumina (γ-Al_2O_3) and a low concentration of a lead salt. Define X_T = total number of adsorption sites on alumina (expressed here in mol liter^{-1}; X_T is obtained from the solid mass by multiplying by the specific site density, approximately 1.3×10^{-4} mol g^{-1} for γ-Al_2O_3), Pb_T = total concentration of lead in the system, and assume that the surface is not close to saturation ($Pb_T \ll X_T$). The surface reactions to be considered are

$$XOH_2^+ = H^+ + XOH; \qquad K_{a1}^{app} = 10^{-6.0} \qquad (21)$$

$$XOH = H^+ + XO^-; \qquad K_{a2}^{app} = 10^{-7.7} \qquad (22)$$

$$XOPb^+ = XO^- + Pb^{2+}; \quad K_1^{app} = 10^{-6.1} \qquad (23)$$

These constants have been chosen from the literature (Hohl and Stumm, 1976) as being applicable in this type of background electrolyte and in the pH range 5 to 6, which is of most interest to us as we shall see. They are *not* universal constants; they depend not only on the pH (see Section 5) and on the background electrolyte but also on the mode of preparation of the aluminum oxide. Since we are not considering here explicitly any electrostatic interactions between the surface and the solutes, the constants are adjusted to reflect some average degree of electrostatic repulsion and attraction.

In addition to surface reactions we have to consider the usual reactions in solution, in particular, the hydrolysis of lead:

$$Pb^{2+} + H_2O = PbOH^+ + H^+; \quad K_{OH} = 10^{-7.7} \tag{24}$$

Other hydrolysis species are not important below pH 7.

Choosing H^+, Pb^{2+}, and XOH_2^+ as principal components (see Tableau 8.1),

TABLEAU 8.1

	H^+	Pb^{2+}	XOH_2^+
H^+	1		
OH^-	-1		
Pb^{2+}		1	
$PbOH^+$	-1	1	
XOH_2^+			1
XOH	-1		1
XO^-	-2		1
$XOPb^+$	-2	1	1

the governing equations are

$$TOTXOH_2 = (XOH_2^+) + (XOH) + (XO^-) + (XOPb^+) = X_T \tag{25}$$

$$TOTPb = (Pb^{2+}) + (PbOH^+) + (XOPb^+) = Pb_T \tag{26}$$

$$(XOH) = 10^{-6.0}(H^+)^{-1}(XOH_2^+) \tag{27}$$

$$(PbOH^+) = 10^{-7.7}(H^+)^{-1}(Pb^{2+}) \tag{28}$$

$$(XOPb^+) = 10^{-7.6}(H^+)^{-2}(XOH_2^+)(Pb^{2+}) \tag{29}$$

Neglecting the last two terms of Equation 25 and substituting Equation 27 yields

$$(XOH_2^+) = X_T[1 + 10^{-6.0}(H^+)^{-1}]^{-1} \tag{30}$$

Introducing Equations 28 and 29 in Equation 26 and solving for (Pb^{2+}),

$$(Pb^{2+}) = Pb_T[1 + 10^{-7.7}(H^+)^{-1} + 10^{-7.6}(H^+)^{-2}(XOH_2^+)]^{-1} \tag{31}$$

Combining Equations 29 and 31,

$$(XOPb^+) = \frac{10^{-7.6}Pb_T(XOH_2^+)}{(H^+)^2 + 10^{-7.7}(H^+) + 10^{-7.6}(XOH_2^+)} \tag{32}$$

Finally an expression for the fraction of Pb adsorbed can be derived by introducing Equation 30:

Percent Pb adsorbed

$$= \frac{(XOPb^+)}{Pb_T} = \frac{X_T}{10^{7.6}(H^+)^2 + 10^{1.6}(H^+) + 10^{-6.1} + X_T} \tag{33}$$

The principal features of this formula are illustrated in Figure 8.5. For a fixed adsorbent concentration the percentage of lead adsorbed increases sharply above a critical pH (ca. 5 to 6 for the conditions of the example), and this "adsorption edge" occurs at lower pH's with increasing solid concentrations (Figure 8.5a). At constant pH the percentage is simply a hyperbolic function of the total adsorbent concentration, as shown in Fig. 8.5b. Because this example is calculated for low adsorption densities ($X_T \gg Pb_T$), the percentage of adsorbed metal is independent of the total metal concentration; otherwise, a typical Langmuir isotherm would be calculated at constant pH.

A simple approximate formula for the point of 50% adsorption is obtained from Equation 33:

$$X_T = 10^{7.6}(H^+)^2 + 10^{1.6}(H^+) + 10^{-6.1}$$

For high enough values of X_T, only the first term of the right-hand side is important: therefore

$$(H^+) = (10^{-7.6}X_T)^{1/2}$$

or, more generally

$$\left[\frac{X_T K_{a1} K_{a2}}{K_1}\right]^{1/2} \tag{34}$$

This approximate formula holds in general if neither the hydrolysis species of the adsorbing metal nor other metal surface species are important under the conditions of interest.

Example 1 illustrates the principal features of metal adsorption on oxides: adsorption increases sharply with pH and is significant even below the first pK_a of the surface acid groups. This is to say that the surface is positively charged when adsorption commences (XOH_2^+ is the dominant surface species) and that

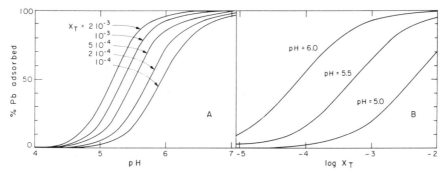

Figure 8.5 Adsorption of lead on alumina as calculated in Example 1: variation in Pb adsorption (A) as a function of pH for various surface site concentrations, and (B) as a function of surface sites concentration at various pH's. Note the effect of surface site concentrations on the "adsorption edge" in panel A.

adsorption takes place *against* an electrostatic repulsion. The extent of this electrostatic repulsion is of course important to assess; in our example it is accounted for by the choice of effective constants.

Even without considering the difficult issue of the coulombic term, an adsorption problem like that of Example 1 can be quite complicated. If adsorption takes place in a high enough pH range, several hydrolysis species of the metal may have to be considered both on the surface and in the bulk solution. In general, one should also include the soluble species of the metal that is in the solid phase (e.g., $Al(OH)_2^+$). Finally, the presence of complexing ligands and competing metals may render the problem too cumbersome for hand calculations.

Example 2. Adsorption of Phosphate on Alumina

As a contrast to Example 1, consider the adsorption of phosphate on aluminum hydroxide. The pH range of interest is somewhat higher than for lead adsorption (ca. 5 to 9 pH), and, as will be seen later, the effective acidity constants are consequently larger:

$$XOH_2^+ = H^+ + XOH; \quad K_{a1}^{app} = 10^{-6.8} \tag{21}$$

$$XOH = H^+ + XO^-; \quad K_{a2}^{app} = 10^{-8.7} \tag{22}$$

In order for this model system to exhibit the appropriate adsorption properties, several phosphate surface species need to be considered:[14]

$$XOH_2PO_3^0 + H_2O = XOH_2^+ + H_2PO_4^-; \quad K^{app} = 10^{-4.5} \tag{35}$$

$$XOHPO_3^- + H^+ + H_2O = XOH_2^+ + H_2PO_4^-; \quad K^{app} = 10^{+1.6} \tag{36}$$

$$(XO)_2HPO_2^0 + H^+ + 2H_2O = 2XOH_2^+ + H_2PO_4^-; \quad K^{app} = 10^{+1.8} \tag{37}$$

$$(XO)_2PO_2^- + 2H^+ + 2H_2O = 2XOH_2^+ + H_2PO_4^-; \quad K^{app} = 10^{+9.2} \tag{38}$$

These surface reactions and their apparent equilibrium constants are chosen so as to obtain the same general dependency of adsorption on concentration and pH, as is observed experimentally. Neither the species nor the constants should be taken literally.

Finally, the acid-base properties of phosphate have to be considered. In the pH range 5 to 9 only the second acidity constant is important:

$$H_2PO_4^- = H^+ + HPO_4^{2-}; \quad K_a = 10^{-7.0} \tag{39}$$

With $H_2PO_4^-$ and XOH_2^+ as principal components (see Tableau 8.2), the two relevant mole balance equations are written

$$TOTX = (XOH_2^+) + (XOH) + (XO^-) + (XOH_2PO_3) + (XOHPO_3^-)$$
$$+ 2[(XO)_2HPO_2] + 2[(XO)_2PO_2^-] = X_T \tag{40}$$

$$TOTP = (H_2PO_4^-) + (HPO_4^{2-}) + (XOH_2PO_3) + (XOHPO_3^-)$$
$$+ [(XO)_2HPO_2] + [(XO)_2PO_2^-] = P_T \tag{41}$$

TABLEAU 8.2

	H^+	$H_2PO_4^-$	XOH_2^+
H^+	1		
OH^-	-1		
$H_2PO_4^-$		1	
HPO_4^{2-}	-1	1	
XOH_2^+			1
XOH	-1		1
XO^-	-2		1
XOH_2PO_3		1	1
$XOHPO_3^-$	-1	1	1
$(XO)_2HPO_2$	-1	1	$1, 2^a$
$(XO)_2PO_2^-$	-2	1	$1, 2^a$

^{a} The coefficient 1 is utilized for mass laws, and 2 for the mole balance.

In the concentration ranges where X_T and P_T are similar—as is typical of most experimental studies of anion adsorption which, by contrast to metal adsorption, are usually performed at high surface coverage—no term in these two equations can be neglected, and the algebra of the problem is a bit unwieldy. Nonetheless, since there is no high exponent in any of the mass laws corresponding to Reactions 35 through 39, for any given set of conditions (pH, X_T, P_T), the problem is amenable to an explicit, if messy, solution:

1. Replace all species in the mole balances given by Equations 40 and 41 by their mass law expressions as a function of $H_2PO_4^-$ and XOH_2^+:

$$P_T = (H_2PO_4^-)[a + b(XOH_2^+)] \qquad (42)$$

$$X_T = (XOH_2^+)[c + d(H_2PO_4^-)] \qquad (43)$$

where

$$a = 1 + \frac{10^{-7}}{(H^+)} \qquad (44)$$

$$b = 10^{4.5} + \frac{10^{-1.6}}{(H^+)} + \frac{10^{-1.8}}{(H^+)} + \frac{10^{-9.2}}{(H^+)^2} \qquad (45)$$

$$c = 1 + \frac{10^{-6.8}}{(H^+)} + \frac{10^{-15.5}}{(H^+)^2} \qquad (46)$$

$$d = 10^{4.5} + \frac{10^{-1.6}}{(H^+)} + \frac{2 \times 10^{-1.8}}{(H^+)} + \frac{2 \times 10^{-8.7}}{(H^+)^2} \qquad (47)$$

2. Solve Equation 43 for (XOH_2^+), and substitute into Equation 42 to obtain the quadratic:

$$A(H_2PO_4^-)^2 + B(H_2PO_4^-) - C = 0 \qquad (48)$$

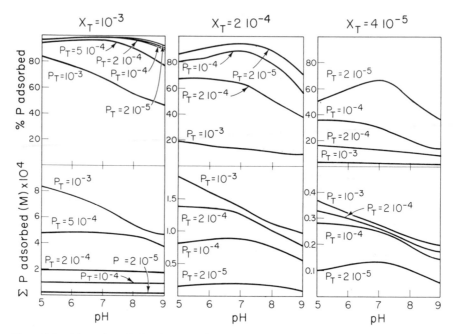

Figure 8.6 Adsorption of phosphate on alumina as calculated in Example 2. Note the general decrease in adsorption at high pH (in contrast to Figure 8.5) and the complex effects of the concentrations of total phosphate, P_T, and total surface sites, X_T, on the pH dependency of the fractional P adsorbed.

where

$$A = ad \tag{49}$$

$$B = b \cdot X_T - d \cdot P_T + ac \tag{50}$$

$$C = c \cdot P_T \tag{51}$$

3. Solve Equation 48 for $(H_2PO_4^-)$, and obtain the adsorbed phosphate concentration by substituting into Equation 52:

$$P_{ads} = \frac{b(H_2PO_4^-)}{c + d(H_2PO_4^-)} X_T \tag{52}$$

The principal feature of the results shown on Figure 8.6 for various total phosphate and surface site concentrations is a decrease in phosphate adsorption with pH. This is due in part to the acid-base properties of the surface and the ligand; for example, at pH just above 7, the principal surface reaction consumes one proton:

$$XOH + HPO_4^{2-} + H^+ = XOHPO_3^- + H_2O$$

It is also due in part to a shift from the monodentate to the bidentate surface species, which results effectively in a halfing of the available surface sites. This

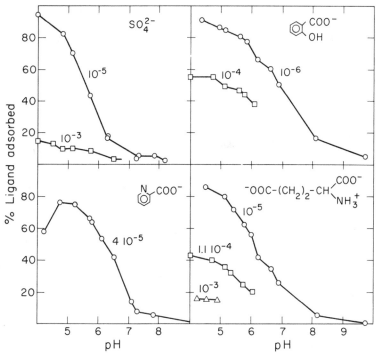

Figure 8.7 Adsorption of sulfate, salicylate, picolinate, and glutamate on amorphous iron oxide $(Fe(OH)_3) = 10^{-3}$ M, except 6.9×10^{-4} M for picolinate). From Davis and Leckie, 1978.[15]

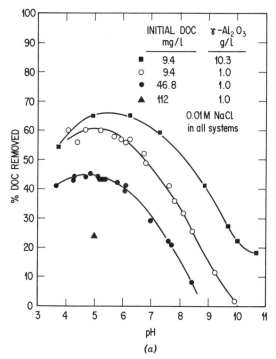

Figure 8.8 Adsorption of natural organic matter on alumina: (*a*) percent DOC adsorbed as a function of pH. (*continued*)

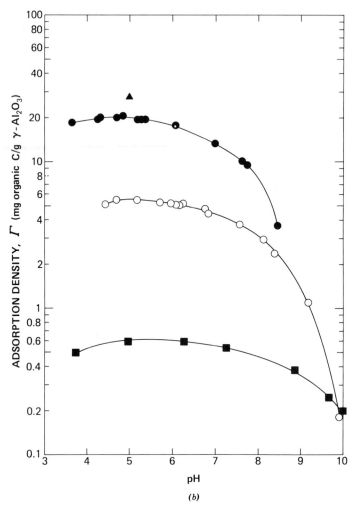

Figure 8.8 (b) adsorption density (DOC adsorbed per mass of solid) as a function of pH, (continued)

second effect is seen even in cases where the phosphate is in large excess of the reactive sites and the surface is saturated (e.g., $P_T = 10^{-3}$ M, $X_T = 2 \times 10^{-4}$ M). At low pH the adsorption often reaches a maximum or even decreases, due to either exhaustion of the phosphate from solution (e.g., $X_T = 2 \times 10^{-4}$ M, $P_T = 10^{-4}$ M) or a proton release according to the reaction

$$XOH_2^+ + H_2PO_4^- = XOHPO_3^- + H^+ + H_2O$$

In this situation the position of the adsorption maximum is controlled by the relative acidities of the surface hydroxyl, the ligand in solution, and the ligand bound to the surface.

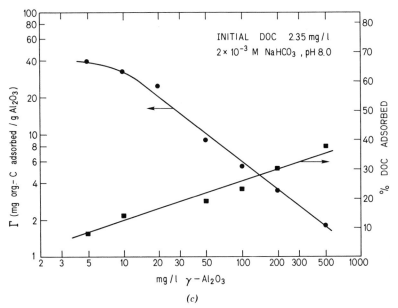

Figure 8.8 (*c*) percent DOC adsorbed and adsorption density as a function of solid concentration. From Davis, 1982.[16]

The characteristics of adsorption illustrated in Example 2 are general for all types of ligands and oxide surfaces. For example, Figure 8.7 shows how the adsorption of sulfate, salicylate, picolinate, and glutamate on amorphous iron oxide all exhibit pronounced maxima at low pH.[15] Adsorption of natural organic material on γ-Al$_2$O$_3$ (Figure 8.8) also exhibits the same characteristic pH dependency, and can be modeled quantitatively by assuming the same type of surface complexes with the surface aluminum ion.[16]

If we consider the adsorption of natural organic matter as a function of suspended solid concentration (Figure 8.8*b*), we find that even at the relatively high pH of 8, the typical DOC concentration of a few milligrams per liter is sufficient to saturate the available inorganic surface on a few milligrams per liter of suspended solid. (Note that the value of 40 mg org·C g^{-1} Al$_2$O$_3$ is consistent with a site density of about 10^{-4} mol g^{-1} Al$_2$O$_3$ and a concentration of functional groups of about 10 meq g^{-1} org.C. Thus in this case approximately one-quarter of the organic functional groups would have reacted with the surface groups.) This result suggests that, while only a fraction of the organic matter in aquatic systems might be adsorbed on suspended solids, the solid surface may be entirely covered by organic material and hence exhibit physical and chemical properties characteristic of organic matter.

3.2 Organic Surfaces

Although oxides are important constituents of the suspended material in natural waters and they provide convenient model systems for laboratory studies, bare (i.e., hydrated and hydrolyzed) oxide surfaces probably do not constitute the principal adsorption sites for trace aquatic constituents. It is now thought that organic surfaces normally contribute most of the functional groups for metal adsorption in aquatic systems,[2,17,18] either in the form of organic particles proper (live organisms, dead remains, condensed humic material) or in the form of organic coatings on inorganic (oxide) particles, as discussed earlier. The predominance of organic surfaces is most simply evidenced by the negative electrostatic charge of aquatic particles which is characteristic of relatively strong organic acid surface groups while (bare) oxides—except for SiO_2—are positively charged near neutral pH. However, at present much less is known about adsorption on organic surfaces than on metal oxides, and a discussion of this subject is necessarily tentative and not easily amenable to meaningful numerical examples.

In the case of organic particles proper, the surface coordination of metals by surface ligands such as carboxyl, phenolic, or sulfhydryl groups is conceptually similar to the equivalent process in solution; the essential difference is that brought about by the long-range coulombic interactions, a subject treated in the next section. The situation is quite different for organic coatings on inorganic surfaces, for the organic ligands themselves may react reversibly with the surface. The system, inorganic surface-organic ligand-metal ion (X-Y-M), may thus yield four general types of surface reactions:

1   Metal adsorption
$$X + M = XM$$

2   Ligand adsorption
$$X + Y = XY$$

3   Metal complexation by the adsorbed ligand, [i.e., formation of ternary surface complex II]
$$XY + M = XYM$$

4   Ligand coordination to the adsorbed metal ion [i.e., formation of ternary surface complex I]
$$XM + Y = XMY$$

In addition the metal and the organic ligand may react in solution according to the usual complexation reactions:

$$M + Y = MY$$

For each of these types of reactions several different species may be formed with various stoichiometries and acid-base chemistries.

If we compare the adsorption of metal ions on oxide surface, in the presence and in the absence of natural organic compounds, the net metal adsorption may then be increased by the organic ligands due to a predominant formation of ternary surface complexes, or it may be decreased due to competition of the ligand for surface sites or, more important, due to complexation in solution

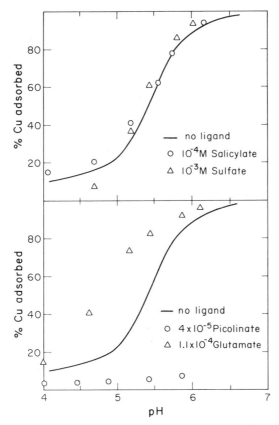

Figure 8.9   Adsorption of copper on alumina in the presence of various ligands. (The adsorption of the ligands is shown in Figure 8.7.) The decrease in copper adsorption in the presence of picolinate is explained by the formation of the nonadsorbable Cu-picolinate complex in solution. The increase in copper adsorption in the presence of glutamate is explained as the formation of a type II ternary complex, X-L-M, on the surface. From Davis and Leckie, 1978.[15]

(competition between the adsorbing surface and the organic ligand). In model systems both of these alternatives are in fact observed, as is the implicit third one: no effect on metal adsorption.

For example, Davis and Leckie (1978)[15] showed that adsorption of copper on amorphous iron oxide is unaffected by sulfate or salicylate, prevented by the presence of picolinic acid, and increased at low pH by the presence of glutamate (Figure 8.9). These results can be rationalized quantitatively by considering the principal copper complexation reactions for these systems (mixed constants corrected for $I = 0.1\ M$):

$$Cu^{2+} + \equiv FeOH_2^+ = \equiv FeO \overset{H}{\underset{Cu}{\diagup}}{}^{2+} + H^+; \quad {}^*K_{ads} = 10^{-2.4}$$

This reaction is chosen to reproduce approximately the graph of Cu^{2+} adsorption in the absence of ligand; the effective constant $*K_{ads}$ includes the specific area of the solid so that the mass law equation can be written with $(Fe(OH)_3 \cdot s)$ instead of $(\equiv FeOH_2^+)$.

$$Cu^{2+} + SO_4^{2-} = CuSO_4; \qquad\qquad K = 10^{1.5}$$

$$Cu^{2+} + H\ Sal^- = Cu\ Sal + H^+; \qquad *K' = 10^{-2.8}$$

$$Cu^{2+} + H\ Pic = Cu\ Pic^+ + H^+; \qquad *K' = 10^{+2.7} \quad \text{(below pH = 5.4)}$$

$$Cu^{2+} + H\ Glut^- = Cu\ Glut + H^+; \quad *K' = 10^{-1.7} \quad \text{(above pH = 4.4)}$$

With the ligand concentrations present in these experiments $[(Fe(OH)_3 \cdot s) = 10^{-3}\ M; (SO_4^{2-}) = 10^{-3}\ M; (H\ Sal^-) = 10^{-3}\ M; (H\ Pic) = 10^{-4.4}\ M; (H\ Glut^-) = 10^{-4}\ M]$, we expect the iron oxide surfaces to outcompete sulfate and salicylate but not picolinate for complexation of the cupric ion. This is indeed observed in Figure 8.9. Note that the concentration of copper ($10^{-6}\ M$) is much smaller than that of any of the ligands and thus cannot affect ligand speciation. Although sulfate, salicylate, and picolinate are largely adsorbed at low pH, their functionalities apparently do not allow formation of type II ternary surface complexes (XYM). The acid functional groups of these molecules are bound to the surface hydroxyl groups. On the contrary, glutamate has coordinative functional groups at both ends of the molecules, and formation of a type II ternary surface complex is evident by the increased Cu adsorption at low pH. At higher pH, when glutamate desorbs, its affinity for copper is not quite high

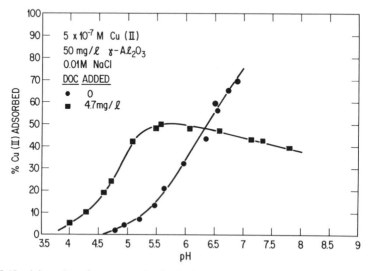

Figure 8.10   Adsorption of copper on alumina in the presence of natural organic matter. The addition of dissolved organic carbon to the system enhances adsorption at low pH (presumably due to type II ternary surface complexes) and decreases it at high pH (presumably due to formation of nonadsorbed complexes). From Davis, 1983.[19]

enough to outcompete the oxide surface. The results of Figure 8.9 provide no evidence for or against formation of type I ternary surface complexes (XMY) whose existence is supported by other data.

In the absence of sufficient information to develop a meaningful theoretical model for the effect of natural organic material on the adsorption of metal ions in natural water, consider the data of Davis on the adsorption of copper on γ-Al$_2$O$_3$ (Figure 8.10).[19] At low pH the presence of natural organic matter drastically increases the copper adsorption, while it actually decreases it at higher pH. These observed effects can be rationalized by the formation of type II ternary surface complexes at low pH where the organic is largely adsorbed (see Figure 8.8) and by complexation of Cu by the desorbed organic ligand at high pH. (The same effect might have been observed with glutamate in Figure 8.9 if the ligand concentration had been higher.)

A study of Cu complexation with the same natural organic matter in solution showed very high affinity of the ligand for copper (similar to Example 8 in Chapter 6), so that above pH $= 5$ all the copper should be complexed (note again the large excess of the ligand over the metal in this example: $Y_T = 5 \times 10^{-5}$ M versus Cu$_T = 5 \times 10^{-7}$ M). The initial increase in Cu adsorption with pH in the presence of the organic matter is thus interpreted as the increase in complexation of the metal by the adsorbed ligand due to the usual acid-base effects (increase in the effective complexation constant). To account quantitatively for the eventual plateau in adsorption, it is necessary to estimate the relative contribution of surface reactions **1**, **3**, and **4** to the overall adsorption process. This is not possible in the absence of further information since the presence of the metal must affect the ligand adsorption in type II ternary surface complexes, and vice versa for type I complexes.

4.   ELECTROSTATIC INTERACTIONS ON SURFACES

In this chapter we have repeatedly pointed out the importance of long-range electrostatic interactions in the adsorption process. In order to obtain a quantitative description of such interactions—which are most responsible for making coordination at surfaces different from coordination among solutes—we start with a system in which only electrostatic forces are at play, by considering an ideal (nonreactive) electrolyte solution containing a charged surface. By using the principle of additivity of energies, we can then superimpose the results with any chosen set-of surface complexation reactions and thus obtain a complete adsorption model that considers both coulombic and chemical interactions. The calculation of the electrostatic interactions—the Gouy-Chapman theory—and its superposition on the chemical interactions—the Stern theory—are formally and conceptually similar to the Debye-Hückel theory of nonideal behavior of electrolytes: the same long-range electrostatic interactions are considered; the same mixture of electrostatic and thermodynamics is used for the derivation; the same approximations (point charges in a continuous uniform medium) are involved.

4.1   Distribution of Ions near a Charged Surface:
The Gouy-Chapman Theory

For simplicity let us consider a one-dimensional system consisting of an inert electrolyte solution neighboring an infinite flat surface at a given electrical potential ψ_d as compared to a reference potential in the bulk solution, $\psi_\infty = 0$. Qualitatively, we know that the charged surface will attract ions of opposite charge and repulse ions of like charge, resulting in a nonuniform equilibrium distribution of ions in the vicinity of the surface. Our objective is to obtain a quantitative description of this distribution.

Following the general methodology of Chapter 2, we can express that, at equilibrium, there should be no work gained or spent in transporting an ion from infinity to any distance x near the surface; the electrostatic work must be balanced exactly by the "concentration" work. This is done most readily by using the condition of uniformity of the molar free energy of the ion (in its extended form including the electrostatic term; see Section 1.91 in Chapter 2):

$$\mu_i = \mu_i^0 + RT \ln [S_i(x)] + Z_i F \psi(x) = \text{constant} = \mu_i(x) \tag{53}$$

The term $[S_i(x)]$ represents the concentration of i (at the distance x from the surface) used here instead of the activity; the activity coefficient is taken to be constant and included in the standard free energy μ_i^0. Z_i is the charge number of i; R, T, and F have their usual meaning (gas constant; absolute temperature; Faraday constant). $\psi(x)$ is the electrical potential at the distance x from the surface. With the condition $\psi(\infty) = 0$, Equation 53 can be rearranged to provide an exponential expression of the ion concentrations as a function of the electrical potential:

$$[S_i(x)] = [S_i(\infty)] \exp\left[-Z_i \frac{F}{RT} \psi(x) \right] \tag{54}$$

If the potential is small (i.e., $\psi_d \ll RT/F = 25$ mV), this exponential expression can be linearized by using the Taylor expansion:

$$[S_i(x)] \cong [S_i(\infty)]\left[1 - Z_i \frac{F}{RT} \psi(x) \right] \tag{55}$$

The problem now—and the only difficulty in this derivation—is to obtain an expression of the electrical potential as a function of distance from the surface. This is achieved by introducing the fundamental law of electrostatics that relates the variation in electrical potential with the charge density; i.e., Poisson's equation which, in one dimension, is written

$$\frac{d^2 \psi(x)}{dx^2} = -\frac{\rho(x)}{\varepsilon} \tag{56}$$

where ρ is the charge density (in Coulombs cm^{-3}) and ε is the electrical permittivity of the medium ($\varepsilon \cong 7 \times 10^{-12}$ Coulombs V^{-1} cm^{-1} in water). The charge density at any point x is obtained by summing up the ion concentrations at that point:

$$\rho(x) = \sum_i Z_i F[S_i(x)] \tag{57}$$

therefore

$$\frac{d^2\psi(x)}{dx^2} = -\frac{F}{\varepsilon} \sum_i Z_i[S_i(x)] \tag{58}$$

While a general solution can be obtained for a symmetrical electrolyte ($Z_{\text{cation}} = -Z_{\text{anion}}$) by introducing Equation 54 into 58 and integrating, let us consider instead the simple case of a small surface potential and introduce the linear expression given by Equation 55 into 58:

$$\frac{d^2\psi(x)}{dx^2} = -\frac{F}{\varepsilon} \left\{ \sum_i Z_i[S_i(\infty)] - \sum_i Z_i^2 \frac{F}{RT} \psi(x)[S_i(\infty)] \right\} \tag{59}$$

The term $\sum_i Z_i[S_i(\infty)]$ must be null as a condition of electroneutrality of the bulk solution, and we are left with an ordinary second-order differential equation:

$$\frac{d^2\psi(x)}{dx^2} = \kappa^2 \psi(x) \tag{60}$$

where

$$\kappa^2 = \frac{F^2}{\varepsilon RT} \sum_i Z_i^2[S_i(\infty)] \tag{61}$$

and therefore

$$\kappa^2 = 2 \frac{F^2}{\varepsilon RT} I \quad (I = \text{ionic strength of bulk solution}) \tag{62}$$

$$\frac{1}{\kappa} = 0.28 I^{-1/2} \quad \left(\frac{1}{\kappa} \text{ in nm and } I \text{ in } M \right) \tag{63}$$

With the boundary conditions $\psi(0) = \psi_d$ and $\psi(\infty) = 0$, the solution of Equation 60 is a simple exponential decrease of the potential as a function of distance from the surface:

$$\psi(x) = \psi_d \exp(-\kappa x) \tag{64}$$

therefore

$$[S_i(x)] = [S_i(\infty)] \left[1 - Z_i \frac{F}{RT} \psi_d \exp(-\kappa x) \right] \tag{65}$$

As shown on Figure 8.11, the distance $1/\kappa$ indicates the extent of the electrostatic influence of the surface in the solution, and it is called the double-layer thickness. (The Gouy-Chapman theory is also called the double-layer theory because it refers to the surface layer of charges and the bulk solution layer of

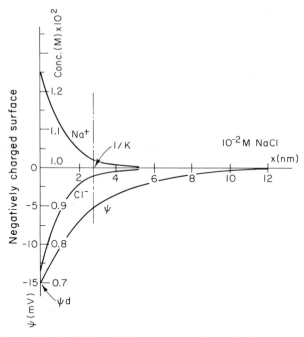

Figure 8.11 Theoretical distributions of ion concentrations and electrical potential near a charged surface. (Calculated for the small potential approximation.) $1/\kappa$ is the thickness of the double layer.

counter charges.) In dilute solution, the double-layer thickness can be relatively large ($1/\kappa = 2.8$ nm for $I = 10^{-2}$ M), but it is "compressed" in high ionic strength systems ($1/\kappa = 0.4$ nm for $I = 0.5$ M).

Although we chose for simplicity in the derivation to characterize the surface by its electrical potential ψ_d, it is sometimes more convenient to characterize it by its surface charge σ_d. The expression relating surface charge and surface potential is obtained by considering that the charge on any area of the surface must be balanced by the charge in the corresponding prismatic volume in solution:

$$\sigma_d = -\int_0^\infty \rho(x)\, dx \tag{66}$$

therefore

$$\sigma_d = \int_0^\infty \varepsilon \frac{d^2\psi(x)}{dx^2}\, dx = -\varepsilon \left[\frac{d\psi(x)}{dx}\right]_{x=0} \tag{67}$$

For small potentials the derivative of the potential at the surface is obtained from Equation 64:

$$\left[\frac{d\psi(x)}{dx}\right]_{x=0} = -\kappa\psi_d$$

therefore

$$\sigma_d = \varepsilon\kappa\psi_d \tag{68}$$

$$\sigma_d = 2.5I^{1/2}\psi_d \quad (\sigma_d \text{ in Coulombs m}^{-2}; I \text{ in } M; \psi_d \text{ in V}) \tag{69}$$

The surface charge necessary to generate a given surface potential increases with the square root of the ionic strength. Equation 69 is a linear relationship between surface and surface potential; the coefficient of proportionality thus has the dimensions of a capacitance:

$$C_{GC} = \varepsilon\kappa = 2.5I^{1/2} \quad (C_{GC} \text{ in farads m}^{-2} \text{ and } I \text{ in } M) \tag{70}$$

This Gouy-Chapman capacitance is on the order of 1 to 0.03 farad m^{-2}, depending on the ionic strength of the solution ($I = 10^{-1}$ to 10^{-4} M). Note that the simple linear relationship between σ_d and ψ_d is due to the approximations introduced for small surface potentials ($\psi_d < 25$ mV). For higher values of the potential (and for a symmetrical electrolyte) the relationship involves a hyperbolic sine function, and the apparent capacitance of σ_d/ψ_d varies with the potential.

4.2 Electrostatic Effects on Adsorption: Sternian Adsorption Models

In Section 4.1 we saw how purely electrostatic effects result in the adsorption (positive and negative) of ions in the "diffuse layer" (also called the Gouy layer) near a charged surface. In fact such an effect is relatively unimportant since ions are affected strictly on the basis of their charge and major electrolyte ions thus contribute practically all the counter charge excess near the surface. In other words, unless surface potentials are very high and available surface areas very large, electrostatic adsorption in the diffuse layer results simply in a minute spatial rearrangement of the major ions.

What is vastly more interesting to us is the influence of the electrostatic effect on the specific (i.e., "chemical") adsorption of trace solutes. This problem is apparently easily solved by considering that the total adsorption energy is the sum of the chemical (intrinsic) and the electrostatic (coulombic) energy terms, as originally defined by the complexation model (see Section 2.3):

$$\Delta G^0_{ads} = \Delta G^0_{int} + \Delta G^0_{coul}$$

Recall from Example 1 that adsorption of cations on oxides usually takes place against an electrostatic repulsion; the electrostatic contribution to the total energy of adsorption can thus often be negative ($\Delta G_{coul} > 0$).

The coulombic term is then calculated by considering the electrostatic work involved in transporting the soluble reactants from the bulk solution to the surface and the soluble products from the surface to the bulk solution:

$$\Delta G^0_{coul} = F\Delta Z\psi_0 \tag{71}$$

where ψ_0 is the surface potential (compared to a reference potential of 0 in the bulk solution) and $F\Delta Z$ is the change in charge of the surface species due to the adsorption reaction. For example,

$$XOH_2^+ = H^+ + XOH; \qquad \Delta Z = -1; \quad \Delta G_{coul}^0 = -F\psi_0$$

$$XOPb^+ = XO^- + Pb^{2+}; \qquad \Delta Z = -2; \quad \Delta G_{coul}^0 = -2F\psi_0$$

$$XOHPO_3^- + H^+ + H_2O$$
$$= XOH_2^+ + H_2PO_4^-; \quad \Delta Z = +2; \quad \Delta G_{coul}^0 = +2F\psi_0$$

What we have done is to consider that the general adsorption reaction

$$XA^{m+} + B^{n+} = XB^{p+} + A^{q+}; \quad \Delta G_{ads}^0 = \Delta G_{int}^0 + \Delta G_{coul}^0$$

is effectively the sum of three reactions:

1   First the soluble reactant moves to the surface

$$B_{bulk}^{n+} = B_{surface}^{n+}; \quad \Delta G_1^0 = \Delta G_{coul,B}^0 = +nF\psi^0$$

2   Then the reaction itself proceeds at the surface

$$XA^{m+} + B_{surface}^{n+} = XB^{p+} + A_{surface}^{q+}; \quad \Delta G_2^0 = \Delta G_{int}^0$$

3   Finally, the soluble product moves to the bulk solution

$$A_{surface}^{q+} = A_{bulk}^{q+}; \quad \Delta G_3^0 = \Delta G_{coul,A}^0 = -qF\psi^0$$

and
$$\Delta G_{coul}^0 = \Delta G_{coul,A}^0 + \Delta G_{coul,B}^0 = (p - m)F\psi^0 = (n - q)F\psi^0$$

Although so far this is all reasonably straightforward, and elegant, we have not yet faced the critical question: What is the surface potential ψ_0 that is applicable to specifically adsorbing ions? This is a much more difficult question than it may seem at first because, according to the Gouy-Chapman theory, the potential near a charged surface decreases over distances that are comparable to the diameter of ions $(1 - 10 \text{ nm})$. Under such conditions we have to decide how close adsorbed ions can get to the adsorbing surface (Are they hydrated or not?); where the ionic charge is in fact located; how planar or irregular, at *the molecular scale*, the surface really is; and more fundamentally, whether or not electrostatic principles that are applicable to continuous media are applicable at the molecular scale where discontinuity would seem to be the rule. Note that all these difficulties are applicable to the first layer of adsorbed ions near the surface, whether they are specifically adsorbed or simply attracted by coulombic forces. The subscript d that we used for the surface charge and surface potential in our derivation of the Gouy-Chapman theory may thus indicate a certain distance from the actual solid surface rather than the surface itself.

A very compelling reason for paying attention to these theoretical difficulties in applying the Gouy-Chapman theory to specifically sorbed ions is that the corresponding equations simply do not fit the available data. Arguably one could use the potential ψ_d as calculated by Equation 69 from the surface charge σ_d (itself calculated from the charge of the adsorbed species; see Section 5) as providing an estimate of the coulombic term that is applicable to specifically adsorbing species. The resulting adsorption model, with *no* fitting parameter for the coulombic term, cannot generally be made to agree with experimental data.

The Stern theory consists basically of taking Alexander's solution to this Gordian knot by considering that (1) the Gouy-Chapman theory is applicable at some distance d from the surface to all ions that interact with the surface through purely electrostatic effects, and (2) in the layer between the surface and this distance d, the potential drops linearly. The existence of this compact or Stern layer near the surface is justified on the basis of the size of the specifically adsorbing molecules (although d is a fitting parameter, not one calculated from ionic radii), and the assumption of a linear potential drop is made for simplicity sake.

In order for this simple idea to be translated into equations, some physical model of the adsorption process is necessary: the location of the charge of the specifically adsorbing ions—those that have some intrinsic chemical affinity for the surface—must be precisely defined. This is where the confusion begins. Both the original literature on this topic[20-22] and recent discussions of adsorption models[23] provide cartoons of adsorption layers, potential planes, equivalent flat capacitors, and so on, that often do not correspond to the descriptive prose or to the resulting equations, and sometimes are even inconsistent with fundamental physical laws. Interestingly enough the mathematical results are almost always equivalent; when they are not, they are usually numerically indistinguishable from each other. This is undoubtedly a reflection of the fact that all reasonable one-parameter models for the electrostatics of adsorption necessarily end up with a simple proportionality between surface charge and surface potential. In this sense all such models are in fact equivalent, "Sternian," and equally valid or invalid regardless of the propriety or incongruity of the physical image of adsorption that they attempt to capture. With this sobering thought in mind, let us describe such a "Sternian" model, one that is in some way iconoclastic because it places the compact layer on the opposite side of the mean plane of specific adsorption, compared to Stern's own model.

In accordance with our general coordination approach to surface adsorption, we first consider that the surface charge is the result of specific surface adsorption—including that of H^+ and OH^-—and that we cannot distinguish, in principle, between some inherent surface charge and that applicable to specifically sorbed ion. In other words, to use the traditional language of surface chemistry, we are considering here that all specifically sorbed ions are potential determining. The key to the model is then to decide on the spatial distribution of this surface charge and that of the balancing diffuse layer. As illustrated in Figure 8.12, three reasonable choices are:

1 That each specifically sorbed ion is positioned at a mean adsorption plane (the mean position of the ion's charge to which the potential is applied, as if it were a point charge) characteristic of that ion. In its strict

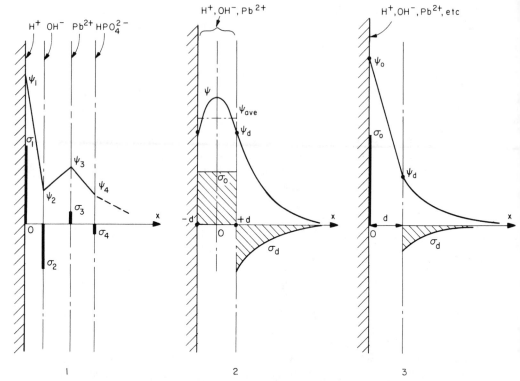

Figure 8.12 Three hypothetical distributions of ions near an adsorbing surface: (1) discrete mean planes of adsorption characteristic of each ion, (2) uniform distribution of ions in a layer adjacent to the surface, and (3) unique mean adsorption plane for all ions (thus defined as the surface itself) whose size is accounted for by the distance that marks the edge of the diffuse layer. All three distributions yield a proportionality between surface charge and surface potential.

interpretation this model is clearly too unwieldly to be useful. Multilayer models provide various degrees of simplification to this approach.

2 That the charge of specifically sorbed ions is evenly distributed in some compact layer, at the outer edge of which the diffuse layer begins. Such a continuous surface layer separating pure phases is the traditional view in the study of the physics and thermodynamics of surfaces.

3 That the charges of specifically sorbed ions (=the surface charge) can be considered as point charges located in a unique plane. In this case, as in the first, the diffuse layer is considered to begin at some distance d from the specific adsorption plane; d is a fitting parameter which is rationalized by the size of the specifically adsorbed ions and facultatively also that of the first layer of the hydrated ions in the diffuse layer.

It is rather intuitive that the last two options must give a very similar result, the single plane of charges simply representing an averaging process for the continuous distribution of charges. We can show that indeed both yield a linear relationship between surface potential and surface charge

$$\psi_0 = \frac{\sigma_0}{C} \tag{72}$$

as would most simple variations on the same theme.

□ □ □

First recall that the variation in potential with charge density is given by Poisson's equation

$$\frac{d^2\psi}{dx^2} = -\frac{\rho}{\varepsilon} \tag{56}$$

In Option **3**, where the charge density is zero between the effective surface plane and the edge of the diffuse layer, the potential must thus vary linearly with distance. As in a flat capacitor the linear coefficient is given by the expression σ_0/ε_s:

$$\psi = \psi_0 - \frac{\sigma_0 x}{\varepsilon_s} \tag{73}$$

where ε_s is the permittivity coefficient applicable in the surface layer.

Expressing the potential at the edge of the diffuse layer as a function of the charge according to Equation 69 applicable at small potentials, we obtain an expression for ψ_0:

$$\psi_0 = \psi_d + \frac{\sigma_0 d}{\varepsilon_s} \tag{74}$$

$$\psi_0 = \frac{\sigma_d}{\varepsilon\kappa} + \frac{\sigma_0 d}{\varepsilon_s} \tag{75}$$

For reasons of electroneutrality, the charge in the diffuse layer is equal and opposite to the surface charge

$$\sigma_d + \sigma_0 = 0 \tag{76}$$

therefore

$$\psi_0 = \sigma_0 \left(\frac{d}{\varepsilon_s} - \frac{1}{\varepsilon\kappa} \right) \qquad \text{Q.E.D.} \tag{77}$$

For Option 2, Poisson's equation is written

$$\frac{d^2\psi}{dx^2} = -\frac{\sigma_0}{2d\varepsilon_s} \tag{78}$$

which, with the boundary condition $\psi(x = d) = \psi_d$, yields the expression:

$$\psi = -\frac{\sigma_0}{4d\varepsilon_s} x^2 + \frac{\sigma_0 d}{4\varepsilon_s} \kappa + \psi_d \tag{79}$$

The average potential applicable to the uniformly distributed charges in the compact layer is then obtained:

$$\psi_0 = \psi_{ave} = \psi_d + \frac{\sigma_0 d}{12\varepsilon_s} \tag{80}$$

Proceeding as above

$$\psi_0 = \sigma_0 \left(\frac{d}{12\varepsilon_s} - \frac{1}{\varepsilon\kappa} \right) \qquad \text{Q.E.D.} \tag{81}$$

Because the coefficients d and ε_s are little more than fitting parameters, all such expressions are equivalently described with the single fitting parameter C, as shown in Equation 72. Note that the final expression would be somewhat more complicated if we did not make the assumption of small potentials in the diffuse layer. However, it is generally found experimentally that diffuse layer potentials (approximated by various electrokinetic methods) are small indeed. Because it is a ratio of electrical charge and potential the coefficient, C, like C_{GC}, has the dimensions of a capacitance and is often referred to as the capacitance of the compact layer.

The simple linear form of Equation 72 is the one that we wish to use initially to study how the coulombic term is included in the surface complexation model. Note that in the classical Stern theory only one type of surface reaction is considered, resulting in an expression of the surface coverage (in the compact layer) that has the general form of a Langmuir isotherm:

$$\Gamma = \Gamma_{max} \frac{K_{ads}(C)}{1 + K_{ads}(C)}$$

But K_{ads} now includes a variable coulombic term:

$$K_{ads} = \exp\left(-\frac{\Delta G_{ads}^0}{RT} \right)$$

$$K_{ads} = \exp\left(-\frac{\Delta G_{int}^0}{RT} \right) \times \exp\left(-\frac{\Delta G_{coul}^0}{RT} \right)$$

$$\Delta G_{coul}^0 = FZ\psi_0 = \frac{FZ\sigma_0}{C} = \frac{F^2 Z^2 \Gamma}{aC}$$

In the last equation Z is the charge of the adsorbed ion and a (m^2 g^{-1}) is the specific surface area, included here because we have chosen to define Γ (mol g^{-1}) as the adsorption density per unit *mass* of solid. As seen in the last expression, the free energy of adsorption and thus the adsorption constant are a function of the adsorption density Γ.

□ □ □

5. INCLUSION OF THE ELECTROSTATIC TERM IN THE SURFACE COMPLEXATION MODEL

Before considering the adsorption of trace constituents let us examine simply the question of the acid-base chemistry of an oxide surface.

Example 3.   Acid-Base Chemistry of γ-Al_2O_3

Consider the acid-base reactions at the alumina surface. The reactions themselves are written as previously, and the usual tableau can be developed with H^+ and XOH_2^+ as components (see Tableau 8.3). In order to include explicitly

TABLEAU 8.3

	H^+	XOH_2^+	$P = e^{-F\psi_0/RT}$
H^+	1		
OH^-	-1		
XOH_2^+		1	0, 1^a
XOH	-1	1	-1, 0^a
XO^-	-2	1	-2, -1^a

^{a} The first coefficient is utilized for mass laws; the second for the "mole balance."

the coulombic interaction term, we must add one unknown (the potential in the adsorption plane) and one equation (the potential-surface charge relation) to our usual system of mole balance and mass law equations. To do this most conveniently, let us take the exponential of the dimensionless potential

$$P = \exp\left(-\frac{F\psi_0}{RT}\right) \qquad (82)$$

as a principal variable and add it to the list components[24] (see Tableau 8.3). By taking the electric charge number of each surface species (relative to that of the chosen surface component species, here XOH_2^+) to be the "stoichiometric" coefficient in the corresponding column of the tableau, the expressions of the mass laws, including explicitly the coulombic term, are written in the usual way from the lines of the tableau:

$$(XOH) = K_{a1}(H^+)^{-1}(XOH_2^+)P^{-1} = K_{a1} \times \exp\left(\frac{F\psi_0}{RT}\right)(XOH_2^+)(H^+)^{-1} \quad (83)$$

$$(XO^-) = K_{a1}K_{a2}(H^+)^{-2}(XOH_2^+)P^{-2} = K_{a1}K_{a2}$$
$$\times \exp\left(\frac{2F\psi_0}{RT}\right)(H^+)^{-2}(XOH_2^+) \qquad (84)$$

The mole balance equations are obtained from the first two columns of the tableau:

$$TOTH = (H^+) - (OH^-) - (XOH) - 2(XO^-) = -X_T + \text{acid} - \text{base} \quad (85)$$

$$TOTX = (XOH_2^+) + (XOH) + (XO^-) = X_T \quad (86)$$

The relationship between surface charge and surface potential is obtained very similarly from the last column of the tableau (using now the actual charge number of the surface species not the value relative to the charge of the component XOH_2^+, as the stoichiometric coefficient). The only difference with other mole balance equations is that the right-hand side must be calculated from the surface potential:

$$\text{``}TOTP\text{''} = \text{surface charge} = (XOH_2^+) - (XO^-) = \frac{A}{F}\sigma_0 = \frac{AC}{F}\psi_0 \quad (87)$$

where A is the area of adsorbing surface per liter of solution. For simplification we define the parameter

$$\lambda = 2.3 \frac{AC}{F} \frac{RT}{F} \text{ mol liter}^{-1} \quad (88)$$

and use it in Equations 82 and 87 to yield the mole balance equation

$$TOTP = (XOH_2^+) - (XO^-) = -\lambda \log P \quad (89)$$

(Note that the appropriate adsorbing surface area A is difficult to measure so that λ can be considered a unique fitting parameter.) The new principal unknown P is thus introduced into the system of equations in a form very similar to the principal component concentrations—except for the logarithm in the "mole balance" given by Equation 89.

Given values for the instrinsic constants K_{a1} and K_{a2}, the total surface site concentration X_T, and the coefficient λ, we can solve this system of two mass law and three mole balance equations to obtain a theoretical acidimetric and alkalimetric titration curve for an oxide suspension. Vice versa, from an experimental titration curve, K_{a1}, K_{a2}, X_T, and λ can be calculated. To get some insight into this process, let us assume that we have an independent estimate of X_T, obtained, for example, from the total number of exchangeable protons. We can simplify the system of equations, considering first that at low pH, XOH_2^+ is the dominant surface species and

$$(XOH_2^+) \quad \text{and} \quad (XOH) \gg (XO^-) \quad (90)$$

Let us define the "excess acid" q as

$$q = \text{acid} - \text{base} - (H^+) + (OH^-) \quad (91)$$

(q is an experimentally measurable parameter). The $TOTH$ equation then reduces to

$$(XOH) = X_T - q \quad (92)$$

and by difference between Equations 92 and 86, we obtain the approximate equality:

$$(XOH_2^+) = q \tag{93}$$

The third "mole balance" given by Equation 89 thus becomes

$$q = \lambda \log P \tag{94}$$

By defining the parameter Q^+,

$$Q^+ = \frac{(X_T - q)(H^+)}{q} \quad \text{(another measurable parameter)} \tag{95}$$

and substituting into the mass law of Equation 83, we obtain the relation

$$\log \frac{(XOH)(H^+)}{(XOH_2)} = \log Q^+ = \log K_{a1} - \log P \tag{96}$$

therefore

$$\log Q^+ = \log K_{a1} + \frac{q}{\lambda} \tag{97}$$

A plot of $\log Q^+$ versus q thus should yield a straight line of slope $1/\lambda$ with an intercept $\log K_{a1}$ on the $\log Q^+$ axis. Similarly at high pH, using the inequality

$$(XO^-) \quad \text{and} \quad (XOH) \gg (XOH_2^+) \tag{98}$$

and defining the parameter

$$Q^- = -\frac{q(H^+)}{(X_T + q)} \tag{99}$$

we obtain the approximate equalities $(XO^-) = -q = \lambda \log P$ and $(XOH) = X_T + q$, leading to the simplified relationship

$$\log Q^- = \log K_{a2} + \frac{q}{\lambda} \tag{100}$$

The second intrinsic acidity constant K_{a2} can thus also be obtained by a linear extrapolation (with the same slope $1/\lambda$) of a $\log Q^-$ versus q plot onto the vertical axis.

The experimental data of Figure 8.13 describing the acid-base titration of 8.174 g liter^{-1} γ-Al$_2$O$_3$[13] ($X_T = 1.12$ mM) has been subjected to these operations, and the results are shown in the lower part of the figure. As can be seen, at low absolute values of the excess acid q, both $\log Q^+$ and $\log Q^-$ strongly deviate from the predicted straight line. This is because, in that region, inequalities given by Equations 90 and 98 do not hold, all three surface acid-base species must be considered simultaneously, and no simplified derivation is possible.

The three parameters

$$pK_{a1} = 7.5$$

$$pK_{a2} = 9.35$$

$$\lambda = 0.63 \; mM$$

obtained from Figure 8.13 have been used in the original equations (Equations 83, 84, 85, 86, and 89) to generate the acid-base titration curve shown in the

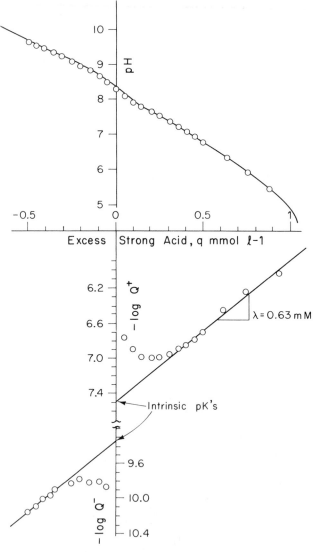

Figure 8.13 Acidimetric titration of an alumina surface (data from Hohl and Stumm, 1976[13]) and graphical treatment of the data illustrating the calculation of intrinsic acidity constants.

upper part of the figure. The fit with the experimental data is excellent, as may be expected since three parameters were adjusted for this purpose—even four if we include X_T. Note that the smeared shape of this titration curve is similar to that shown in Chapter 6 for a fulvic acid sample (Figure 6.8). Similar physical processes are at play, and a similar mathematical treatment can in principle be used to analyze and describe the acid-base chemistry of polymeric humic substances.

Although it does not appear very complicated, the system of Equations 83, 84, 85, 86, and 89 is unwieldy because of the logarithmic term in Equation 89 and a computer algorithm (part of the MINEQL program)[25] was in fact used to generate the titration curve of Figure 8.13. An illustration of the predominant influence of the coulombic term on the acid-base chemistry of surfaces (which is the reason for the numerical difficulties) is shown on the log concentration versus pH diagram of Figure 8.14 which compares the distribution of the acid-base surface species as calculated here for γ-Al_2O_3 to what they would be if there was no coulombic interaction (and the pK_a's were fixed at their intrinsic values). Because of electrostatic repulsion, which makes gain or loss of a proton increasingly unfavorable at low or high pH, the domain of importance of the neutral surface species is greatly enlarged. The equivalence points $(XOH_2^+) = (XOH^0)$ and $(XOH^0) = (XO^-)$ are displaced by approximately one pH unit. A plot of the apparent pK_a's (including the coulombic term) shows practically a linear dependency on pH, with a slope of about 0.5, in the pH range 6 to 11. This reflects of course the variation in surface charge (see Equation 96) which is also linear in that pH range, and goes through zero at $pH_{IEP} = \frac{1}{2}(pK_{a1} + pK_{a2})$ (isoelectric point). Note that the choice of acidity constants in Examples 1 and 2 correspond to the calculated apparent pK_a's for pH's of 5.5 and 7, respectively.

If we consider the estimated specific surface area of 117 m^2 g^{-1} for γ-Al_2O_3 ($\times 8.174$ g liter^{-1}), we can calculate from λ the value of the surface capacitance C according to Equation 88:

$$C = \frac{F}{2.3RT} \times F \times \frac{\lambda}{A}$$

therefore

$$C \cong 1.6 \times 10^6 \frac{\lambda}{A} \text{ Coulombs V}^{-1} \text{ m}^{-2} \quad (\lambda \text{ in } M \text{ and } A \text{ in } m^2 \text{ liter}^{-1})$$

$$C \cong 1.06 \text{ farads m}^{-2}$$

This capacitance is much higher than the compact layer capacitance measured at the surface of mercury or silver iodide electrodes (about 0.02 farads m^{-2}) whose electrical properties have been the object of most classical double layer studies. Such a high capacitance appears characteristic of oxide surfaces, and it corresponds to a large difference between the surface charge per se (as measured chemically in titration experiments) and the charge at the edge of the

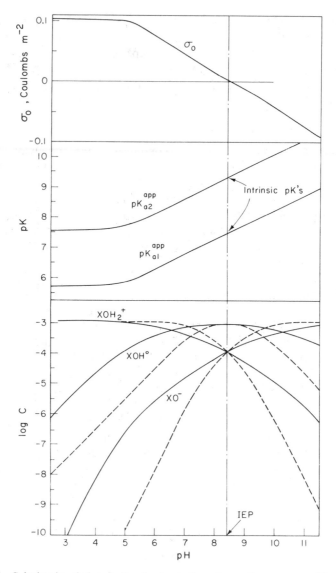

Figure 8.14 Calculated variations in the alumina surface charge, the apparent acidity constants, and the surface species distribution as a function of pH. The effects of the electrostatic energy term on the species distribution is seen by comparison with the distribution that would exist in the absence of the coulombic term (dashed line). At the isoelectric point [IEP $= \frac{1}{2}(pK_{a1}^{int} + pK_{a2}^{int})$] the surface is uncharged, and the apparent acidity constants equal to the intrinsic ones.

diffuse layer (e.g., as approximated by electrophoretic measurements). Note, however, that in this case the ionic strength is high (0.1 M NaClO$_4$) and that the capacitance is not much larger than the diffuse layer capacitance calculated on the basis of the Gouy-Chapman theory (Equation 70 yields $C_{GC} = 0.8$ farads m^{-2} for I = 0.1 M).

Example 4.   Influence of Coulombic Effects on the Adsorption of Pb^{2+} on γ-Al$_2$O$_3$

If we compare the results of Example 1 with experimental data (Figure 8.15), we see that the increase in lead adsorption with pH is experimentally more gradual than it is calculated to be on the basis of the simple surface complexation reaction considered in that example. Can this "smearing" of the adsorption graph be accounted for by long-range coulombic interations?

The system is defined by

$$Pb_T = 2.94 \times 10^{-4} \ M$$

$$X_T = 0.137 \ mmol \ g^{-1} \times 11.72 \ g \ liter^{-1} = 1.61 \ mM$$

and

$$A = 117 \ m^2 \ g^{-1} \times 11.72 \ g \ liter^{-1} = 1.37 \times 10^3 \ m^2 \ g^{-1}$$

The relevant mole balance and mass law equations are written according to our adsorption model (see Tableau 8.4):

TABLEAU 8.4

	H$^+$	Pb^{2+}	XOH$_2^+$	$P = e^{-F\psi_0/RT}$
H$^+$	1			
OH$^-$	-1			
Pb^{2+}		1		
PbOH$^+$	-1	1		
XOH$_2^+$			1	0, 1^a
XOH	-1		1	-1, 0^a
XO$^-$	-2		1	-2, -1^a
XOPb$^+$	-2	1	1	0, 1^a

^{a} The first coefficient is utilized for mass laws; the second for the mole balance.

$$TOTH = (H^+) - (OH^-) - (PbOH^+) - (XOH)$$
$$- 2(XO^-) - 2(XOPb^+) \tag{101}$$

$$TOTPb = (Pb^{2+}) + (PbOH^+) + (XOPb^+) = Pb_T = 2.94 \times 10^{-4} \tag{102}$$

$$TOTX = (XOH_2^+) + (XOH) + (XO^-) + (XOPb^+)$$
$$= X_T = 1.61 \times 10^{-3} \tag{103}$$

$$TOTP = (XOH_2^+) - (XO^-) + (XOPb^+) = -\lambda \log P \tag{104}$$

$$(PbOH^+) = 10^{-7.7}(Pb^{2+})(H^+)^{-1} \tag{105}$$

$$(XOH) = 10^{-7.5}(XOH_2^+)(H^+)^{-1}P^{-1} \tag{106}$$

$$(XO^-) = 10^{16.85}(XOH_2^+)(H^+)^{-2}P^{-2} \tag{107}$$

$$(XOPb^+) = 10^{-7.5}(Pb^{2+})(XOH_2^+)(H^+)^{-2} \tag{108}$$

The value of the parameter λ is obtained from the previous example:

$$\lambda = \frac{AC}{1.6 \times 10^6} = 9.1 \; mM$$

Although this system of equations could be solved iteratively for any given value of pH, the process is tedious, so this problem has again been solved by computer to obtain the curve of Figure 8.15. Although the value of the lead adsorption constant has been chosen to give a reasonable fit of the upper portion of the data, we see that the results are practically indistinguishable from those of Example 1 and do not fit the data any better. This is explained by the fact that the principal reaction controlling lead adsorption at pH below 6 according to this model is

$$XOH_2^+ + Pb^{2+} = XOPb^+ + 2H^+ \tag{109}$$

Such a reaction involves no change of electrical charge for the surface species and consequently is not affected by coulombic interactions. As long as XOH_2^+ is the principal surface species $((XOH_2^+) \cong X_T)$, the models of Examples 1 and 4 are equivalent. (Note that they include practically the same constant for Reaction 109.) In order to obtain from the calculation the smearing effect observed in the data, it is then necessary to consider the formation of other

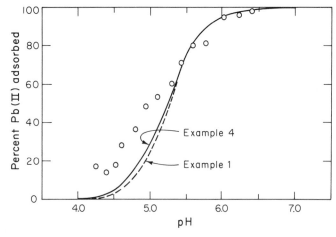

Figure 8.15 Model calculations for lead adsorption on alumina with (Example 4) and without (Example 1) consideration of the electrostatic energy term. In neither case do the results of the simple model fit the experimental data.[13]

surface species such as $X\text{OHPb}^{2+}$, $X\text{OPbOH}$, and $(X\text{O})_2\text{Pb}$, which provide a different pH dependency for lead adsorption. Alternatively, one could consider a different adsorption model in which Pb^{2+} and H^+ would react at different surface planes, and hence Reaction 109 would include a coulombic term. In either case an excellent fit of the data is possible, but this example illustrates the data-fitting nature of the surface species considered in adsorption models.

Fitting of Experimental Data: Model Dependency of Adsorption Constants

Experimental data on which one may base an adsorption model and from which one may calculate the necessary stoichiometric and thermodynamic constants are principally of two kinds: acid-base titrations of suspended solids in various background electrolytes and partition (solute vs. solid) of adsorbing species as a function of system composition, primarily pH and total concentration of the adsorbate. At present there is no possible direct determination of the chemical nature of the surface species so that chemical reasonableness is the only guide for choosing species that may fit experimental data.

Because the experimental data set is usually not very large, there are many different combinations of surface species and surface parameters [site density and surface capacitance(s)] that can fit the data quite closely. This is illustrated in Figure 8.16 which shows how large variations in the values of the model

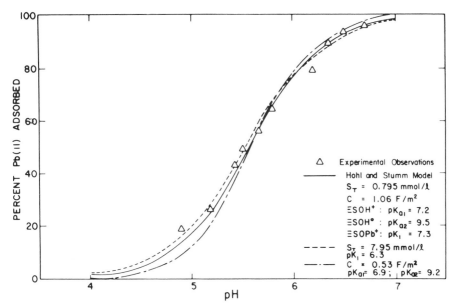

Figure 8.16 Model calculations for lead adsorption on alumina. The experimental data can be fitted with various sets of model parameters, for example, by compensating a higher surface site concentration or a lower capacitance with different choices of surface acidity constants. From Morel *et al.*, 1981.[26]

parameters can be made to compensate for each other and provide an equally good fit of the data. For a given adsorption model, the problem of parameter determination can be better constrained by acquisition of additional data. However, different models that include different basic assumptions must *necessarily* consider different surface species and include different constants. For example, the presence or absence of surface complexes with major electrolyte ions that may contribute to the surface charge or the possible subdivision of the adsorption layer lead to widely different expressions of the coulombic term and hence widely different choices of surface species to account for the data. Even such simple parameters as the intrinsic acidities of the surface groups are estimated differently in different models depending on the role attributed to surface species of major electrolyte ions in controlling the surface charge. The point here is that the inclusion of a given set of species in an adsorption model is a data-fitting procedure and not necessarily a description of chemical reality. While excellent predictive capability may be obtained by using consistently a given model, one should never use the species and the parameters of one model in another.

Because not enough attention has been paid to this problem, existing compilations of adsorption constants are not very useful, and they may be misleading in some cases. In addition the original experimental data from which the constants were derived are often not reproducible with the type of equilibrium calculations shown in the preceding examples, because indispensible parameters such as surface capacitance or site densities, consistent with the reported constants, are not reported. In an effort not to confuse this issue any further, we do not provide here a table of surface acidity and complexation constants and suggest that the original literature always be utilized for the purpose of making adsorption calculations for natural waters (until such time that all data are coherently interpreted with a single, uniformly accepted, adsorption model).

6. PARTITIONING OF SOLUTES INTO BULK PARTICULATE PHASES

A number of aquatic processes that remove solutes from solution and incorporate them into particles are no more adequately described by surface adsorption than they are by precipitation of pure solid phases. Aside from the important phenomenon of uptake by living organisms, these processes, which can be described by the generic term "sorption," involve the partitioning of solutes between the solution and the whole of a particulate phase. The particle-solution interface is not considered to control such processes, which are dominated by the bulk properties of the particles themselves. The word particle here signifies a phase that is hydrodynamically and thermodynamically distinct from the bulk solution; it may be an aqueous solution, a solid, or an organic phase, as we now discuss briefly in the cases of ion exchange, solid solution, and organic film solvation.

6.1 Ion Exchange

The general model of an ion exchanger consists of a porous lattice containing fixed charges.[27] Hydrated clays and condensed humic material (see Chapter 6) are the most common aquatic particles approximating this model. The water included in the lattice defines an aqueous solution distinct from the bulk solution in which electroneutrality is maintained by inclusion of an excess of counter ions (ions of opposite charge) from the bulk solution. (Our previous model of a surface phase in which charges are evenly distributed over a finite thickness is in effect that of an ion exchanger.)

Using, for example, x to refer to a cation exchanger, we can write the exchange reaction for sodium and calcium:

$$2Na_x^+ + Ca^{2+} = Ca_x^{2+} + 2Na^+ \tag{110}$$

As is the case for surface adsorption, the energy that attracts ions in the exchanger can be purely electrostatic or partly chemical. Even when the energy of attraction is purely electrostatic, high specificity may be exhibited due to matching of the pore or mesh size of the exchanger to that of the hydrated ions. On the basis of a largely electrostatic interaction, a preference for multivalent ions is generally expected and observed in ion exchangers; that is, considering ion partitioning as caused purely by coulombic interactions, we can write

$$Na_x^+ = Na^+; \quad \frac{\{Na_x^+\}}{\{Na^+\}} = \exp\left(\frac{F\Delta\psi}{RT}\right) \tag{111}$$

$$Ca_x^{2+} = Ca^{2+}; \quad \frac{\{Ca_x^{2+}\}}{\{Ca^{2+}\}} = \exp\left(\frac{2F\Delta\psi}{RT}\right) \tag{112}$$

Such equations describe the partitioning of ions in relatively dilute solutions on both sides of a semipermeable membrane—the Donnan equilibrium—and they explain qualitatively the role of electrical charge in ion exchange. The equations are, however, of little quantitative use for most ion exchangers because activity coefficients are all but unknown in the ion-exchanger phase, a medium of effective ionic strength in excess of 0.3 M in which some of the charges are fixed. The same difficulty applies to the use of the equilibrium constant for Reaction 110:

$$K = \frac{\{Ca_x^{2+}\}\{Na^+\}^2}{\{Na_x^+\}^2\{Ca^{2+}\}} \tag{113}$$

The corresponding "selectivity coefficient"

$$K_c = \frac{(Ca_x^{2+})(Na^+)^2}{(Na_x^+)^2(Ca^{2+})} \tag{114}$$

is far from constant over a range of sodium and calcium concentrations, and there is no simple functionality to relate it to the equilibrium constant.

Although a number of theories have been developed to predict or at least rationalize the behavior of ion exchangers, empirical information on the ion exchange properties of clays or condensed humic material is probably most useful to aquatic chemists. In classical ion-exchange literature this empirical information usually takes the form of an ion-exchange isotherm where the mole fraction of a given ion in the exchanger is plotted against the mole fraction of the same ion in solution. More directly pertinent to our general formulation of equilibrium is to plot the (log of) selectivity coefficient as the ordinate, as we showed in Chapter 6 for the ion exchange of sodium and calcium in humic acid

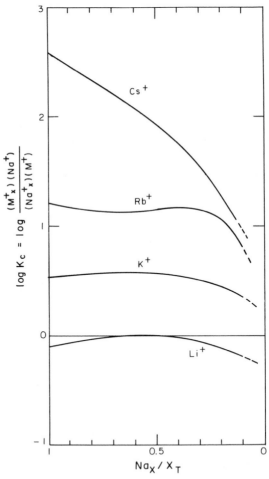

Figure 8.17 Selectivity coefficients for the exchange of monovalent cations in a bentonite clay. The smaller hydrated ions are exchanged preferentially over the larger ones, in accordance with a principally electrostatic interaction energy. Adapted from Gast, 1969.[28] (With permission from Soil Science Society of America.)

(Figure 6.9). This is the representation that is given in Figure 8.17 for the exchange selectivity of a bentonite clay for monovalent cations.[28] The selectivity sequence, $Cs^- > Rb^+ > K^+ > Na^+ > Li^+$, is opposite to that of the hydrated radii. A selectivity for smaller hydrated ions is expected on the basis of a principally electrostatic sorption energy. Note that a strong preference for retention of potassium over sodium in clays is consistent with the relative sodium impoverishment in sedimentary rock compared with igneous rock (see Chapter 5).

6.2 Solid Solution

When a solid phase precipitates in a multicomponent system, foreign ions, metals or ligands, are often incorporated into the solid matrix. Although the first step in such a process may be that of surface adsorption (as it is for precipitation of the pure solid), the actual inclusion of a foreign ion in the bulk of the solid phase is restricted to ions whose valences and interatomic bonding distances are a close match for those of the precipitating ions. For example, in aquatic systems, Sr^{2+}, Mg^{2+}, and Mn^{2+} are often incorporated in calcium carbonate.[29] Many metal oxides and sulfides show various degree of enrichment in other trace metals, as is conspicuously exemplified in ferromanganese nodules.

A thermodynamic description of this process is obtained by considering the solid phase as a solution of one pure solid in the other and attributing to each of these solids an activity that is not unity. Consider, for example, the substitution of Mn^{2+} for Ca^{2+} in calcite:

$$Mn^{2+} + CaCO_3(s) = MnCO_3(s) + Ca^{2+} \tag{115}$$

The apparent equilibrium constant for this reaction—now called a distribution coefficient—expressed as

$$D = \frac{\{MnCO_3 \cdot s\}}{\{CaCO_3 \cdot s\}} \times \frac{\{Ca^{2+}\}}{\{Mn^{2+}\}} \tag{116}$$

is equal to the ratio of the solubility products of calcium and manganese carbonates:

$$K_{CaCO_3} = \frac{\{Ca^{2+}\}\{CO_3^{2-}\}}{\{CaCO_3 \cdot s\}} \tag{117}$$

$$K_{MnCO_3} = \frac{\{Mn^{2+}\}\{CO_3^{2-}\}}{\{MnCO_3 \cdot s\}} \tag{118}$$

therefore

$$D = \frac{K_{CaCO_3}}{K_{MnCO_3}} \cong \frac{10^{-8.3}}{10^{-10.4}} = 10^{2.1} \tag{119}$$

The difficulty with this approach of course is to evaluate the activities of the solids in the solid solution. (This can be done empirically from the equilibrium partitioning of Mn^{2+} and Ca^{2+}, thus making the whole approach an empirical rather than a predictive one.) If we make the gross assumption that the solid solution is ideal, that is, if we equate the activities of the solids to their mole fractions, we obtain a qualitative description of the role of solid solutions in the solubility of ions.

Consider, for example, the addition of manganese ion to a system defined by a fixed free carbonate ion concentration and a given total concentration of calcium; $(CO_3^{2-}) = 10^{-5.3}$ M; $TOTCa = 10^{-2}$ M. By defining the total solid concentration

$$S_T = (CaCO_3 \cdot s) + (MnCO_3 \cdot s) \tag{120}$$

the activities of the solid phases can be written:

$$\{CaCO_3 \cdot s\} = \frac{(CaCO_3 \cdot s)}{S_T} = \frac{(Ca^{2+})(CO_3^{2-})}{K_{CaCO_3}} = 10^3(Ca^{2+}) \tag{121}$$

$$\{MnCO_3 \cdot s\} = \frac{(MnCO_3 \cdot s)}{S_T} = \frac{(Mn^{2+})(CO_3^{2-})}{K_{MnCO_3}} = 10^{5.1}(Mn^{2+}) \tag{122}$$

The three mole balance expressions

$$TOTCa = (Ca^{2+}) + (CaCO_3 \cdot s) = (Ca^{2+})(1 + 10^3 S_T) = 10^{-2} \tag{123}$$

$$TOTMn = (Mn^{2+}) + (MnCO_3 \cdot s) = (Mn^{2+})(1 + 10^{5.1} S_T) \tag{124}$$

$$S_T = (CaCO_3 \cdot s) + (MnCO_3 \cdot s) = S_T[10^3(Ca^{2+}) + 10^{5.1}(Mn^{2+})] \tag{125}$$

complete the mathematical description of the system. Algebraic manipulation of these equations leads to the approximate solution

$$(Mn^{2+}) = \frac{TOTMn}{TOTMn + TOTCa} \times \frac{K_{MnCO_3}}{(CO_3^{2-})} \tag{126}$$

which is plotted on Figure 8.18. The principal feature of this result—and the major effect of solid solution formation—is the decrease in the soluble manganese (Mn^{2+}) at low total manganese concentration in the presence of calcite. With the assumption of ideality of the solid solution, this decrease is a factor of more than one thousand. This large effect is due in part to the greater insolubility of $MnCO_3$ compared to $CaCO_3$, and in part to the large concentration of solid compared to soluble calcium, as seen in the formula

$$TOTMn = (Mn^{2+})\left[1 + \frac{K_{CaCO_3}}{K_{MnCO_3}} \times \frac{S_T}{(Ca^{2+})}\right] \tag{127}$$

Note that a decrease in the solubility of calcium due to a decrease in the activity of $CaCO_3(s)$ is only seen when the total manganese is in excess of the total calcium, not a very probable circumstance in natural waters. However, such a decrease in the soluble concentrations of a precipitating metal due to solid solution with another may occur in the case of metal oxides, when the metals are both present at trace concentrations.

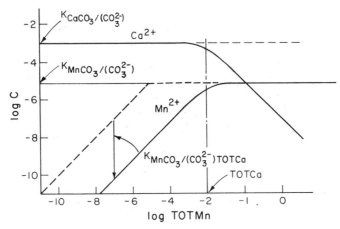

Figure 8.18 Variations in soluble calcium and manganese ions concentrations as a function of total manganese in the presence of an ideal $CaCO_3$-$MnCO_3$ solid solution [(CO_3^{2-}) is fixed at $10^{-5.3}$ M]. Note the large decrease in soluble Mn compared to total manganese at low concentrations.

6.3 Organic Film Solvation

Many trace organic pollutants are hydrophobic, and their fate in aquatic systems is largely governed by their partitioning among gas, water, and solid phases. A widely accepted and reasonably successful predictive model for the water-solid partitioning of such hydrophobic trace compounds is to consider a two-phase equilibrium between the water and what is imagined to be an organic film coating the sediment or soil particles.[30,31] Three major empirical observations are at the core of such a simple model:

1. Sorption isotherms for hydrophobic compounds are typically linear over the range of concentration of interest.

2. For a given compound, sorption is proportional to the total organic content of the solid (the coefficient of proportionality is somewhat affected by the size class of the sediments).

3. For a given sediment, sorption is proportional to the octanol-water partition coefficient of the organic compound. (Note also that octanol-water partition coefficients are inversely correlated with water solubilities.)

Although these observations may support other models, the image of a bulk organic phase in which the hydrophobic compound is dissolved—effectively a solvent extraction procedure—is particularly simple and appealing even if it does not have a strong conceptual or theoretical backing. The general formula for such organic film solvation of trace compounds is written

$$C_x = \alpha \times K_{ow}(OC) \times C \qquad (128)$$

where C_x is the concentration of the compound in the solid phase, α is an empirical coefficient of proportionality ($\alpha \cong 0.63$) representing the solvent properties of condensed aquatic organic matter compared to octanol, K_{ow} is the octanol water partition coefficient, (OC) is the fraction of organic carbon in the solid, and C is the concentration of the compound in water. α, K_{ow}, and (OC) are dimensionless, and both C_x and C are expressed in the same units: C_x is in g or mol kg^{-1} of dry solid and C in g or mol kg^{-1} of solution. The wide applicability of this formula and its very small data requirements make it a very valuable tool for predicting the fate of organic pollutants in aquatic systems.

7. KINETIC CONSIDERATIONS

In the introduction to this chapter we noted that the geochemical fate of trace elements and compounds was often controlled by (ad)sorption on a solid surface and eventual settling and incorporation in sediments. This overall phenomenon includes not only the two obvious steps of sorption and settling but also the process of particle aggregation through which colloidal material (<10 μm) is transformed into particles that have nonnegligible settling velocities (>10 μm). Any of the three processes, sorption, aggregation, and settling, may control the overall rate of elimination of a conservative chemical compound from the water column. We now discuss each of these briefly.

7.1   Adsorption Kinetics

Most of the experimental adsorption data that are shown or alluded to in this chapter were obtained with equilibration times of the order of minutes to hours and were usually observed to be reversible on the same time scale. When the experiments were continued for longer times, often a second, slower, and typically irreversible adsorption process was observed with a characteristic equilibration time of the order of days to months. This second process is usually thought to correspond to a slow change in the nature of the solid itself (adsorption data are often obtained with metastable solids) or to a diffusion of the adsorbate toward the interior of the solid which may be porous. Indeed, many theoretical treatments of adsorption kinetics focus on diffusion and often consider a two-step process: diffusion to the particle in the boundary layer and diffusion within the particle.[32] The expression for boundary layer diffusion has the usual form introduced in Chapters 4 and 5, while that for pore diffusion is complicated by considerations of pore size and retardation by reaction of the sorbate with the pore surface. Other general rate equations simply describe the kinetics of the adsorption process as a bimolecular reversible chemical reaction:

$$X + C \rightleftarrows XC \tag{2}$$

therefore

$$\frac{d(XC)}{dt} = k_f(X)(C) - k_b(XC)$$

In this formulation, which is consistent with a Langmuirian description of adsorption equilibrium, the law of microscopic reversibility is usually considered to hold, and only one independent kinetic constant must be obtained:

$$\frac{d(XC)}{dt} = k\left[(X)(C) - \frac{(XC)}{K_{ads}}\right]$$

As noted in Chapter 5, reaction and diffusion control of solute-solid reactions are often difficult to distinguish from each other, and applicability of a given rate law is rarely sufficient proof of the underlying mechanism.

7.2 Settling

The settling velocity of aquatic particles is generally assumed to follow Stokes law:

$$V_s = \frac{g}{18\mu}(\rho_s - \rho)d^2$$

where V_s is the settling velocity, g is the acceleration of gravity, μ is the dynamic viscosity, ρ_s is the density of the particle, ρ is the density of water, and d is the diameter of the particle. While we shall say little of the density of aquatic particles (some value, slightly in excess of that of water and increasing with the inorganic content of the particle), we wish to focus our attention on particle size to which, because of the square term, the settling velocity is most sensitive. The most important notion to be conveyed here is that a particle diameter is really not an inherent, conservative property of a given particle. In the size range below 10 μm, particles continuously aggregate, and the balance of aggregation and settling processes results in a characteristic size distribution that is a property of the suspension, not that of the individual particles. Thus, in order to describe quantitatively the fate of particles (and of their load of sorbed material), we must understand the mechanisms of particle aggregation in natural waters, coagulation, and ingestion by organisms.

7.3 Particle Aggregation

From a thermodynamic point of view suspensions of small particles are unstable since the total surface energy is greater in the dispersed state than it is in the aggregated state.* The classical treatment of the energetics of particle

---

* Surface energies are in fact sufficiently large to affect the solubility of solid phases. See, for example, discussion in Stumm and Morgan (1981, 295–299).

interactions, the DLVO theory (Derjaguin, Landau, Verwey, and Overbeek), considers the balance of short-range attractive Van der Waals forces and the longer range repulsive electrostatic forces.[33] Because the Van der Waals forces vary with (the inverse of) the fourth power of the interparticle distance (x^{-4}) and the electrostatic forces with x^{-2}, a plot of interaction energy versus x yields the typical graph of Figure 8.19. When the particles are very close together, attractive forces dominate, so the particles are coagulated. When the particles are farther apart, electrostatic repulsion creates an energy barrier that prevents coagulation and "stabilizes" the suspension. The large effect that increasing ionic strength has on lowering the coagulation energy barrier illustrated in the figure is simply due to compression of the double layer and decrease in the electrostatic repulsion among particles.

There are three principal mechanisms by which particles can overcome their electrostatic energy barrier, collide, and coagulate:[35] Brownian motion as determined by the thermal energy of the suspension (perikinetic coagulation), laminar or turbulent fluid shear controlled by the kinetic energy of the fluid (orthokinetic coagulation), and differential settling which depends on the gravitational (and hence kinetic) energy of the particles. These processes are very much dependent on the sizes of the coagulating particles as well as on the

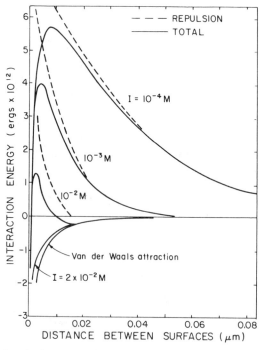

Figure 8.19 Variations in the energy of attraction (Van der Waals) and repulsion (electrostatic) as a function of distance between two charged spheres ($d = 0.2$ μm, $\psi_0 = 50$mV) in a medium of low ionic strength ($2 \times 10^{-2}\,M$ to $10^{-4}\,M$). From Kasper. 1971.[34]

temperature and mixing regime of the water. Under the conditions prevalent in natural water, it is principally the relative motions imparted to the particles by the turbulence of the water[10]—orthokinetic coagulation—that is thought to control the collision process. The particle size distribution observed in aquatic systems (see Figure 8.1 and Table 8.1) can be rationalized as a steady state between coagulation of the smaller particles and settling of the larger ones, and values, $p \cong 4$, of the exponent of the power law in Equation 1 are predicted by such theory. In situations where coagulation is the step limiting the overall sedimentation process, the rate of elimination of particles from the water column may be dependent on the square of the particle concentration (second-order process) rather than being proportional to the particle concentration, as it is when settling is the rate limiting step.

Although the subject of particle ingestion by aquatic organisms is beyond the scope of this chapter, it should be strongly emphasized that zooplankton feeding, shedding of outer layers, and packaging of feces are generally considered to be important, perhaps dominant, factors in the sedimentation of aquatic particles, particularly in the sea.[7,8,36] Although zooplankton certainly exhibit some taste and discrimination in their feeding habits, one may hope that on a larger scale their particle aggregating activities are describable by coagulation equations. Although this "biological coagulation" mechanism may be sometimes more effective than orthokinetic or perikinetic coagulation and promote a faster sedimentation rate, in some places over some size range it may result in the same type of particle size distribution in the water column.

REFERENCES

1 P. W. Schindler, *Thalassia Yugoslavia*, **11**, 101–111 (1975).

2 L. Balistrieri, P. G. Brewer, and J. W. Murray, *Deep-Sea Res.*, **28A**, 101–121 (1981).

3 M. C. Kavanaugh and J. O. Leckie, Eds., *Particulates in Water*, Advances in Chemistry Series 183, ACS, Washington, D.C., 1980.

4 M. A. Anderson and A. J. Rubin, Eds., *Adsorption of Inorganics at Solid-Liquid Interfaces*, Ann Arbor Science, Ann Arbor, 1981.

5 W. Stumm and J. J. Morgan, *Aquatic Chemistry*, 2nd ed., Wiley, New York, 1981, pp. 600–601.

6 R. J. Gibbs, Ed., *Suspended Solids in Water*, Plenum, New York, 1974.

7 J. K. B. Bishop, J. E. Edmond, D. R. Ketten, M. P. Bacon, and W. B. Silker, *Deep Sea Res.*, **24**, 511–548 (1977).

8 A. Lerman, *Geochemical Processes, Water and Sediments Environments*, Wiley, New York, 1979.

9 A. Lerman, K. L. Calder, and P. R. Betzer, *Earth Planet. Sci. Letters*, **37**, 61–70 (1977).

10 J. R. Hunt, in *Particulates in Water*, M. C. Kavanaugh and J. O. Leckie, Eds., Advances in Chemistry Series 183, ACS, Washington, D.C., 1980.

11 A. W. Adamson, *Physical Chemistry of Surfaces*, 2nd ed., Wiley, New York, 1967.

12 P. W. Schindler, in *Adsorption of Inorganics at Solid-Liquid Interfaces*, M. A. Anderson and A. J. Rubin, Eds., Ann Arbor Science, Ann Arbor, 1981.

13 H. Hohl and W. Stumm, *J. Colloid. Interf. Sci.*, **55**, 281 (1976).

14 J. G. Yeasted, "The Modeling of Lake Response to Phosphorus Loadings: Empirical, Chemical and Hydrodynamic Aspects," Ph.D. Dissertation, Department of Civil Engineering, Massachusetts Institute of Technology, Cambridge, 1978.

15   J. A. Davis and J. O. Leckie, *Env. Sci. Technol.*, **12**, 1309–1315 (1978).

16   J. A. Davis, *Geochim. Cosmochim. Acta*, **46**, 2381–2393 (1982).

17   K. A. Hunter and P. S. Liss, *Nature*, **282**, 823–825 (1979).

18   E. Tipping and D. Cooke, *Geochim. Chim. Acta*, **46**, 75–80 (1982).

19   J. A. Davis, submitted for publication.

20   D. C. Grahame, *Chem. Rev.*, **41**, 441–501 (1947).

21   O. Stern, *Z. Elektrochem.*, **30**, 508–516 (1924).

22   J. Th. G. Overbeek, in *Colloid Science*, vol. 1, H. R. Kruyt, Ed., 1952, ch. 4.

23   J. C. Westall and H. Hohl, *Adv. Colloid Interf. Sci.*, **12**, 265 (1980).

24   J. C. Westall, in *Particulates in Water*, M. C. Kavanaugh and J. O. Leckie, Eds., Advances in Chemistry Series 183, ACS, Washington, D.C., 1980.

25   J. C. Westall, J. L. Zachary, and F. M. M. Morel, Technical Note 18, R. M. Parsons Laboratory, Massachusetts Institute of Technology, Cambridge, 1976.

26   F. M. M. Morel, J. G. Yeasted, and J. C. Westall, in *Adsorption of Inorganics at Solid-Liquid Interfaces*, M. A. Anderson and A. J. Rubin, Eds., Ann Arbor Science, Ann Arbor, 1981.

27   F. Hellferich, *Ion Exchange*, McGraw-Hill, New York, 1962.

28   R. G. Gast, *Soil Sci. Soc. Amer. Proc.*, **33**, 37–41 (1969).

29   W. Stumm and J. J. Morgan, *Aquatic Chemistry*, 2nd ed., Wiley, New York, 1981, pp. 287–291.

30   C. T. Chiou, L. J. Peters, and V. H. Freed, *Science*, **206**, 831 (1979).

31   S. W. Karickhoff, B. J. Brown, and T. A. Scott, *Water Res.*, **13**, 241–248 (1979).

32   D. W. Hardwicks and L. G. Kuratti, *Water Res.*, **16**, 829–837 (1982).

33   W. Stumm and J. J. Morgan, *Aquatic Chemistry*, 2nd ed., Wiley, New York, 1981, p. 656.

34   D. Kasper, "Theoretical and Experimental Investigations of the Flocculation of Charged Particles in Aqueous Solutions by Polyelectrolytes of Opposite Charge," Ph.D. Dissertation, California Institute of Technology, Pasadena, 1971.

35   S. Friedlander, *Smoke, Dust and Haze: Fundamentals of Aerosol Behavior*, Wiley, New York, 1977.

36   M. M. Mullin, in *Particulates in Water*, M. C. Kavanaugh and J. O. Leckie, Eds., Advances in Chemistry Series 183, ACS, Washington, D. C., 1980.

PROBLEMS

8.1   a. Assuming a solid density $\cong 1$, and a power law size distribution with an exponent of -4 between radii of 0.1 and 10 μm, what is the total surface area of 5 mg liter^{-1} of suspended solids?

   b. Using a cross-sectional area $\cong 10^{-15}$ cm^2 for typical ionic groups, estimate an upper limit on the adsorption capacity (X_T) of such a suspension. Compare your result with the values given in Example 1 and Example 4 (specific site density and specific surface area per solid mass).

8.2   Consider a solid surface characterized by the reactions

$$XOH_2^+ = H^+ + XOH; \quad pK_{a_1} = 5.0$$

$$XOH = H^+ + XO^-; \quad pK_{a_2} = 7.0$$

$$XO^- + Cu^{2+} = XOCu^+; \quad pK = -8.0$$

and for which electrostatic effects are considered unimportant.

   a. Calculate the adsorption isotherm at pH $= 7$, $X_T = 10^{-6}$ M.

b. Assuming no other solution species than Cu^{2+}, calculate the fraction of Cu adsorbed as a function of pH ($X_T = 10^{-6}$ M, $Cu_T = 10^{-8}$ M).

c. Assuming the same inorganic species and the same water composition as in Example 3 of Chapter 6, calculate the speciation of copper at pH $= 7$.

8.3 a. Calculate the double-layer thickness in a solution of 5×10^{-4} M NaCl.

b. In such a medium what is the charge density of a surface at a potential of -20 mV?

c. Calculate the effects of the adsorption in the diffuse layer on the bulk ion composition (Na^+ and Cl^-), if the solid is present at the high experimental density of 10 g liter^{-1} (specific surface area $= 100$ m^2 g^{-1}).

d. Assuming that the surface charge has been developed by removal of acidic protons from the surface and that the solution is not buffered at all, what is the pH of the solution?

e. What is the pH at the surface?

8.4 Consider a 10 mg liter^{-1} solid suspension (specific site density $= 10^{-4}$ mol g^{-1}; specific surface area $= 100$ m^2 g^{-1}) and a 10^{-4} M solution of an organic ligand, Y, both characterized as monoprotic acids:

$$XOH = H^+ + XO^-; \quad pK_{ax} = 6.0$$

$$HY = H^+ + Y^-; \quad pK_{ay} = 4.5$$

The adsorption reaction is written:

$$XOH + Y^- = XOHY^-; \quad \log K = 5.0$$

a. Neglecting coulombic effects, calculate the speciation of the surface and of the organic ligand as a function of pH.

b. Assuming an inner capacitance of 1 farad m^{-2}, redo part a according to the electrostatic model presented in the chapter (e.g., Example 4).

INDEX